Healthy at Work

Markus Wiencke • Mirella Cacace •
Sebastian Fischer
Editors

Healthy at Work

Interdisciplinary Perspectives

Springer

Editors
Markus Wiencke
Federal Institute for Occupational Safety and Health (BAuA)
Berlin, Germany

Mirella Cacace
SOCIUM Research Center on Inequality & Social Policy
University of Bremen
Bremen, Germany

Sebastian Fischer
Leuphana University
Lüneburg, Germany

ISBN 978-3-319-32329-9 ISBN 978-3-319-32331-2 (eBook)
DOI 10.1007/978-3-319-32331-2

Library of Congress Control Number: 2016950620

Printed on acid-free paper

This Springer imprint is published by Springer Nature
The registered company is Springer International Publishing AG Switzerland

Foreword

In a changing working world with its demands for increased flexibility and blurred boundaries, occupational health management needs are becoming increasingly heterogeneous. Urgent questions concerning the sustainability of health and safety measures in the everyday work environment, how to deal with increased stress, and how to negotiate the gap between health and safety requirements and the autonomy of individuals can only be answered on the basis of an interdisciplinary approach.

Our knowledge about the factors that influence the effectiveness of occupational health and safety measures has grown steadily in recent years. However, how and under what organizational conditions these factors come together remains relatively unclear. Occupational health and safety appears to be inextricably linked to aspects of culture, organizational processes, and leadership and thus stands in a context of change management and private and public policies.

With this in mind, a group of scholars from disciplines as disparate as psychology, economics, management, medicine, and sociology present here a framework for the design of occupational health management. They orient the reader on how to identify and manage the factors that influence health in organizations. Thus, health management emerges as the effective interplay of organizational culture, leadership, policies, and evidence-based interventions.

The book provides a fascinating introduction to the current scholarly discourses on occupational health and provides a variety of suggestions for its transfer in practice.

European Agency for Safety and Health at Work Christa Sedlatschek
Bilbao, Spain

Contents

List of Contributors

Mats Alvesson works at Lund University, Sweden, the University of Queensland, and Cass Business School. He has written extensively on organizational culture, leadership, identity issues at work, critical theory, and qualitative method. His publications include *The Triumph of Emptiness* (2013), *Understanding Organizational Culture* (2013), and *Reflexive Methodology,* with Kaj Sköldberg (2009), as well as thirty other books.

Peter Angerer, M.D., Ph.D. is professor and director of the Institute of Occupational and Social Medicine, Centre for Health and Society at the University of Düsseldorf, Germany. His primary interests in the field of occupational medicine are the effect of work stress on health and the effect of stress management interventions focused on working conditions and individual stress management.

Robert Persson Asplund, M.Sc. is Ph.D. candidate and researcher in clinical psychology at Linköping University, Sweden. His clinical work is mainly devoted to rehabilitation and occupational health at an occupational healthcare clinic in Stockholm. His research is focused on internet-based screening and healthcare interventions.

Sophie Baeriswyl has studied work and organizational psychology at the University of Bern, Switzerland, and currently works as a research scientist at the University of Applied Sciences and Arts Northwestern Switzerland. Her area of interest and research focuses on the relationships among working conditions, well-being, and health, with a special emphasis on health-related behaviors as intervening variables.

Miriam K. Baumgärtner works as a project leader and postdoctoral researcher at the Center for Disability and Integration (CDI-HSG), at the University of St. Gallen, Switzerland. She earned her Ph.D. in management from the University of St. Gallen and her M.Sc. in psychology from the University of Mannheim, Germany. For several years, she has worked for GESIS–Leibniz Institute for the Social Sciences and gained practical experience as a consultant at Kenexa in Munich. Her research focuses on human resource management, leadership, and workplace diversity. She has a special interest in the inclusion of people with disabilities in the workplace and health-focused leadership.

Wilhelm Beckmann is a research associate at the Dresden University of Applied Sciences, where he received his bachelor's degree equivalent diploma (FH) in business administration and engineering in 2011. His field of activity is in work analysis and design, in particular job autonomy as well as conceptualizing and facilitating the health circle in the workplace.

Martial Berset is a senior researcher at the University of Applied Sciences and Arts Northwestern Switzerland. He received his Ph.D. in work and organizational psychology from the University of Bern, Switzerland. His main area of interest is the relationship between working conditions and health-related behaviors.

Stephan Böhm is associate professor of organizational behavior and director of the Center for Disability and Integration (CDI-HSG) at the University of St. Gallen, Switzerland. He received his M.Sc. and Ph.D. from the University of St. Gallen. From 2008 to 2009, he worked as a visiting research fellow at the Oxford Institute of Ageing at the University of Oxford, Great Britain. His work focuses on leadership, human resource management, diversity, and change management. He has a special interest in the vocational inclusion of employees with disabilities, health-focused leadership, and management of demographic change. His research has been published in top-tier journals including *The Academy of Management Journal*, *Personnel Psychology*, *The Leadership Quarterly*, *Journal of Management Studies*, *Journal of Organizational Behavior*, and *Human Resource Management*. In 2014, he and Miriam Baumgärtner won the Swiss Public Health Award for the concept of health-focused leadership.

Daniel Braun-Beustrin holds a diploma in business economics and an M.A. in sustainable economics. As a business development agent at the Leuphana University of Lüneburg, Germany, he consults and supports innovation-incubator project teams by creating new business models. The innovation incubator is an EU project that focuses on supporting local economies by combining science and economics.

Mirella Cacace holds a Ph.D. in economics and currently works as postdoctoral researcher at the Research Center on Inequality and Social Policy (SOCIUM) at the University of Bremen, Germany. In her previous position as visiting professor at the Leuphana University of Lüneburg, Germany she was head of research of the Healthy at Work (GiB) team dealing with occupational health in small- and medium-sized enterprises. In 2008 and 2009, she was a Harkness Fellow in International Health Policy and Practice, sponsored by the Commonwealth Fund, and hosted by Columbia University in New York City. Her main areas of research concern institutional economics and international healthcare systems. In autumn 2016 she will join the APOLLON University of Applied Sciences in Bremen, Germany, as professor of health economics.

Anca Chiriac, M.D., Ph.D. Apollonia University, Department of Dermatology, Nicolina Medical Center, P. Poni Research Institute of Macromolecular Chemistry, Romanian Academy, Iaşi, Romania.

Harald Christa received his master's degree equivalent diploma in sociology from the University of Bamberg, Germany, in 1988. His Ph.D. thesis was on social marketing strategies for NPOs in the field of elder care (1996). Since 1998, he has been a professor of social management at the University of Applied Sciences for Social Work, Education and Nursing in Dresden, Germany. His research is focused on marketing of nonprofit and public organizations, quality management, and networking management and evaluation. He also works as a consultant in the fields of elder care and social work with ex-offenders. He has published four scientific books and 86 papers in German journals and publications. He works as a scientific reviewer for a research journal and a website on social management publications.

Jessica de Bloom, Ph.D. works as a research fellow at the Institute of Advanced Social Research at the University of Tampere, Finland. Her field of expertise concerns longitudinal and intervention research on work stress, recovery, and the effects of leisure time on employee health, well-being, and job performance. She is currently developing a Smartphone application to improve and prolong the beneficial effects of vacation on well-being.

Nicole Deci holds an M.Sc. in psychology. She is a research associate in the Department for Work and Organizational Psychology at the University of Hamburg, Germany, and also works as a consultant and lecturer in the field of social psychology. Her research focuses on work stress and coping, as well as employee health.

Jan Dettmers holds a Ph.D. in psychology. Since 2015, he has been professor for work and organizational psychology at the Medical School Hamburg (MSH), Germany. Previously, he has been an assistant and/or visiting professor at the Universities of Hamburg and Leipzig. His main research interests concern work and health in the domain of flexible and boundary-less forms of work and new approaches to prevention.

Laura Diedrich received an M.Sc. in psychology from the Freie Universität Berlin in 2013. She also studied psychology at the Universität Hamburg. From 2013 to 2014, she was a research assistant at the Lüneburg Innovation Incubator. She has worked as a personnel developer at an IT company in Hamburg, Germany, since 2014.

David Daniel Ebert, Ph.D. focuses on the development and evaluation of internet- and mobile-based interventions for the prevention of common mental health disorders. Previously, he was operational head of the Division of Online Health Training at Leuphana University of Lüneburg, Germany, and now leads the unit for e-mental health research at the University of Nuremberg-Erlangen, Germany. He is a member of the board of directors of the International and the European Society for Research on Internet Interventions, and member of the editorial board of *Internet Interventions,* published by Elsevier.

Hanna Lisa Eckerth is a research assistant at the Institute for Performance Management at the Leuphana University of Lüneburg, Germany. Currently, she coordinates the research network "organizational health." She holds an M.Sc. in public health from the University of Applied Science in Fulda, Germany.

Fokko-Jelto Eller holds an M.A. in management and business development from the Leuphana University of Lüneburg, Germany (2014). He studied management and entrepreneurship at the Leuphana University of Lüneburg and international business at the University of Applied Sciences and Arts, Dortmund, Germany. He worked as a graduate assistant at the Lüneburg Innovation Incubator. Since 2015, he has been a Ph.D. student and research assistant at the Leuphana University of Lüneburg. His field of research is sustainable entrepreneurship.

Astrid Emmerich works as researcher at the University of Leipzig in work and organizational psychology. Her primary research focus is authenticity as a health-promoting factor for individuals, teams, and leaders. She is a qualified systemic consultant, works as a freelancer in training and occupational health management, and has international teaching experience.

Sebastian Fischer has been a postdoctoral researcher at the Leuphana University of Lüneburg, Germany, where he also received his Ph.D. in organizational psychology. He researched the influence of situational characteristics on employees with different personalities. Situational characteristics ranged from a groups' shared climate and culture perceptions to occupational characteristics or the immigrant status of particular groups. His research focuses on explaining employee behaviors and feelings ranging from stress and withdrawal from work to innovative and proactive behaviors. Recently, he joined market research focusing on outcome perspectives of innovation rather than on processes. E-mail: Sebastian.fsc@gmail.com.

Ingrid Franz holds an M.Sc. in nutrition sciences and food economics. In her position as research associate at Leuphana University of Lüneburg, Germany, she did quantitative analysis for the Healthy at Work (GiB) project. Her primary research interest is health economics, especially international healthcare systems. She is currently writing her doctoral thesis on healthcare system performance.

Benjamin Fuchs graduated with a diploma in social sciences at the University of Erlangen, Germany (2011). He is currently working as a scientific researcher and policy consultant at the Institute for Employment Research of the German Federal Employment Agency (IAB). His primary fields of research are fertility, the sociology of health, and social policy. He uses a rigorous, quantitative approach in his research.

Burkhardt Funk is a full professor of information systems at Leuphana University of Lüneburg, Germany, and the chief information officer of the presidential board. His research interests include building decision models and support systems based on methods from machine learning, in a variety of application areas such as e-mental health and e-business. He was co-PI of the EU-funded e-mental-health project, Online Health Training, in which he was responsible for the implementation of web and mobile applications. As a regular advisor to leading European firms, he has a wealth of practical data-mining experience.

Petia Genkova is a researcher and professor of business psychology at Osnabrück University of Applied Sciences. She researches and teaches in the areas of intercultural communication and cross-cultural psychology, media psychology, lifelong learning, competence advancement and measurement, and diversity, among others. E-mail: Petia@genkova.de.

Anna Geraedts is a clinical psychologist who received her Ph.D. in psychology at Vrije University Amsterdam, the Netherlands, on the topic of e-mental health in occupational health care. She is a research consultant in e-health at ArboNed, the leading occupational health and safety service in the Netherlands. Her work focuses on the development, research, and implementation of occupational e-mental health interventions.

Ben Godde is professor of neuroscience at Jacobs University Bremen, Germany. His research focuses on the plasticity of brain and behavior over the life span. He also investigates environments and interventions that maintain and improve perceptual, motor, and cognitive functioning in older adults and which promote lifelong learning.

Anne-Katrin Haubold is professor of human resources management at the Dresden University of Applied Sciences (HTW Dresden). She holds a Ph.D. in business administration, as well as graduate degrees in social and applied psychology and business administration. She has worked as a human resources specialist in a multinational firm and as an international business consultant. Her research interest is focused on sustainable human resources practice in times of demographic change.

Elena Heber holds a diploma in psychology and works as a research associate at the Division of Online Health Training, Leuphana University of Lüneburg, Germany. Her field of expertise is research in occupational health interventions, in particular internet- and mobile-based training for work-related stress and depression. She has a particular interest in the role of health promotion in the workplace, especially in the design and delivery of health and well-being interventions for employees.

Torsten J. Holstad holds a Ph.D. in psychology and is currently a personnel development and selection consultant for the Hay Group in Frankfurt, Germany. He has worked as a researcher in the Department of Work and Organizational Psychology at the University of Leipzig. His main area of research is leadership, especially the relationship between leadership and well-being.

David Horstmann works as a research assistant and is a Ph.D. student at the Institute for Performance Management at the Leuphana University of Lüneburg, Germany. His research interests are organizational health and health-oriented leadership. He has a degree in psychology from the Universität Hamburg, with a focus on organizational psychology and personnel development.

Ariunaa Ishig graduated as a work and organizational psychologist and works on the Silqua-joint-project, "Healthy aging in the work on care of the elderly," on a salutogenic coaching program for management staff. She is a freelance behavioral coach for vocational training, a systemic consultant, and a systemic therapist.

Swen Malte John, M.D., Ph.D. is professor and the chairman of the Department of Dermatology, Environmental Medicine, at the University of Osnabrück, Germany, and of the Lower-Saxonian Institute of Occupational Dermatology. He is a dermatologist with a special interest in occupational dermatology and allergology. His research interests include the implementation and scientific evaluation of systematic, interdisciplinary approaches to the prevention of occupational skin diseases, with an emphasis on worker-education strategies ("Osnabrück model"). He is inter alia coordinator of the EADV Healthy Skin@Work europrevention campaign, an official partner of EU-OSHA's Healthy Workplaces. He has been a WHO and ILO expert in dermatological issues for several years.

Nadine Kakarot is an organizational psychologist specializing in the analysis of mental workload. She also works on the development and implementation of sustainable solutions for reducing work-related stresses and strains, e.g., by using employee-oriented methods.

Zarnie Khadjesari is a mixed-methods researcher working in the Addictions Department at King's College London. She has a background in health services research, holds B.Sc. and M.Sc. degrees from the University of York, and a Ph.D. in population health from University College London. She has worked in the field of electronic screening and brief alcohol intervention (eSBI) for 9 years and is interested in the development, evaluation, and implementation of eSBI as a means of broadening access to SBI in non-healthcare settings.

Kai-Michael Kleinlercher studied business psychology at the Leuphana University of Lüneburg, Germany. After receiving his B.A., he worked as a research assistant at the Lüneburg Innovation Incubator. He has worked on the project team for Healthy at Work (GIB), in which he worked on the development and validation of a Health Risk Assessment that aimed to focus psychological demands and resources in the organizational context.

Sabine Korek earned her Ph.D. in work and organizational psychology at the University Leipzig, Germany, where she works as a research assistant. Her research interests include leadership, work and health, gender, and science management.

Andreas Krause is professor of applied psychology at the Institute of Humans in Complex Systems at the University of Applied Sciences and Arts Northwestern Switzerland. Previously, he worked on the occupational and organizational psychology research team at the University of Freiburg. He has had a postdoctoral fellowship for Leading Early Career Researchers from Baden-Württemberg Stiftung. He completed his Ph.D. in occupational and organizational psychology at the Europa-Universität Flensburg's International Institute of Management and Economic Education (2002). Currently, his research team is studying the ambivalent effects of outcome-oriented management methods.

Lars Martin Kreissner holds a diploma in psychology from the University of Constance and currently works as a research associate at the Center for Disability and Integration (CDI-HSG) at the University of St. Gallen, Switzerland. His former positions include occupations in psychotherapeutic and consultative areas at different clinical facilities. His research focuses on workplace inclusion of people with disabilities and workplace retention of employees suffering from mental health issues.

Dany Kuntzsch received her B.A. in business administration from the Dresden University of Applied Sciences (HTW Dresden) in 2012, specializing in human resources management. Since 2013, she has been working as a research associate in the field of salutogenic-oriented HRM at the University of Applied Sciences for Social Work, Education and Nursing in Dresden. In addition, she works as a lecturer in human resources management and ergonomics at the Dresden University of Applied Sciences. Her research interests include aspects of occupational health management, in particular reintegration of employees with long-term illnesses.

Mary Kuntzsch works for the chair of Human Resources Management of Professor Haubold at the Dresden University of Applied Sciences (HTW Dresden). She holds a B.A. in business administration, with a concentration in human resources management. Kuntzsch's core competence is the conception and execution of health-oriented employee training programs. In addition, she is actively involved in planning and coordinating international M.B.A. programs at the university. Her current research focuses on coping strategies and the individual factors that govern the well-being of employees in geriatric nursing.

Dirk Lehr is a full professor of health psychology and applied biological psychology at Leuphana University of Lüneburg, Germany. He is responsible for the Division of Online Health Training at Leuphana. He holds a Ph.D. in medical psychology from Philipp University of Marburg, Germany, and he is a licensed psychotherapist. Occupational health is his main research interest, especially internet interventions for workers experiencing work-related stress, as well as traditional face-to-face stress management training.

Valentina Leier holds a B.A. in public health from the University of Bremen, Germany. Currently, she works as a researcher at University Medical Center Hamburg-Eppendorf, with the intention of pursuing her doctorate. In her former position as research associate at the Leuphana University of Lüneburg, Germany, she was an analyst in the Healthy at Work (GiB) project. Her main areas of interest concern prevention research and institutional economics.

Sina Lürßen holds an M.A. in public health, with a focus on occupational health management, from the University of Bremen, Germany (2014). Since graduation, she has worked as a research assistant in the field of occupational health management and organizational development at the Centre for Social Policy (ZeS) on the projects FUTURE: CARE (ZUKUNFT: PFLEGE) and DEMOKabine.

Friedrich Müller is professor of industrial, environmental, and perception psychology. He is an experimenter with a passion to make mental phenomena measurable: "The most important challenge for psychologists is to develop measures and methods suitable to transform so-called soft factors and phenomena into hard data."

Beate Muschalla studied social anthropology and psychology. She holds a diploma (2006) and a Ph.D. (2008) in psychology. She is trained as a behavior therapist in a rehabilitation clinic for patients with mental disorders and became a state-licensed psychotherapist in 2010. As a work-clinical scientist-practitioner, she has focused on work anxieties, vocational reintegration, and work ability since 2005. She has been working on several research projects at the Research Group Psychosomatic Rehabilitation at Charité Berlin, the Work and Organizational Psychology Department in Potsdam, and the Department of Clinical Psychology of the Freie Universität Berlin. From 2012 to 2014, she conducted her own therapy study on work anxieties in a German rehabilitation clinic.

Sabine Nitsche studied psychology at the Technical University of Berlin and specialized in organizational behavior at the University of California at Berkeley. After graduation, she pursued a master's degree in industrial relations and personnel management at the London School of Economics. She has broad international experience in human resource management, having spent several years in London, Switzerland, and Brussels. She has worked in the human resources (HR) departments of multinational organizations like the Royal Bank of Scotland, Merrill Lynch, and Toyota Motor Europe. She was also HR consultant at Hudson and an associate senior lecturer at the Business School of The Open University. She completed her Ph.D. at St. Gallen University, Switzerland, investigating the need for standardization vs. differentiation in HR policies and practices for multinational organizations operating in Europe. She was granted her first professorship in 2008 and appointed professor for HR and organization management at the University of Applied Sciences–HTW Berlin in 2010. Since 2011, she has been leading the research project InnoGema (www.innogema.de), developing an occupational health management network for small and medium-sized enterprises.

Jan Oltmanns completed his Ph.D. in psychology at Jacobs University Bremen, Germany, in 2015. Under the supervision of Ursula M. Staudinger and Ben Godde, he investigated the effects of high (versus low) work-task mobility as a dynamic characteristic of the work setting that may buffer the long-term negative effects of low-complexity work in middle-aged production workers.

Dieter Ratz holds a degree in business administration from the University of Nuremberg. He is a certified market researcher and an expert in empirical social research with broad experience in quantitative data analysis, in particular multivariate methods. Since 2007, he has been the head of M-Forschung, Hamburg, providing market research and consulting services for leading market research institutes such as GfK and Ipsos. Further, he has completed many academic projects and studies in the field of market research.

Götz Richter holds a Ph.D. in social science and works for the Federal Institute for Occupational Safety and Health (Bundesanstalt für Arbeitsschutz und Arbeitsmedizin) in Dortmund and Berlin, Germany. His major scientific responsibility is demographic change and change of work and strategic management of the Initiative New Quality of Work (INQA).

Bettina Riegel holds a B.A. in pedagogy and sports science from the University of Oldenburg, Germany, and an M.A. in sports science and nutrition from the University of Halle, Germany. She has worked on various projects in the field of occupational health management over the course of her career. Currently, she is completing a part-time master's program in general management and is employed by a consultancy that provides specialized services for healthcare management systems.

Wolfgang Ritter graduated from the University of Bielefeld, Germany, in organizational sociology and worked there for more than 5 years at the Department of Public Health. With Bernhard Badura, he developed a guideline for workplace health promotion. After completing his Ph.D., he worked at the University of Bremen as a senior researcher focusing on occupational safety and health and developing and researching organizational networks. He currently works as a manager of a social insurance association.

Wulf Rössler was full professor of clinical and social psychiatry at the University of Zurich, Switzerland, and clinical director at the Psychiatric University Hospital Zurich from 1996 to 2013. Since 2013, he has been senior professor at the Leuphana University of Lüneburg, Germany, and a professor of postgraduation at the University of São Paulo, Brazil. He is the author of about 550 publications and editor of various textbooks on emergency psychiatry, psychiatric rehabilitation, social psychiatry, prevention of mental disorders, and art therapy. He is on the scientific board of many scientific journals and the chief editor of *Frontiers in Public Mental Health.*

Silke Rupprecht is passionate about bringing mindfulness to workplace settings. For her dissertation, she researched the impact of mindfulness training on school teachers' health and quality of instruction. She delivered mindfulness therapy in a clinical setting for people suffering from depression and anxiety. At Leuphana University of Lüneburg, Germany, she was the program manager of a Germany-wide research initiative aimed at improving school quality through enhancing teacher and student health. She currently lives and works in London.

Isabella Schimitzek has worked at the APOLLON University of Applied Sciences, Bremen, Germany, as a scientific employee since April 2014. Previously, she worked as a scientific employee at the Centre for Social Policy Research (ZeS) at the University of Bremen on the project, "Network structures in ambulant care." Since 2014, she has been working on her Ph.D. dissertation entitled, "Care and supply structures in ambulant care," at the University of Bremen.

Britta Schmitt-Howe, Ph.D. is a sociologist working in the fields of organizational sociology and international relations. She is a quality-management consultant and expert in evaluation. Since 2008, after years of working on integration of quality and OSH management systems at the statutory accident insurance and prevention association for the German health and welfare sector (BGW), she has been providing global evaluation for the Joint German OSH Strategy. Currently, she conducts the research project, "Variables that promote and impede implementation of OSH measures and their relationships with risk-related characteristics of enterprises," for the Federal Institute for Occupational Safety and Health (BAuA) in Berlin.

Caius Solovan, M.D., Ph.D. is board certified in dermatology-venereology and head of pathology at the University Clinic for Dermatology and Venereology, Timisoara, Romania, with a major interest in immuno-inflammatory skin disease (psoriasis, bullous dermatosis) and skin cancer (cutaneous lymphoma).

Erika Spieß, Ph.D. (1987), has been associate professor at the chair of Economic and Organizational Psychology of Professor Brodbeck since 2007. From 1996 to 2006, she held various assistant and visiting professorships; from 1981 to 1996, she was on the research staff at the chair of organizational and industrial psychology

(Professor Lutz von Rosenstiel) at LMU Munich. Her Habilitation was in 1996. From 1973 to 1980, she studied psychology at the University of Heidelberg and LMU Munich. Her research focuses on cultures of learning and cooperation, intercultural behavior in business contexts, well-being-oriented leadership, and social support in organizations.

Peter Stadler, Ph.D. has been working at the Bavarian Health and Food Safety Authority (LGL) since 2001. His main emphasis is on work-related psychological stress. From 1993 to 2001, he was the manager of the Institute of Applied Psychology D & T. From 1985 to 1993, he was an operative in the organizational department of a Bavarian bank. He studied psychology at the University of Regensburg and LMU Munich and holds a Ph.D. from the University of Darmstadt. His research focuses on health-related topics like psychological workload, health-promoting leadership, burnout, bullying in the workplace, psychological load and back pain, and the psychology of work safety.

Ursula M. Staudinger is the founding director of the Robert N. Butler Columbia Aging Center and professor at Columbia University in New York City. She is also vice president of the German National Academy of Sciences Leopoldina. She is a psychologist, aging researcher, and a leader in the field of life span psychology. Her work focuses on the positive plasticity of human aging, in particular in cognitive and personality functioning. She investigates the cumulative effects of work on human aging as well as the development of life insight and wisdom.

Markus Steiner holds a Ph.D. in occupational and environmental medicine and joined GO Health Service, the occupational health service of the NHS Grampian, part-time in 2010. He also works as a postdoctoral clinical research fellow at the University of Aberdeen Department of Child Health on projects which use linked, routinely collected health data. He studied medicine in Homburg/Saar and Hamburg, where he worked for many years for NORDIG, Institute for Health Research and Prevention in occupational and environmental epidemiology. His other work has included clinical and research positions at the University of Hamburg, Institute for Occupational Medicine, project management at the medical advisory board of statutory health insurances (MDK) Berlin-Brandenburg, and researcher at the German Environmental Agency (UBA) and the University of Edinburgh, where he also completed an M.Sc. in epidemiology. Further, he was visiting research fellow and scientific lead for the Healthy at Work (GiB) project at the Leuphana University of Lüneburg, Germany.

Christiane Stempel holds a diploma in psychology and currently works as a lecturer and Ph.D. researcher at the Department of Work and Organizational Psychology at Leipzig University, Germany. Her former projects include the international research project, Rewarding and Sustainable Health Promoting Leadership (ReSuLead), and the educational Erasmus Intensive Programme: Shift-it!

Shaping Healthy Work in Future. Her main areas of interest are leadership, gender, and health.

Katherine Targett is a consulting occupational physician with GO Health Services, NHS Grampian, and honorary lecturer with the Department of Medicine and Therapeutics, University of Aberdeen. She studied biochemistry at Leeds, medicine at Oxford University, and trained in respiratory medicine before commencing specialist training in occupational medicine. Her research interests include the impact of chronic diseases on employment. She has an active teaching portfolio, is associate editor for *Occupational Medicine* and *Journal of the Society of Occupational Medicine*, and sits on the faculty of the Occupational Medicine Workplace-based Assessments Advisory group.

Rüdiger von der Weth received his master's degree equivalent diploma in psychology from the Technical University Berlin in 1985. His Ph.D. thesis was on goal conflicts in complex management tasks (1989). He became a full professor of applied psychology at the University of Applied Sciences Stuttgart (HTW Stuttgart) in 2000 and professor of human resource management and ergonomics at the Dresden University of Applied Sciences (HTW Dresden) in 2003. His research focuses on the role of motivational and emotional aspects in technical and organizational innovation processes, planning and decision making in professional contexts (industry, public service, health care, agro forestry), development of methods for participation in industrial, public, and political decision making, training for project managers, and evaluation of the effects of training. He also works as a consultant and management trainer. He has published three scientific books and 76 papers in international and German journals and publications. He works as a scientific reviewer for several international research journals, for example, *Research in Engineering Design* and *Learning and Education,* and was a member of the management committee of the research network, Human and Organizational Factors in Planning and Scheduling (HOPS).

Harald Walach is professor and the director of the Institute of Transcultural Health Sciences at the European University Viadrina in Frankfurt (Oder), Germany, where he also coordinates a postgraduate training program for doctors in complementary medicine and cultural studies. He holds a double Ph.D. in clinical psychology and the history and theory of science. He has authored more than 100 peer-reviewed papers, several books, and many book chapters. From 2005 to 2009, he worked with the University of Northampton, UK, where he established and directed the M.S. program in transpersonal psychology and consciousness studies. His research interests are in the methodology and evaluation of complementary medicine and the impact of consciousness.

Patricia Weinert, M.Sc. is a social scientist who has been working for many years as senior project manager at the International Social Security Association (ISSA) in Geneva, Switzerland. She was responsible for ISSA's worldwide prevention program and acted as ISSA's coordinator of the World Congress on Safety and Health at Work, which is organized every 3 years together with the International Labor Organization (ILO). She is presently working as project coordinator at the Institute for Interdisciplinary Dermatological Prevention and Rehabilitation (iDerm), which is one of the biggest research institutions for the prevention of occupational skin diseases worldwide. She is responsible for the coordination of European and international projects.

Markus Wiencke holds a Ph.D. in psychology and an M.A. in social and cultural anthropology from Freie Universität Berlin. He has training as a systemic therapist and counselor. In his work at the Federal Institute for Occupational Safety and Health (BAuA) in Berlin, he focuses on structures and strategies for health and safety at work. Previously, he was visiting research fellow for the Healthy at Work (GiB) project at the Leuphana University of Lüneburg, Germany. Further, he has worked as the therapeutic head of ward for a psychiatric clinic and as a management consultant. He has written on the social and cultural dimensions of health (promotion), based on his fieldwork in Germany, Brazil, Chile, India, and Tanzania. E-mail: markuswiencke@gmx.de.

Michaela Wieser holds a B.A. in media and communication from the University of Passau, Germany.

on output but can also lead to long absences and even early retirement. Muschalla identifies work-related anxieties among people participating in medical or vocational rehabilitation programs whose return to the working world is particularly difficult. For the author, the concept of work-related anxiety provides a tool for developing a preventative design focused on matching personal skills and job requirements in the workplace.

Oltmanns, Richter, Godde, and Staudinger expand on these considerations in their chapter on the consequences of shifting demographics for the labor market. Drawing on the results of two research projects, the authors introduce the concepts of *person–environment fit* and *work–task mobility*. Person–environment fit refers to the fit between people (i.e., their individual skills, behaviors, goals, and attitudes) and the demands of a workplace environment (e.g., job profiles, requirements, support structures, and organizational culture). Work–task mobility refers to the repeated shifts in tasks an employee must complete while maintaining a constant level of quality. The authors discuss the two concepts as important management tools in developing systemic and dynamic workplace health management structures that preserve an employee's health and ability to work throughout the span of their working lives.

2.2 Part II: The Importance of Leadership for Occupational Health

Workplace health is a resource—a valuable asset—and therefore warrants the highest concern in the top organizational levels of an enterprise, the stewardship or leadership level. Leadership affects employee health and well-being in many ways, ranging from physical and mental health and health behavior to job satisfaction and well-being. In short, as Boehm, Baumgärtner, and Kreissner state in their opening contribution, *"good" leadership* keeps employees healthy and happy. This statement, however, prompts the following questions: What is good leadership? What responsibilities do supervisors have for the health of their employees? How can health-promoting leadership be conceptualized and implemented? What criteria might the scientific community use to measure health-promoting leadership behavior?

Starting with what leaders should not do, Boehm and co-authors emphasize avoiding harmful influences on employees' health, such as constant work overload, frequent overtime, and a paucity of opportunities to recover. Boehm et al. compare traditional and *health-oriented leadership styles* and consider which leadership style is best for a health-promoting environment. From a theoretical perspective, the authors hypothesize that it is beneficial if health-focused leadership styles amend or even replace established approaches. However, caveats are that the extent to which health-oriented leadership styles contribute to the desired outcome is unproven and the debate over whether health or well-being is the most suitable outcome indicator is still unsettled. The authors thus suggest more research to establish sound relationships between leadership styles and desired outcomes.

According to Spiess and Stadler, it is commonly accepted that an organization cannot achieve long-term success without healthy employees. Leaders are in charge of providing job resources like autonomy, social support, meaning of work, a climate of fairness, and appreciation, but health promotion has not yet been fully recognized as management's responsibility. As a consequence, such promotion has not been widely implemented. Building on an established *model of organizational health*, Spiess and Stadler base expectations on leadership and develop a four-level model to answer the question of what makes employees healthy. Their answer comprises several elements, such as participatory models of work, appreciation by leaders, and a deep anchoring of health promotion in the work process.

Although employees give rather mediocre ratings to leadership behavior related to health and safety (DGUV 2015),[1] good news comes from the contribution authored by Holstad, Emmerich, Stempel, and Korek. They posit that *health-promoting leadership* can be taught. Learning is possible if the top management provides supplementary resources and support. Interestingly, Holstad et al. emphasize that health-oriented leadership is not a one-way road; leaders may promote followers' well-being and *vice versa*. However, the authors' contribution highlights a data gap that needs to be addressed. On the practical side, the authors state that workplace health needs to be firmly integrated into work routines and requires support from the top organizational level.

Leadership in times of rapidly changing economic environments is the focus of Eller, Fischer, and Diedrich's contribution. Uncertainty caused by a rapidly changing work environment has a major, negative impact on employees' health, and under these circumstances, another facet of leadership quality comes to the fore. Providing stability in turbulent times promotes well-being and contributes to maximizing employee performance. This chapter also presents the idea that leadership and *sensemaking*—as the authors call this particular leadership skill—can be taught in order to foster this core competency in an organization.

Focusing on nursing homes, Horstmann and Eckerth shed light on a sector in which leaders face extreme challenges. While providing health care themselves, employees in this sector are under particular strains caused by heavy physical workloads, legal requests, and compliance with documentation requirements. In addition, nursing homes are frequently SMEs dealing with skilled labor shortages, limited personnel and financial resources, and little capacity to develop leadership skills. According to the authors, leadership styles need to address *sector-specific* challenges by creating customized concepts, if possible within a broader network. Interestingly, the authors also emphasize that leaders are role models for their subordinates and thus call for them to be aware of their own resources and needs.

[1] DGUV (=Deutsche Gesetzliche Unfallversicherung) 2015: Employees give a 3+ in leadership behavior and working atmosphere. The survey shows deficits in occupational safety and health. http://www.dguv.de/de/mediencenter/pm/Pressearchiv/2015/quartal_3/details_q3_113985.jsp. Accessed 10 October 2015 (in German).

2.3 Part III: Public and Private Policies Promoting Occupational Health

John and Weinert open the next part of the volume with their perspective on innovation through cooperation between researchers and policy-makers: the creation of prevention awareness and the correlation of policies and programs to specific organizational contexts. Their case study dealing with occupational skin diseases provides valuable general insights into the *development and implementation of prevention programs*. The authors show how scientific knowledge can be used to raise awareness for prevention. Cooperation between important stakeholders and policy-makers make it possible to design preventive measures and better context-specific dermatological interventions.

A core motif in this part of the book is the area of conflict between public and private occupational safety and health policies and the dynamic occupational contexts, in which they are implemented. The chapters provide insights into the logic and methods of contextual adaptation processes of health policies and the possibilities for influencing them. In doing so, different forms of cooperation are described through which new occupational safety and health measures can be integrated into existing practices.

Steiner and Targett describe how occupational health and safety has been integrated into the British public health system. The example of a regional provider in Scotland can be read as a heuristic model for how an *integrated prevention and care package* might be structured. New forms of access for occupational health services thus emerge, particularly for SMEs.

In her contribution, Nitsche presents a private *regional network that aids SMEs* in implementing occupational health management structures. She illustrates the potential benefits for sustainable health promotion when the synergies created by cooperating stakeholders are accessed through a moderating support system. In addition, the chapter shows how the challenges of transferring a network model from a densely populated city (Berlin) to a rural region (Brandenburg) might be met.

Ritter, Schimitzek, and Lürssen present the project Zukunft:Pflege (Future:Care) in their article. The network, which ties together SMEs in the fields of care, education providers, and outpatient care associations, illustrates how the challenges of demographic change can be met on a regional level. The authors describe how the cooperation arrangement can stimulate *interactive learning processes among network partners*. From the perspective of the coordination team they show how—despite competition between participating SMEs—interactive learning processes are enacted based on the common goal of maintaining the long-term employability of the workforce.

The issue of what SMEs want in terms of occupational health and safety support is addressed by Cacace, Franz, Braun-Beustrin, and Ratz from a health economics perspective. Via an *adaptive choice-based conjoint analysis model* combined with a short willingness-to-pay questionnaire, they determine the preferences of SMEs in a rural region. The authors find that occupational health services should ideally be individually tailored to companies and billed on a pay-per-use basis. Cooperation with social insurance providers appears to make sense for ensuring access, especially for SMEs.

With Schmitt-Howe, we delve deeper into the perspectives of the enterprise. From a sociological perspective and with a mixed-method approach, she examines the *attitudes of entrepreneurs* towards occupational safety and health. The point of departure for the study is the discrepancy found elsewhere between public policies related to occupational safety and health and their actual implementation. Against this background, Schmitt-Howe examines *attitudes and communication* patterns in enterprises and how they might affect occupational safety and health outcomes.

In his contribution, Fuchs examines the tension between public and private policies and business practices. He begins with the current scientific discourse on the links between *parenting and health in the working population* in the United States and then introduces a sociological model for examining the correlations in more detail. A variety of recommendations emerge for how scientific approaches might benefit the development of policy.

2.4 Part IV: Evidence-Based Studies and Interventions Promoting Occupational Health

A plethora of ways exist through which health at work can be examined and explained. In the last part of the book, we present a number of approaches for assessing employee health and reducing illness.

Lehr, Geraedts, Asplund, Khadjesari, Heber, de Bloom, Ebert, Angerer, and Funk summarize the growing area of *internet-based interventions* related to occupational health and present some of their own interventions. Internet interventions may provide accessible low-threshold help for those in need when they need it. The authors especially elucidate differences between traditional stress management training and internet interventions and provide insight on the strengths and weaknesses of internet-based interventions. From this foundation, the authors develop an agenda for future research on internet interventions, which calls for expanding these interventions in the field of mental health and incorporating additional concepts as they become established in occupational health programs.

Fischer, Kleinlercher, and Rössler develop and test an *online tool for assessing work-related antecedents for mental health.* Going beyond traditional person-based approaches to mental illness, their model includes organizational influences that may affect the mental health of people at work. The tool can provide valuable insights for reducing mental health problems within organizations, especially for groups of people facing extreme stressors, and for lowering barriers to mental health that exist within the culture of a firm. Thereby, it enables organizations to develop interventions according to their specific needs.

Nurses in the geriatric sector are an employee group that contends with both physical and psychological stressors. Haubold, Kuntzsch, Beckmann Ishig,

Kuntzsch, Christa, and von der Weth describe the development of a *training program* specifically aimed at reducing the strain on these employees, with the *long-term goal of reducing health-related early retirement.* Their approach is based on Antonovsky's salutogenesis concept, an approach that builds on individuals' resources to cope with a stressor, which has only recently been applied to the health of caregivers.

Aside from building the resources necessary to handle stressors at work, mindfulness, the ability to voluntarily direct attention to the present moment in an accepting and nonjudgmental manner, focuses on coping with stress by changing the perception of the stressor. Rupprecht and Walach provide a *map of mindfulness training in the workplace*. They outline various training formats and review recent empirical findings on the effectiveness of mindfulness training for different types of employees. Thereby, they underscore the importance of the mindfulness concept for workplace health interventions.

The next chapters shift focus to interventions for better physical health. Cacace, Riegel, and Draht shed light on *interventions for employees with work-related skin problems*. Employee groups such as hairdressers, florists, and so forth frequently work in wet or damp environments or with strong chemical substances that lead to skin problems. Training on handling such environments or substances can prevent skin problems, but they are difficult or expensive to administer, especially in rural areas and in sectors that are dominated by SMEs. The authors have developed cost-effective training tailored to specific industries and have evaluated their effectiveness, both in terms of reduced skin problems and in terms of costs per participant.

In a similar direction, Chiriac and Solovan focus on who will be affected by skin problems while working in a high-risk industry. They investigate the *prevalence of skin problems in specific occupational groups* to identify associations between the skin problem and the occupation. However, few associations could be found for the skin problems they investigated, leading the authors to conclude that skin diseases may be triggered by occupational factors only for those susceptible to the disease.

Müller and Kakarot explore the challenge of an ageing workforce and keeping older employees in good health, which is difficult in sectors characterized by heavy manual labor. In a sewage company, Müller and Kakarot installed a *program to measure physical work demands* and the physical and mental efforts needed to meet those demands. They then investigated how age influences perceived demands and efforts for various heavy tasks. Through this program, the company was able to align demands of a task with the required effort and ability needed for a person to accomplish it. The program may be used to build a healthy workforce even among older employees.

Finally, Genkova and Wieser shed light on another group under pressure at work, working mothers. In their study they find that a *work–life balance* is valued by working mothers in both the United States and Germany. However, when it comes to the burden of having to work, a mother's age seems to have an impact. The older the mother, the fewer the impacts of the work burden and having a child. The authors discuss the strengths and limitations of their approach.

Acknowledgements This book emerged as a result of the research project, Healthy at Work—Prevention and Treatment of Occupational Diseases of the Innovation Incubator/Leuphana University Lüneburg. The project was generously funded by the European Union and the federal state of Lower Saxony. We thank those who have successfully applied for this funding and provided guidance and advice: Wulf Rössler, Nils Ole Oermann, Matthias Augustin, Björn Maier and Swen Malte John.

We would like to thank Patricia Weinert for heading the conference organizing committee with passion and excellence. We would further like to thank here all our colleagues for their continuous support (in alphabetical order): Kristin Bahr, Daniel Braun-Beustrin, Björn-Helge Busch, Laura Diedrich, Fokko Eller, Ingrid Franz, Sabine Hirth, Debra Hopkins, Wolfram Kawohl, Kai Kleinlercher, Jürgen Klose, Valentina Leier, Hannah Luedtke, Brita Müller-Kanneberg, Bettina Riegel, Lars Rusch, Stefanie Scheske, Markus Steiner, Yvette Tasto, Nazan Ulusoy, Silke Volkhardt, Max von Soest, Patricia Weinert, Anita Wiemer, and Susanne Wöhe.

References

Ahola K, Honkonen T, Virtanen M, Aromaa A, Lönnqvist J (2008) Burnout in relation to age in the adult working population. J Occup Health 50(4):362–365

BAuA (2015) Volkswirtschaftliche Kosten durch Arbeitsunfähigkeit 2013. [Economic cost of absenteeism 2013]. www.baua.de. Accessed 10 Oct 2015 (in German)

DGUV (2015). Online available http://www.dguv.de/de/mediencenter/pm/Pressearchiv/2015/quartal_3/details_q3_113985.jsp. Accessed 10 Oct 2015 (in German)

Eurofound (2010) Absence from work. Eurofound EurWORK European Observatory of Working Life

Hasle P, Limborg HJ (2006) A review of the literature on preventive occupational health and safety activities in small enterprises. Ind Health 44(1):6–12

Lansdown TC, Deighan C, Brotherton C (2007) Health and safety in the small to medium-sized enterprise: psychosocial opportunities for intervention. Health and Safety Executive, London

Maslach C, Schaufeli WB, Leiter MP (2001) Job burnout. Annu Rev Psychol 52(1):397–422. doi:10.1146/annurev.psych.52.1.397

Wachter JK, Yorio PL (2014) A system of safety management practices and worker engagement for reducing and preventing accidents: an empirical and theoretical investigation. Accid Anal Prev 68:117–130. doi:10.1016/j.aap.2013.07.029

Walters DR (1996) Health and safety strategies in Europe. J Loss Prev Process Indust 9 (5):297–308

Wittchen HU, Jacobi F, Rehm J, Gustavsson A, Svensson M, Jönsson B, Olesen J, Allgulander C, Alonso J, Faravelli C, Fratiglioni L, Jennum P, Lieb R, Maercker A, van Os J, Preisig M, Salvador-Carulla L, Simon R, Steinhausen H-C (2011) The size and burden of mental disorders and other disorders of the brain in Europe 2010. Eur Neuropsychopharmacol 21(9):655–679. doi:10.1016/j.euroneuro.2011.07.018

Wulfhorst B, Bock M, Skudlik C, Wigger-Alberti W, John SM (2011) Prevention of hand eczema: gloves, barrier creams and workers' education. In: Johansen JD, Frosch PJ, Lepoittevin J-P (eds) Contact dermatitis. Springer, Berlin, pp 985–1016

Part I

Organizational Culture and Occupational Health

Organizational Culture and Health

Mats Alvesson

Abstract
This chapter provides a brief introduction to the concepts of culture, particularly organizational culture; the latter emphasizes the meanings, orientations, and symbolism shared by people in organizational settings. We argue that many health issues can be productively conceptualized as cultural phenomena. Health is understood in a variety of ways in various societies, organizations, occupations, and other groups, with correspondingly varying implications for action. The cultural meanings of health guide the ways in which organizations, managers, and professionals try to influence people and improve how they deal with health issues. We discuss the expansion of the meaning of the term "health" and a change in its focus to include "right" attitude to work; these changes have resulted in "health" being incorporated into the sphere of well-being by managerial and professional regimes. Here it is important to consider both innocent well-intentioned activities and issues of power and control.

1 Introduction

Organizational culture is one of the major issues in academic research and education for both organization theory and management practice. There are good reasons for this: the cultural dimension is central to all aspects of organizational life. Even in organizations in which cultural issues receive little explicit attention, how people in a company think, feel, value, and act are guided by ideas, meanings and beliefs of

M. Alvesson (✉)
Lund University, Lund, Sweden

University of Queensland, Brisbane, QLD, Australia

Cass Business School, London, UK
e-mail: Mats.Alvesson@fek.lu.se

M. Wiencke et al. (eds.), *Healthy at Work*, DOI 10.1007/978-3-319-32331-2_2

a cultural (socially shared) nature. The significance of culture is not reduced by whether managers think that culture is too soft or too complicated to bother about or whether there is no unique corporate culture. Senior organizational members are always, in one way or another, managing culture; that is, underscoring what is important and what is less so and framing how the corporate world should be understood. Organizations practicing intensive numbers management may develop and reproduce cultures that celebrate performance indicators and rituals around the handling of these. In most contemporary organizations, corporate culture receives a lot of attention and is seen as crucial. A key concern is that "culture management aspires to intervene in and regulate being, so that there is no distance between individuals' purposes and those of the organization for which they work" (Grey 2005, p. 68). However, all people in organizations are involved in the production and reproduction of culture at the level of the work-place.

This chapter provides an introduction to organizational culture and points to its significance for illuminating health at work. It thus does not concern health as such, or health issues as 'objective facts', but rather cultural meanings around health issues, how health is understood and how—if at all—it should be managed and, if so, by whom. The chapter builds on a cultural approach to work and organizations developed in Alvesson (2013a) and takes seriously some key insights of critical studies (e.g., Alvesson and Willmott 2012) to illuminate the sometimes not entirely innocent or unproblematic meanings of "health" at work.

2 The Concept of Culture; in Particular, Organizational Culture

A glance at just a few works that use the term "organizational culture" will reveal enormous variation in the definitions of this term and even more in the use of the term "culture". Even in anthropology, "culture" has no fixed or broadly agreed meaning (Borowsky 1994; Ortner 1984) and there are also variations in its use in published material on organizational culture. These variations are partly related to considerable differences in the purposes and depth of these books and articles. Additionally, the broad variety of scientific disciplines and research orientations involved in organizational culture studies makes the field very heterogeneous. The concept of culture seems to lend itself to very different uses; for example, as collectively shared forms of ideas and cognition, as symbols and meanings, as values and ideologies, as rules and norms, as emotions and expressiveness, as the collective unconscious, as behavior patterns, structures and practices and so on, all of which may be the targets of studies. Of course, culture is not unique in this way; most, if not all, significant concepts in organization studies and social science tend to be associated with a variety of different meanings and definitions (Palmer and Hardy 2000).

Culture is a tricky concept because it can easily be used to cover everything and consequently nothing. The fact that certain researchers are interested in "culture", or at least use this term, does not mean that they have very much in common.

"Culture" frequently seems to refer to little more than a social pattern; that is, it refers to surface phenomena rather than exploring the meanings and ideas behind them. In many cases it would be advisable to abandon the term in favor of terms like "informal behavior patterns", "norm system", or simply "social pattern". Many people referring to culture seem to do so in a very vague way; it is important to use the concept without losing focus, direction, and interpretive depth (Geertz 1973).

Many academics, including the author, view culture broadly as a shared and learned world of experiences, meanings, values, and understandings that inform people and are expressed, reproduced, and communicated partly in symbolic form (Alvesson 2013a). This view is consistent with a variety of approaches to the conduct of concrete studies.

The phrase "organizational culture" as an umbrella term for a way of thinking that takes a serious interest in cultural and symbolic phenomena and focuses on shared meanings. This term directs the spotlight in a particular direction rather than mirroring a concrete reality for possible study. I agree with Frost et al.'s (1985, p. 17) definition of organizational culture; namely, "Talking about organizational culture seems to mean talking about the importance for people of symbolism—of rituals, myths, stories and legends—and about the interpretation of events, ideas, and experiences that are influenced and shaped by the groups within which they live". In this chapter, I will also take organizational culture to include values and assumptions about social reality; however, for me values are less central and less useful than meanings and symbolism in cultural analysis. This position is in line with the view broadly shared by many modern anthropologists (especially Geertz 1973). Culture is thus understood to be a system of common symbols and meanings. It provides "the shared rules governing cognitive and affective aspects of membership in an organization, and the means whereby they are shaped and expressed" (Kunda 1992, p. 8).

Culture is not primarily inside people's heads, but somewhere between the heads of a group of people in settings where symbols and meanings are publicly expressed; for example, in work group interactions and board meetings as well as in material objects. Being the meaning aspect of what is being socially expressed, it is simultaneously visible and invisible. It is the meanings behind what is being expressed (materially, in actions, or in verbal and textual communication).

Culture then, is central to governing an understanding of behavior, social events, institutions, and processes. Culture is the setting in which these phenomena become comprehensible and meaningful. It is important not to overemphasize the static elements of culture: even if tradition, framework, rules, and fairly stable meanings are part of the picture, culture is not best understood as a homogenous, cohesive and causal force, but rather as something that people do, it is emergent, dynamic, situationally adaptive, and co-created in dialogue (Heijes 2011). This is true partly because individuals and groups are meaning-seeking entities and partly because a multiplicity of complex meanings are put into motion in specific settings and interactions. Meanings need to be negotiated and processed, not rigidly applied. For example, the term "managerial authority" does not have a static and uniform meaning in the work-place; rather, the cultural context provides a framework for its negotiation, aiding interactions (Lundholm 2011). Are managers clearly superior figures, people you are supposed to have deep respect for, or are they more accurately perceived as equals? Culture helps to answer this type of question on a general organizational level and offers some framing and reduction of uncertainty regarding specific relationships between individual managers and subordinates (if the term subordinates is now accurate).

3 Key Concepts of Culture: Symbols and Meanings

Of the several concepts significant for cultural understanding—including assumptions, beliefs, ideas, rites, rituals, myths, identity, and values—symbols and meanings are clearly the most significant ones.

Meaning refers to how an object or an utterance is interpreted. It points to what something is seen to be standing for. Meaning has a subjective referent in the sense that it appeals to an expectation, a way of relating to things. Meaning makes an object or phenomenon relevant and meaningful.

In a cultural context it is always socially shared meanings that are of interest, highly personal meanings less so. The scope of the social may vary from a work group to an entire organization or a profession, industry, society or even a civilization (e.g., Muslim world, the West). Individuals may be more or less bound by authority and obey rules or they may dislike and rebel against bureaucracy; they may see rules as indicators of order and rationality or as straitjackets, obstacles to the exercise of judgment and responsibility. Individual meanings are certainly important and may vary considerably within a group; however, cultural understandings concentrate *not* on individual idiosyncrasies but on shared orientations within an organization or other group. Even though people in work and other contexts always have their own idiosyncrasies and, as expressed by Starbuck (2010, p. 1398) "everyone's perceptions blend prior beliefs with new observations", idiosyncrasies are reduced and perceptions and beliefs become more socially homogenous (less heterogeneous) through culture. Thus, through common frameworks, values, and definitions of reality, a shared sense of reality is created.

A *symbol* can be defined as an object—a word or statement, type of action or material phenomenon—that stands ambiguously for something else or something more than the object itself or both (Cohen 1974). A symbol is rich in meaning, condensing a complex set of meanings in a particular object, thus communicating meaning economically. The complexity of a symbol and its meaning occasionally require considerable interpretation and deciphering. People have private symbols; however, in organizational contexts collective symbolism is of greater interest. In this sense "health" is a theme with a strong symbolic meaning. It is not just a matter of the absence of sickness and suffering, but relates to meanings such as care, responsibility, morality, employer's concerns, lifestyle, social status and so on.

When thinking about culture it is important to bear in mind what culture is *not*, that is, what a cultural perspective does not focus on. Making a distinction between culture and social structure is helpful here. Culture is regarded as a more or less cohesive system of meanings and symbols in the context of which social interactions take place, whereas social structure is regarded as the behavioral patterns engendered by social interactions. Thus, in the case of culture, we have a frame of reference of beliefs, expressive symbols, and values, by means of which individuals define their environment, express their feelings and make judgments. At the social structural level, we have a continuous process of interaction. As Geertz (1973, p. 145) states, culture is the creation of meaning through which human beings interpret their experiences and guide their actions, whereas social structure is the form that actions take or the network of social relationships that actually exist.

4 Organizational Culture Between Macro and Micro

The term organizational culture is frequently used to indicate a view of organizations as typically unitary and unique, characterized by stable sets of meanings. Most organizations are then viewed as mini-societies with distinct sets of meanings, values, and symbols that are shared by, and unique for, the majority of the people working in the organization. Culture then has an integrative community-creating function, reducing fragmentation and variation within an organization.

This view fits nicely with interest in using corporate culture as a tool for increasing integration and performance and for promoting leadership ideas and actions that have a broad impact. It is definitely easier to make a strong case for management and leadership putting imprints on the organization as a whole, if the latter is, at least in some key respects, fairly homogenous. The organizational culture concept has a stronger appeal if there is something unique to point at. Being similar to other organizations in the industry is not attractive. Most people, both individually and in organizational contexts, like to emphasize their distinctiveness (identity). However, viewing organizational culture as unitary and unique is problematic in several ways. It can be challenged with arguments both from below and from above. Challenges from below emphasize the pluralism of organizations: different groups develop different outlooks on the world. This is often referred to as organizational subcultures. Challenges from above point to the powerfulness of ideas, values, and symbolism that are shared by broader groups of people, associated with civilizations, nations, regions, industries, and occupations. Taken together, this means that both local and more macro contexts need to be considered to understand cultural manifestations at the organizational level.

A strong case can be made that societies—nations or groups of nations with similar characteristics—put strong imprints on organizational cultures. The concept of an *industrial subculture* draws attention to the fact that culture corresponds most fully to a society and that the sphere of industry includes "a distinctive set of meanings shared by a group of people whose forms of behavior differ to some extent from those of the wider society" (Turner 1971, p. 1). Turner's point of departure is his experience that when "moving from one industrial organization to another, it is possible to observe certain similarities" that differ from behavior elsewhere in society. Here, the entire industry is conceptualized as a subculture.

Individual organizations then may appear as sub-sub-cultures. Drawing attention to the cultural context of the focal object encourages a broader view of it.

Much interest has focused on national cultures, about which there is a rich body of studies that are beyond the scope of this chapter. It is important to bear in mind that there are problems associated with studying something as vague, large-scale, and perhaps only superficially homogenous as national cultures. Additionally, the idea that these are unitary, static and measurable is questionable. See, for example, McSweeney's critique (2002) of the highly influential work of Hofstede (1980). Other macro-level foci may include societal trends and dynamics, including the effects of consumerism, hedonism, narcissism, changes in work morality, and disinclinations to submit to authority in many countries (Alvesson 2013b). These are not merely external, society-level phenomena but are also (re-) produced at work-place levels. Rather than, or at least in addition to studying culture as such on different, neatly separated levels, it is preferable to look at the interplay between overall changes and what seems to happen at the local, organizational level, for example, in terms of reception, transformation, and resistance.

The so-called institutional theory school emphasizes isomorphism, a trend implying that organizations are becoming increasingly alike. Patterns of meanings are imported from various entities outside individual organizations, having broad and fairly homogenous impact across organizations. For reasons of legitimacy and adapting a cognitive view of the social world as ordered and comprehensible, people in organizations are sensitive to the meanings, ideas, and definitions of what is natural, rational, and good developed by various institutions, such as professions, state agencies, science, management consultants and so on (Scott 1995). Creation of local meaning is thus seen as less relevant, as it is assumed to exhibit fairly little variation. Instead, the overall level driving organizations to conformity is viewed as more crucial to understanding organizational reality. From a cultural point of view—at least as it is used in the present book—institutional theory tends to embrace a rather crude idea of meaning. Institutional theory illuminates structural arrangements (organizational forms, techniques, policies) associated with fairly standardized meanings and constructions rather than the more nuanced and specific meaning-creating processes and symbolism. The latter are observed much more closely and richly by organizational cultural studies. Institutional theory does not emphasize interpretive and cultural depth and, as a consequence, may overemphasize homogeneity and conflate meaning. Nevertheless, it draws attention to valuable macro aspects of meaning creation and the numerous studies concerning this clearly indicate a need to go outside the individual organizational level and consider macro aspects in operation.

In contrast to the macro view, a micro or *local* perspective typically views an organization as the macro context and cultures within it as the more important phenomena. Although such a local perspective may use the term "subculture", its proponents frequently refrain from explicitly or implicitly giving priority to the organizational level, that is, from defining "sub" in relationship to that level. Local views are closely associated with an emphasis on work context and social interaction; however, authors also point to other origins or vital mechanisms for the development of cultural orientations within a specific group of people.

In an influential local approach to culture, Van Maanen and Barley (1984, 1985) call attention to the existence within organizations of groups composed of individuals with different backgrounds and professional affiliations yet have high degrees of internal interaction; such groups consequently share very little between them. Van Maanen and Barley (1985) argue that "unitary organizational cultures evolve when all members of an organization face roughly the same problems, when everyone communicates with almost everyone else, and when each member adopts a common set of understandings for enacting proper and consensually approved behavior" (p. 37). These conditions are, of course, rare. These researchers emphasize subcultures created through organizational segmentation (division of labor hierarchically and vertically), importation (through mergers, acquisitions, and the hiring of specific occupational groups), technological innovation (which creates new group formations), ideological differentiation (e.g., when some people adopt a new ideology of work), counter-cultural (oppositional) movements, and career filters (the tendency for people moving to the top to have or develop certain common cultural attributes) (pp. 39–47). When addressing subcultures, it is important not to proceed from "sociological fact sheet" categories like gender, class, and ethnicity, but to be open for what may be relevant in specific organizations. In a case study of organizations, Parker (2000) found three major sources of differentiation: spatial/functional (associated with geographical location and work function), generational (connected to age and length of time in the organization) and occupational/professional.

When considering cultural phenomena in organizations, it is thus important to consider the organizational level, but also seriously consider how macro cultures and internal differentiation create subcultures and sometimes clashes and ambiguities of meanings between various groups.

5 Culture and Health

The relevance of organizational culture is obvious when it comes to a wealth of health and security-related issues. The broad "wellness movement" (Dale and Burrell 2014) is a cultural phenomenon that influences understandings among a variety of groups and organizations. Issues around risk-taking, precautions, support, and recognition of vulnerabilities are very much a matter of cultural understandings that are more or less shared by people at work: how people relate to themselves and their work and what is defined as natural, reasonable and good. This is nicely illustrated by a case study of an oil platform on which the number of accidents and injuries was dramatically reduced over a brief period. This mainly resulted from a cultural change in which the oil workers redefined their views of the work-place and their relations (Ely and Meyerson 2010). The meanings of gender—rather than sex—seem to be vital here (Alvesson and Billing 2009; Gherardi 1995). Prior to the cultural change, the dominant view was that competent workers were masculine, individualistic, tough, efficient, and inclined to get on with their work. Being overly cautious or worrying too much was viewed as bad. Partly as a result of managerial initiatives toward creating a new work-place culture, this view changed in the direction of a stronger orientation toward "feminine" ideals. Workers began to acknowledge and express fears, anxieties and dependencies. They consequently became more safety conscious, both from their own and from others' perspectives, emphasized caring, social relationships, and responsibilities (at work but also in relation to family) and acknowledged rather than minimized or denied vulnerability. Of course, numerous other aspects around safety at work are also a matter of culture, including priorities and attention paid to performance pressure, education and training, the use and meaning of safety-related equipment, formal rules and so on.

However, health is also a cultural phenomenon in itself, not merely something that can explain hazards and accidents at work. There are some objective aspects around health that do not really warrant consideration in terms of culture, such as clearcut, objective, bodily characteristics that indicate disability or sickness; these are only a minor component of the entire field of health at work. Most aspects are ambiguous, open to a variety of meanings, and more or less shared within an organization or another collective (shop floor, a specific group of health professionals). Dale and Burrell (2014) point to some overall changes in the understandings of health issues, which have moved progressively toward a focus on well-being. These changes include increased individualization, the conflation of health with attitudes and satisfaction, and health as an economic resource: fitness for purpose. The extent to which health is being individualized (rather than related to the work environment), turned into well-being (partly a matter of the right attitude) and made an instrument that is a key element in corporate performances (rather than a matter of care for its own sake) differs greatly between organizations and groups. The cultural understandings and consequences of this are crucial.

Based on these overall trends, we point briefly at only a few important themes here in an appeal for critical scrutiny and reflection for anyone addressing the subject matter and to offer potentially valuable topics for study. All point to the open nature of "health" and how a variety of potential cultural meanings are attributed to it, framing interpretation and guiding action for dealing with "health problems". (We sometimes use quotation marks to indicate that what is referred to as "health" is not self-evidently the best or only way of referring to it. Health is a social construct and what the term/representation is supposed to be referring to can sometimes be constructed in other, more insightful, ways.).

5.1 Health, Well-Being and Positive Attitudes to Work and Oneself

Being "healthy" is not a discrete, crystal clear category; contemporary understandings tend to include and/or overlap ideas such as well-being and positive attitudes. This means a broad and ambiguous space for cultural meanings around what is normal and/or to be aspired to. Dale and Burrell (2014) list five positive aspects of health/well-being that are often packaged together: health (physical), contented (emotional), flourishing (personal development), committed (values), and prospering (work/organization). The negative aspects of each of these are sick, distressed, demotivated, disengaged, and failing, respectively. The exact meaning of health/well-being and health-related issues (policies, leadership, human resources and interventions by health professionals) is seldom clear and consistent, but varies between cultural contexts and plays various roles in organizational life. Health thus becomes linked with a variety of management and performance issues that are framed and acted upon in different ways depending on their "healthiness" or well-being meanings. Can absence from work, suboptimal performance, social conflicts, lack of the right attitude, evidence of little commitment, resistance to management and work conditions, and so on be viewed as a matter of health (psychic illness, obesity, lack of self-care, insufficient exercise, stressful work, attitude problems . . .)? Orientations here vary with cultural contexts.

5.2 Health as an Object/Value in Itself Versus Health as a Means of Improving Performance

In both a narrow and broad sense, health can be viewed as more or less related to various performance- and effectiveness-related issues. Health is sometimes closely linked with absence. The latter is often interpreted or categorized or both as indicating sickness; however, absence may be viewed as an outcome of a variety of issues besides pure ill-health. Dealing with health with the aim of increasing presence at work is somewhat different from dealing with health for its own sake. In practice, health can sometimes be promoted by absence from work, whereas efforts to reduce absence through focusing on "health" may increase presence, but not necessary "health" as such. (Health-related interventions and monitoring may lead to increased surveillance, making disinclined to stay at home for fear that employees absence from work will mean that they are perceived as being unable to manage their own health or suspected of having difficulties in achieving a home/work balance). Thus, viewing health as of value in itself or more or less directly related to work performance are two different cultural text ways of viewing it. Groups differ, often having mixed and/or unstable sets of meanings; for example, absence, poor performance and so on may be greater or lesser key concerns, which can lead to targeting health issues or indicators of health problems without any direct cultural understanding of implications for performance. Human Resources departments, health units, and managers may be humanistic, client-centered, or more strictly focused on corporate affairs and priorities. Ideas and priorities around links may also differ; for example, increased "health" may be seen as directly leading to reduced absence, more energy/commitment, and better performance. Alternatively, "health" may be viewed as being good for an organization in more indirect ways; for example, by providing evidence of a caring employer and increasing identification and loyalty, which may then improve performance.

5.3 Health Promotion as a Means of Nurturing the Strategic Value of Employees

An even more ambitious view of "health" than simply seeing it simply as a means of reducing certain costs and producing a more reliable and efficient labor force is to mobilize it in an overall strategy designed to increase human capital and nurture the value of employees through various procedures. The aim here is to make it possible to place higher demands on workers. Through massive investments in promoting and encouraging being healthy, eating well, exercising and learning to cope with stress and psychological problems, a mentally and physically fit work-force is created that is able to cope with more demanding tasks, for example, dealing with potentially stressful demands for flexibility. This strategy was used by the lorry manufacturer Scania to supplement and support their design of the labor process. The company has a very large department of occupational health specialists, employs physicians, psychologists, trainers, physiotherapists, and nutrition experts, and offers a wealth of courses, assessments, and support for healthy life styles (Maravelias et al. 2013). Here, health monitoring, education, and therapy are seen as a way of making workers fit for and accepting of production regimes and able to adapt to the system. "Health" as a management tactic for optimizing use of labor represents a very ambitious form of health/employment management that is central to organizational culture.

5.4 Health as Image and a Component of Elitism

When the workers at Scania were targeted for ambitious improvements in "health", this was integrated with a design for production and their tasks. It was thus an internal process that aimed to improve productivity and achieve flexible forms of production. However, health can also have clearer external or image-related aspects, signaling cultural ideals to various audiences and being part of impression management. With the increase in service and some knowledge-intensive, professional work, appearance and image matters increasingly more for a range of organizations (Alvesson 2013b). We can talk about the increase in pink and gold workers (service and professional service work), which to some extent has occurred at the expense of white and blue collar occupations and tasks. Here, looking good and appearing to be fit is important. Health then becomes a matter of controlling body weight and presenting an appearance that is associated with healthiness, which can mean looking youthful or, in some cases, strong. For the business elite, at least in those up to middle-aged, it appears to be increasingly important to appear athletic. "Elite health" then becomes an important theme and ideal. Many executives, perhaps particularly in large cities, spend considerable time in fitness centers and emphasize their good physiques. An interesting case was that of the executives at Pepsi, who described themselves as "the Marine Corps of the Business World" and put considerable energy into being in good mental and physical shape and communicating this fact (Sculley 1987). Rather than seeing this focus on being in good shape as marginal and potentially (if time-consuming) a waste of time or at least much less important than other virtues (like intellectual and cultural activities), many have emphasized that this is a key quality for strong and competent executives, perhaps being more into body and energy than brain and reflection.

5.5 Who Is Responsible for Health?

Most people would probably believe that those responsible for health comprise a broad set of actors: authorities, various health professionals, employers, managers, unions, employees themselves, and so on. In this chapter, we are not interested in coming up with an objective or policy-oriented answer, but rather describing the variety of cultural meanings around this question. Within a work-place context, organizational culture may emphasize *individuals* and target them as the key agents responsible for dealing with health issues. Through self-care and self-management, exercising and eating appropriately, getting enough sleep, handling stress at work through planning, working efficiently, and managing demands and responses thoughtfully, individuals are considered responsible for their work situations and themselves. Another possibility is that a *group* or set of peers is emphasized—such a group or team addresses health issues through teamwork, collegial relationships that offers support, or balancing of work tasks and pressures. Alternatively, in a managerially focused culture *leadership* is viewed as key. The leader (supervisor) is in charge of health issues and accomplishes a healthy work-place through planning and allocating work, providing support, and helping with problem-solving. Then again, work conditions associated with *work organization*, performance targets, division of labor, resource allocation, organizational structure providing the right mix of autonomy and supportive-constraining structures (giving guidelines) health at work can be accomplished. Yet another possibility is to view the entire *corporation* (or top management and Human Resources staff) as key, and focus on the policies and resources put into health issues. Are there hazard prevention measures at work, investments in a clean and pure physical environment, allocation of resources for recreation and healthy life styles, and employment of health professionals, work environment experts and other such specialists?

Understanding health responsibility can thus circle around a variety of meanings of responsibilities and agency. Individualistic, collectivistic, managerialistic, and expert-focused cultural understandings point in somewhat different directions. Is health a medical, psychological or managerial concern? Different groups and organizations vary in how they understand these matters. For example, is the function of occupational health units understood to be to control and treat clear examples of physical unhealthiness, or are they supposed to check and work with everything that may have health implications, including family problems, conflicts at work, and experiences of stress? Is the company a source of patriarchal care or are employers supposed to respect and not interfere with their employees' lifestyles until those lifestyles directly and visibly appear to affect their work? A range of meanings may here guide the work and interventions of various health professionals and units and mandate, in narrow or broad terms, communicating specific notions of health and healthiness in different ways, shaping cultural understandings of these issues in the work-place.

6 Summary

This chapter provides a brief introduction to the concepts of culture, particularly organizational culture; the latter emphasizes the meanings, orientations, and symbolism shared by collectives in organizational settings. The focus on the structuring of experience and guiding of action by in-depth meanings creates a different approach to accounting for objective facts or trying to point out behaviors and structures. Culture indicates the collective meanings behind, and expressed in, surface manifestations.

Many health issues can be productively conceptualized as cultural phenomena. Health is understood in a variety of ways in various societies, organizations, occupations, and other groups, with correspondingly varying implications for action. The cultural meanings of health guide how organizations, managers, and professionals try to influence and improve how people deal with health issues.

There seems to have been an expansion and change of focus to health being about "right" attitude to work that has resulted in "health" being incorporated into the sphere of well-being in managerial and professional regimes. Here it is important to consider both innocent well-intentioned activities and issues of power and control. Power is partly exercised through cultural control; that is, domination over meanings affecting how people relate to themselves and work. Whether health problems are understood as a consequence of organizational control or of individual shortcomings in self-management makes a considerable difference. The same is true if certain orientations and resistances are viewed as indicators of a problematic work situation or are attributed to individuals' psychological states.

Although, almost by definition, "health" is a good thing, over-expanding its meaning such that "health" includes well-being, satisfaction, and a positive attitude may result in its exploitation by certain forms of managerial and professional power, leading to intensification of domination of the work force, stigmatization of "unhealthy" workers and targeting them for interventions based on subtle control techniques. Given that much work seems meaningless (repetitive, serving no clear social purpose), the "wrong" attitude, for example cynicism and/or resistance, may be "healthy" (Paulsen 2014). Health themes are not necessarily good or innocent but call for careful scrutiny regarding non-obvious functions and effects on power (Foucault 1980).

To sum up, a qualified understanding of health at work must consider cultural meanings (not only or mainly objective facts) among various health expert communities, society, organizational (sub)cultures, managers, and workers. Rather than simply aiming to improve "health", policies and interventions need to consider the cultural context carefully and be mindful of unintended adverse effects.

References

Alvesson M (2013a) Understanding organizational culture. Sage, London
Alvesson M (2013b) The triumph of emptiness. Oxford University Press, Oxford
Alvesson M, Billing YD (2009) Understanding gender and organization, 2nd edn. Sage, London
Alvesson M, Willmott H (2012) Making sense of management. Sage, London
Borowsky R (ed) (1994) Assessing cultural anthropology. McGraw-Hill, New York
Cohen A (1974) Two-dimensional man. An essay in the anthropology of power and symbolism in complex society. Routledge & Kegan Paul, London
Dale K, Burrell G (2014) Being occupied: an embodied re-reading of organizational 'well-ness'. Organization 21:159–177
Ely R, Meyerson D (2010) An organizational approach to undoing gender: the unlikely case of offshore oil platforms. Res Organ Behav 30:3–34
Foucault M (1980) Power/knowledge. Pantheon, New York
Frost PJ, Moore LF, Louis MR et al (eds) (1985) Organizational culture. Sage, Beverly Hills
Geertz C (1973) The interpretation of culture. Basic, New York
Gherardi S (1995) Gender, symbolism and organizational cultures. Sage, London
Grey C (2005) A Very Short, Fairly Interesting and Reasonably Cheap Book about Studying Organizations. Sage, London
Heijes C (2011) Cross-cultural perception and power dynamics across changing organizational and national contexts: Curaçao and the Netherlands. Hum Relat 64(5):653–674
Hofstede G (1980) Motivation, leadership and organization: do American theories apply abroad? Organ Dyn (Summer):42–63
Kunda G (1992) Engineering culture: control and commitment in a high-tech corporation. Temple University Press, Philadelphia
Lundholm S (2011) An act of balance—hierarchy in contemporary work. PhD thesis, Lund Business Press, Lund
Maravelias C, Thanem T, Holmqvist M (2013) March meets Marx: the politics of exploitation and exploration in the management of life and labour. Res Sociol Organ 37:129–160
McSweeney B (2002) Hofstede's model of national cultural differences: a triumph of faith—a failure of analysis. Hum Relat 55(1):89–118
Ortner S (1984) Theory in anthropology since the sixties. Comp Stud Soc Hist 26:126–166
Palmer I, Hardy C (2000) Thinking about management. Sage, London
Parker M (2000) Organizational culture and identity. Sage, London
Paulsen R (2014) Empty labor. Cambridge University Press, Cambridge
Scott R (1995) Institutions and organizations. Sage, Thousand Oaks
Sculley J (1987) Odyssey: Pepsi to Apple. Harper & Row, New York
Starbuck B (2010) What makes a paper influential and frequently cited? J Manage Stud 47:1394–1404
Turner B (1971) Exploring the industrial subculture. Macmillan, London
Van Maanen J, Barley SR (1984) Occupational communities: culture and control in organizations. Res Organ Behav 6:287–365
Van Maanen J, Barley SR (1985) Cultural organization. Fragments of a theory. In: Frost PJ, Moore LF, Louis MR, Lundberg CC, Martin J (eds) Organizational culture. Sage, Beverly Hills, pp 31–53

Effects of Practices of Belonging on Accident and Illness Prevention in Small- and Medium-Sized Enterprises

Markus Wiencke

Abstract

This chapter draws on an empirical study to describe the importance of context in the prevention of accidents and illness in the work-place. The empirical data are based on individual and group interviews with employers and employees of small and medium-sized enterprises as well as with occupational health and safety experts. The analysis shows that cultural aspects and internal communication processes have a major impact on how prevention is understood and implemented. The results can be condensed into the proposition that prevention may come into conflict with the practices that generate belonging in the work-place. Questions that arise from this include "Which communication processes and emotional meanings are realized with belonging?" and "How do these meanings interact with rules concerning prevention?" Further, it is suggested to describe the concept of belonging in terms of "stereotypes", "degree of empathy", "coherence", and "continuity". These considerations result in suggestions on how inhibiting and facilitating factors for illness prevention might be examined in future comprehensive studies of work-place interventions.

1 Introduction

An intervention for promoting work-place health and safety is a planned action in an organization that is often initiated outside that organization with the aim of realizing a predetermined goal within it. This process is affected by a number of inhibiting and facilitating factors within the organization that can be studied independent of concrete interventions. In intervention studies to date, the effects on the behavioral level of explicit rules have been the predominant focus (Elke

M. Wiencke (✉)
Federal Institute for Occupational Safety and Health (BAuA), Berlin, Germany
e-mail: markuswiencke@gmx.de

M. Wiencke et al. (eds.), *Healthy at Work*, DOI 10.1007/978-3-319-32331-2_3

et al. 2015). Thus, a general conceptual framework for investigating inhibiting and facilitating factors in an intervention at the level of the organization has not yet been developed. However, with the increasing flexibility of the labor market, emotional values associated with the work-place context are progressively becoming more influential in determining action (see Breuer 2014). Accordingly, the question I will address here is how the work-place context influences the effectiveness of interventions (cf. Wiencke and Sommer 2015).

To answer this question, a theoretical framework model such as that proposed by Hale and Borys (2013a, b) on explicit and implicit rules might be useful. For Hale and Borys, rules are the core concept of work-place health and safety management. By defining limits, rules coordinate conduct in organizations. Questions then emerge about how those rules are complied with or violated. In response to Dekker's work (2005), Hale and Borys (2013a) see two conceptually different approaches to rules that they call "Model 1" and "Model 2". Model 1 is a rational approach in which rules are static and documented. Following a top-down approach, rule violations are perceived as negative and sanctioned. Conversely, Model 2 follows a constructivist, bottom-up approach in which rules are dynamic and, as socially constructed behavior patterns, evolve in parallel with the experience of the actors involved. Accordingly, many important, lived rules in an organization are neither explicit nor documented. Violations of these rules are a necessary, positive part of a constant adaptation process. Hale and Borys (2013b) combine both approaches in a behavior control model. Thus, behavior control is described and explained on the one hand by explicit rules and rational behavior (Model 1) and on the other hand by implicit, lived rules (Model 2). Rule management is seen as a dynamic process in the course of which rules are adapted to changing contextual situations. The rules can be adapted in their application to specific contexts in the course of dialogue between experts and users and the concomitant feedback loops.

In their study on the prevention of musculoskeletal disorders in bricklayers, Hasle et al. (2014) provide an example of such feedback loops and the ties between explicit and implicit rules. State regulators, professional associations, employers and the bricklayers themselves contributed jointly to implement the measures. A combination of external pressure from the regulators, normative processes in the relevant professional associations, and mimetic imitation processes among employers and employees transported the measures to the construction site and realized them individually in various specific contexts. New standards were developed in a dialogue between the legal framework and the ever changing contextual situations.

After addressing these considerations, I will discuss factors that play important roles in the process of adapting of rules to work-place contexts. In Sect. 2, I will present the framework for the study, after which I will describe the assessment of the empirical data collected using methods of qualitative social research. In Sect. 4, I will place the findings in a broader context of research on response and effect.

2 Empirical Approach

Our investigation began with questions about the factors that facilitate and inhibit interventions aimed at promoting work-place health and safety. We had no concrete intervention in mind; rather, we wanted to look in general at what factors can influence the effectiveness of interventions in work-place contexts.

Our study being part of the EU project, "Innovation Incubator Lüneburg", we were required to comply with certain criteria. The empirical survey had to take place in small and medium enterprises (SMEs; with no more than 250 employees)[1] in the northern German state of Lower Saxony and focus on how they dealt with work-related illnesses. In selecting our test enterprises, we focused on those in which skin and musculoskeletal disorders are an issue. These included construction and metal-working enterprises, auto garages, and health-care. Of the 250 enterprises contacted by letter and telephone, 28 agreed to take part in the study. In addition, 32 employees, some of whom were working for one of these 28 enterprises and some of whom were not, were interviewed. A total of 14 interviews with work-place physicians, field supervisors from the work-place accident insurance providers, work-place health and safety experts from the health insurance companies, and staff of professional associations were also conducted. Most of the enterprises in the study had between three and 50 employees, the largest had 160 employees. Many of the enterprises were located in rural areas with little connection to urban infrastructures. We generally conducted the interviews, which were recorded and later transcribed, in the premises of the enterprise or the organization in which the interviewee worked. We conducted the individual and group interviews using interview guidelines based on recommendations for developing problem-centered interviews (Witzel 1989) or group discussions (Bohnsack 2003). As described in the following paragraph, the guidelines were modified in the course of the research process. They included, among others, questions regarding the implementation of preventive measures, how accidents and health problems were dealt with, the motivations behind work-place health and safety, social relationships in the work-place, and communication about risk.

The interviews and their evaluation involved a circular process in that we varied the interview guidelines according to the patterns that emerged during analysis of the material. We followed the rules of the coding method of Grounded Theory as described by Strauss and Corbin (2007). Reflecting on internal and external perspectives on work-place dynamics as provided by employers and employees and external experts, respectively, proved highly productive for our analysis. We also tried to introduce external perspectives in the interviews by using the technique of circular questioning as developed in systemic therapy and counseling (cf. Simon and Rech-Simon 2013). For example, interviewees were asked how colleagues

[1] SMEs are of central economic importance in Germany. 99.3% of all German enterprises having been SMEs in 2012. Of the 26.4 million persons employed in the economic sectors covered, more than 60% worked in SMEs in 2012 (Federal Statistical Office 2015).

would behave should a supervisor ask the interviewee to work unsafely, for example without safety glasses, or without ear protection, or carrying heavy loads. By introducing an external perspective, we encouraged the interviewees to reflect on how implicit rules functioned in their everyday working life. In the evaluation process, Clarke's (2005) reflections on the "thick analysis" of situations proved important. Clarke's call to analyze "situations" also opened a perspective on the relationships between social interactions and discourses and how these are determined by various situations. Second, it provided a methodological approach for finding integrating patterns in heterogeneous and contradictory material.

When we focused attention on the work situations in the interviews, a paradoxical picture emerged. The interviewees initially described preventive practices as something apart from their work routines and interactions with colleagues that had little to do with them. Whereas they perceived prevention as static, they considered other aspects of the work context to be dynamic. If, however, we introduced an outside perspective by using circular questioning during reflection processes in discussion groups, a different picture emerged. Participants then described prevention-related practices as a component of social interaction processes. This shows the broken and contradictory nature of prevention and can be condensed into our proposition that prevention may come into conflict with practices that generate belonging in the work-place.

3 Findings

3.1 Implicit Safety Rules

In our interviews, the conflict that Hale and Borys (2013a, b) sketch in their rule management model became apparent. The participants saw the formal rules (as documented risk assessments and instruction) as separate from the dynamic rules of everyday life. That said, they described the results of the interaction between formal and everyday rules as relatively open. Here we found that the interviewees described phenomena that fit with what Hale and Borys called feedback loops between experts and users. An example of this can be seen in the following interview with a work-place physician who was speaking about his experience in a metal-working enterprise with about 150 employees:

> . . .In this factory. . .I campaigned for 15 years; now we do it for five years. . .Accidents had to happen; the occupational accident insurers had to come with mobile vision tests—and 30 % of the people who did the eye test couldn't see properly. . .

In another interview, a work-place physician described how the implementation of formal preventative check-ups enabled her to intervene in a chance encounter:

> . . .as for the issue of skin care, I had another case today. . .I was only there for a work-place safety committee meeting that, in the end, did not take place, and then the chef came in. . .and because I had already examined him once in the past, he had enough confidence in

> me to tell me about an employee who had had a serious rash for a few months...and I spoke to her...After examining 10, 15 people and saying again and again 'It's all confidential, the employer will not be informed', and they saw that everything in fact remained confidential...Now, my place in this company has developed to the point where they will talk to me in such chance encounters. If I had not been here today, nothing would have happened.

In these statements, the experts describe acting as "change agents" who can intervene as a result of being present in the work-place context. Implicit in both statements is that both of these work-place physicians had already acquired a certain degree of knowledge of the work-place context. In the second case, the word "confidence" is used. Their pre-existing knowledge enabled these two work-place physicians to introduce their external perspectives and thus change the perspective in these contexts.

We know that internal perspectives tend to be somewhat resistant to change. In a different part of the interview, the first work-place physician spoke more clearly about the internal perspective. When the interviewer asked about the influence of risk assessment on everyday practice, he said:

> ... very few enterprises have ... a professional sense of health, yes, most say: 'We do not hurt our people, we work as safely as we need to.' But what the...regulators demand...that's all just paper that needs to be filed. And I will show it to anyone who asks to see it. But, in the everyday working situation, it has little meaning.

In the everyday working situation, the rule that "we work as safely as we need to" is subject to a context-sensitive dynamic. This is evidenced by the fact that employers feel responsible for occupational health and safety; they see themselves as having a duty of care for their employees and they link this duty to the provision of safety equipment and instruction. Conversely, they attribute responsibility for the implementation of accident prevention and health care to the employees in the work situation. Thus, a manager of a roofing contractor with about 45 employees said:

> Yes, health, it's like this...: Do I send the boys out or not? ... Like this morning, it started here... it began to snow at a quarter past six. Can they work on a roof? Should they? Will they work in the damp all day or not? These are the questions, and you are stuck in between. On the one hand you say that you want them all to come back healthy tomorrow and not catch a cold, or get problems with their backs. And on the other hand you have your deadlines and customers...And we have ...we have actually discussed this...we leave in the morning and look and see...And then the foreman has to decide and say, 'Look, there's no point' or 'Let's wait an hour, maybe it'll clear up'...A lot of our contractors depend on us... tilers, electricians...they like it warm and dry. So, if we don't finish, they can't start. And then things get tight. And my boys know...that they are expected to avoid risking their health...but they won't go home because of three drops of rain. So there, there is the middle road, and finding it is the responsibility of those on site.

I cite this example because it describes the dilemma that arises when safety rules need to be renegotiated on a daily basis. The manager refers to criteria over which he has no direct influence: the weather and the situation at the construction site, which is the result of the expectations of various stakeholders (his own firm, other contractors, clients). Against this background, his formulation of the responsibility for the health of his employees was very fuzzy. What expectations he actually had in terms of criteria for decision-making on site remained unspoken. He leaves dealing with the contradictions between health considerations and meeting deadlines to his workers.

3.2 Belonging

If we change our perspective, we can see that workers can be exposed to a great deal of ambivalence. The construction trainee cited in the following example does not disclose he has back pain and continues to work despite of it because he does not want his behavior to stand out[2]:

> ...And then you hear: "don't make such a fuss! You're still young!" and so on...and of course, you listen; you have your pride. You're not a girl, right? And of course you do it, take a deep breath even if you feel it in your back...Before I say anything I duck, grit my teeth and do it...

This young man's behavior was influenced by the stereotypical comments of his colleagues. In the interviews with workers, there were many examples of how preventive behavior is influenced by emotionally charged situations. I would like to encapsulate this as a "negotiation of belonging". From this perspective, the young man's coworkers are obviously important for his sense of belonging, which is based on his own stereotypical notions of masculinity. It is a very distant, limited sense of belonging. In other cases we were able to identify other degrees of empathy in belonging. In the same group discussion another trainee described his relationship with his foreman as much more personal:

> So I thought, foremen, well, what they say is right. And if he tells me to put my foot down, then I would think that I'm a bit slow or whatever? Okay, I could say something, like: 'Can't go so fast, my back's playing up' or something like that. But then you think, he's been doing this for so long, and then you want to try to maintain the pace, his pace.

The emotional meaning tied to belonging can be characterized in terms of "stereotypical communication" and "degree of empathy". In the second example, the interviewee attributed more expert knowledge to the foreman than to himself. It was more important for him to match his pace with that of the foreman than to safeguard his back. With more external knowledge, this attitude would have been challenged. Working safely does not necessarily mean working slowly. Rather, stereotypical beliefs that health and safety disrupt the workflow have come into play

[2] The statement was made during a group interview conducted in a public training facility with six trainees from different enterprises of 20–160 employees.

here. The young man's responsibility for his own health has become secondary to his desire to work like his foreman does. Thus, he achieves a form of belonging that is tied to recognition of the foreman's authority.

In the interviews, the phenomenon that we have identified as belonging was tied to certain communication patterns that influence the implicit rules for safety and health-related behavior. From the employee's perspective, the question of how people are charged with emotional meanings in the work environment seemed to be important. In the interviews with employers, this perspective shifted; they often described how they created opportunities for communication patterns. Two aspects that we identified as "coherence" and "continuity" emerged as important. I will now explain these on the basis of interview material. The following statement is taken from an interview with the owner of a home-care enterprise with about 40 employees.

> Well, my goal was then...to get a team set up for good that I could take with me until I retire. I am only turning 40 this year...but I have already thought about this for many years...and I am building up this team, and it's exciting that I will be able to work with it for the long term. That gives me extreme trust in my people and this constant...what you get everywhere in the care services...constant need to control everything, I stopped doing that. Because I know I can trust them. Of course, anybody who comes in new, they will have to earn that trust, they need to get into the team, but the team is so solid in a sense that they can deal with somebody coming in and trying to abuse that...They just teach them the rules of the game. And I don't mean bitching or anything like that, but that people are taken aside and told: 'Look, this is how we do it, so stick with our ways and you will have a great time.'

The employer is providing his staff with a vision of being able to grow old in the company. That employees meet the demands communicated in this quotation may be part of the vision: if they do not, they will have to leave the team. In the interviews with employers, even health is perceived in this context. Thus, health-related recognition processes are linked with more general recognition processes. From the employers' perspective, employees tend to get recognition when they have integrated themselves into work-place communication patterns. The following excerpt from an interview with the manager of an old-age home with about 100 employees shows how, from the employer's perspective, work can contribute to coherency.

> For example, we had a nurse with severe depression. The first thing the doctor did, of course, was to put her on sick leave for four weeks. And she got so depressed that she got suicidal and hurt herself...She broke down completely because she had lost her fixed structures here... She came to us every day and talked...about how she was doing and how we could help her. And then that was gone. And then the doctor said: 'No, I'll cancel the sick leave, go back to work'. And we looked with her, for what we could do for her. So we said, for the moment, no direct contact with the residents because illness had a negative effect on her...If residents were seriously ill or dying...But she did a lot of administrative work...Keeping patient records, structuring things, overseeing shifts...And that has helped her a lot. So sick leave is not always the right way, better to look and see what a person really needs.

This interviewee clearly needed to believe that the causes for the depression were in the private sphere, outside the work-place and the job. This interpretation was important because it allowed her to take on the role of caring employer. Attempts to solve psychological problems in the work-place created a certain kind of belonging and employee loyalty. The essential meaning she wanted to convey was that employees are important to the employer even if they cannot work as they once did. This employer showed considerable flexibility. The authentic connection and recognition the employee received may well have reduced the duration of her medical treatment.

In our interviews, employers described how such forms of continuity in the work-place contribute to coherence. Accordingly, prevention is charged with specific meanings that are commonly associated with coherence and continuity. The causes of disturbances in an employee's sense of coherence must then be sought outside the employer's sphere of influence, such as a relationship with a client or in private life. From this perspective, a positive effect on the health of employees tends to be attributed to social interaction processes in the work-place. I have selected these examples from interviews with employees to show that the normative, emotional and social aspects that prevention acquires can also be a hindrance. Stereotypical perceptions and empathy are two factors that can affect the process of negotiation of safety rules and impact the independent behavior of employees. The associated processes of ambivalence are related to the processes that create coherence and continuity in the work-place. Contrary to what the employer's perspective suggests, prevention is thus embedded in work-place communication patterns and general recognition processes. Accordingly, the rules that govern safety and health behavior are linked to the more general emotional meanings in the work-place. Safety and health at work are social and moral phenomena.

4 Discussion

With these considerations on belonging, I wish to suggest that it is time to more strongly consider the context of work-place interventions. The model introduced in Sect. 1 on rules management (Hale and Borys 2013b) can be used to structure inhibiting and facilitating factors, as Elke et al. (2015) have demonstrated. These authors have provided a systematic review of published discussions and meta-analyses of studies of occupational health and safety interventions. The studies they examined demonstrate a significant relationship between predictors and moderators in occupational health and safety. That said, a criticism made by Elke et al. is that relevant factors were not examined systematically or across interventions. Thus, it remains unclear how and under what contextual conditions these factors function together. Studies of interventions often fail to present the conceptual framework behind the design of the intervention or its implementation. Elke et al. (2015) suggest Hale and Borys' approach for developing a systematic overview of effective factors. Based on Hale and Borys' model, Elke et al. define the identified factors either as aspects of explicit (Model 1) or implicit (Model 2) control of behavior. They then group the factors into three categories of explicit behavior control (organization/structure, systematic approach, and organizational readiness for change) and two categories of implicit behavioral control (activation of human capital, culture/climate, and communication).

Zohar (2010) has made the important suggestion that "social climate" emerges in the context of social interpretation and identification process (cf. Blumer 1969; Weick 1995). "It is apparently this social verification process that motivates the formation or emergence of organizational safety climate" (Zohar 2010, p. 1520). Edwards et al. (2013) have expanded on these considerations to include the structures and processes of an organization: "Safety culture can be viewed as the assembly of underlying assumptions, beliefs, values and attitudes shared by members of an organization which interact with on organization's structures and systems and the broader contextual setting to result in those external, readily-visible, practices that influence safety (p. 77)." There are implicit and explicit categories embedded in this definition: the implicit category "culture/climate" can be found in the concepts "underlying assumptions, beliefs, values and attitudes shared by members of an organization" whereas the explicit category "organization/structures" is self-evident. Implicit and explicit elements are connected via the concept of "practices".

To identify practices (or social interaction processes), I suggest breaking "belonging" down into the aspects of "stereotypes", "degree of empathy", "coherence" and "continuity". Against this background, prevention may come into conflict with practices that generate belonging in the work-place. The questions that follow from this include "Which communication processes and emotional meanings are realized with belonging?" and "How do these meanings interact with prevention rules?" For Elke et al. (2015) most published studies refer to Model 1; none have focused on the possible moderating influences of cultural aspects and communication. These authors thus see a need for more research regarding implicit rules and their interactions with the explicit rules of behavior control. The considerations presented here may inspire ideas about how to investigate the impact of implicit rules in the future.

Acknowledgements This research was part of the project "Healthy at Work" at the Innovation Incubator Luneburg/Leuphana University of Lüneburg, Germany. Research and development within the Innovation Incubator was funded by the European Union and the state of Lower Saxony. Thanks to Silke Volkhardt and Jürgen Klose for their work in conducting the study and to Andreas Hemming for translating the article into English. I would also like to thank all interviewees for participating in the study.

References

Blumer H (1969) Symbolic interactionism: perspective and method. Prentice-Hall, Englewood Cliffs

Bohnsack R (2003) Gruppendiskussion [Group discussion]. In: Flick U, von Kardorff E, Steinke I (eds) Qualitative Forschung. Ein Handbuch [Qualitative Research. A Handbook], 2nd edn. Rowohlt, Reinbek bei Hamburg, pp 369–384 (in German)

Breuer J (2014) Wir brauchen eine echte Kultur der Prävention [We need a real culture of prevention]. DGUV Forum Fachzeitschrift für Prävention, Rehabilitation und Entschädigung 11:14–15 (in German)

Clarke AE (2005) Situational analysis: grounded theory after the postmodern turn. Sage, Thousand Oaks

Dekker SWA (2005) Ten questions about human error: a new view of human factors and system safety. Lawrence Erlbaum, Mahwah

Edwards JR, Davey JD, Armstrong KA (2013) Returning to the roots of culture: a review and re-conceptualisation of safety culture. Saf Sci 55:70–80

Elke G, Gurt J, Möltner H, Externbrink K (2015) Arbeitsschutz und betriebliche Gesundheitsförderung—vergleichende Analyse der Prädiktoren und Moderatoren guter Praxis [Occupational safety and health and workplace health promotion—comparative analysis of predictors and moderators of good practice]. Bundesanstalt für Arbeitsschutz und Arbeitsmedizin, Dortmund (in German)

Federal Statistical Office (2015) https://www.destatis.de/EN/FactsFigures/NationalEconomyEnvironment/EnterprisesCrafts/SmallMediumSizedEnterprises/Current.html. Accessed 6 June 2015

Hale A, Borys D (2013a) Working to rule, or working safely? Part 1: a state of the art review. Saf Sci 55:207–221

Hale A, Borys D (2013b) Working to rule, or working safely? Part 2: The management of safety rules and procedures. Saf Sci 55:222–231

Hasle P, Limborg HJ, Nielsen KT (2014) Working environment interventions—bridging the gap between policy instruments and practice. Saf Sci 68:73–80

Simon FB, Rech-Simon C (2013) Zirkuläres Fragen. Systemische Therapie in Fallbeispielen: ein Lehrbuch [Circular questioning. Systemic therapy in case examples: a textbook], 10th edn. Carl-Auer, Heidelberg (in German)

Strauss AL, Corbin J (2007) Basics of qualitative research: techniques and procedures for developing grounded theory, 3rd edn. Sage, Thousand Oaks

Weick KE (1995) Sensemaking in organizations. Sage, Thousand Oaks

Wiencke M, Sommer S (2015) Arbeitsschutz in der Dienstleistungsgesellschaft: Theoriebasierte Wirkungsforschung [Occupational health and safety in the service economy. Conceptual considerations about how to study the effects of interventions]. Z Arb Wiss 69(3):146–151 (in German)

Witzel A (1989) Das problemzentrierte Interview [The problem-centered interview]. In: Jüttemann G (ed) Qualitative Sozialforschung in der Psychologie [Qualitative Social Research in psychology], 2nd edn. Weinheim, Beltz, pp 227–255 (in German)

Zohar D (2010) Thirty years of safety climate research: reflections and future directions. Accid Anal Prev 42(5):1517–1522

Self-Endangering Work Behavior

Jan Dettmers, Nicole Deci, Sophie Baeriswyl, Martial Berset, and Andreas Krause

Abstract

Confronted with progressively increasing demands for greater flexibility, organizations apply organizational and managerial practices that build on their employees' self-organization and self-discipline, thereby increasing their employees' control over achieving their everyday work tasks. Although these practices offer opportunities for personal growth and coordination of work and private life, the resultant demands for increasing autonomy and self-organization may overtax employees' capacities, leading to stress. Furthermore, employees are increasingly reacting in ways that are self-exploitative and detrimental to their own health in response to large work-loads and strong demands for self-organization. This chapter introduces the concept of *self-endangering work behavior*; namely, behaviors that may be functional with regard to attaining work goals but dysfunctional with regard to health and long-term ability to work. We propose that *self-endangering work behavior* can be viewed as a form of coping reaction when workers are confronted with large work-loads and strong demands for self-organization. Based on the transactional stress model and regulation of behavior theory, we propose that self-endangering work behavior may mediate the effect of large work-loads and strong demands for self-organization on impaired well-being. We summarize empirical findings that confirm

J. Dettmers (✉)
Department of Psychology, Medical School Hamburg (MSH), Hamburg, Germany
e-mail: jan.dettmers@medicalschool-hamburg.de

N. Deci
Work and Organizational Psychology, University of Hamburg, Hamburg, Germany
e-mail: nicole.deci@uni-hamburg.de

S. Baeriswyl • M. Berset • A. Krause
Institute Humans in Complex Systems, School of Applied Psychology, University of Applied Sciences, Northwestern Switzerland, Olten, Switzerland
e-mail: sophie.baeriswyl@fhnw.ch; martial.berset@fhnw.ch; andreas.krause@fhnw.ch

M. Wiencke et al. (eds.), *Healthy at Work*, DOI 10.1007/978-3-319-32331-2_4

the detrimental role of self-endangering behaviors and their capacity to increase the detrimental effects of work demands. Finally, we discuss the benefit of assessing self-endangering behavior with the aim of taking preventive measures.

1 Introduction

Since the late 1980s, organizations and workers in industrialized economies have been confronted with progressively increasing demands for greater flexibility. This has changed the nature of work and employment and the quality of working life in general (Allvin et al. 2011; Felstead and Jewson 1999; Näswall et al. 2008). Flexible forms of work such as telework, flexible work times, and mobile work are becoming increasingly prevalent. It has even been proposed that they are indicators of how modern organizations aim to increase their employees' productivity and work engagement (de Menezes and Kelliher 2011; Posthuma et al. 2013). These changes impact not only on organizational structures and practices but also on the working conditions and everyday lives of employees in general (e.g., Felstead and Jewson 1999; Rousseau 1997).

There are conflicting assumptions about the health-related effects of the development of flexible forms of work (e.g., Badura et al. 2012). On the one hand, there is evidence that new, flexible, and more autonomous forms of work potentially have positive effects on well-being and can promote a good fit between work and private life (Allen et al. 2013; Gajendran and Harrison 2007; Joyce et al. 2010; Pfeiffer 2012; Zok and Dammasch 2012). On the other hand, scholars who base their views on sociological approaches such as the concepts of *boundarylessness* and *subjectivization* (Moldaschl and Voss 2003) are arguing that the increasing demands for flexibility and self-management (Graf 2012) accompanying these developments may place employees at risk. Höge and Hornung (2013) reported that demands for flexibility, including the requirement for employees to structure their own work procedures, are accompanied by increased emotional and cognitive irritation. Other studies have revealed links between flexible-autonomous work and perceived stress, feelings of time pressure (Ducki 2009), reduced sleep quality, and psychosomatic complaints (Janssen and Nachreiner 2004). However, evidence of the negative effects of increasing demands for flexibility is inconsistent. It can be assumed that the way requirements for increasing flexibility and self-management relate to health-related outcomes is affected by the specific behaviors employees use to cope with these demands and the consequences these behaviors have for their health and well-being (Kaur et al. 2010).

In this chapter, we will discuss a specific way in which employees react when confronted with flexible work requirements that may explain the negative effects of flexible and autonomous forms of work on well-being. This reaction, which can be termed *self-endangering work behavior*, includes discrete employee behaviors that assist in coping with excessive demands at work in the short term but have negative adverse effects on health and well-being in the long term. We will conceptualize such behavior in the context of coping behavior and contrast it with existing constructs such as work engagement, overcommitment, and workaholism. Furthermore, we will discuss the specific effects of such behavior on health and well-being.

2 Self-Endangering Work Behavior as a Reaction to Demands for Flexibility

Confronted with new demands for flexibility and continuous change, organizations are increasingly moving away from traditional Tayloristic ways of controlling transformation of employees' latent working capacities into performance (Höge 2011). New managerial practices build on employees' self-organization and self-discipline. Organizational strategies such as reducing hierarchy levels, a matrix structure, and management by objectives increase employees' control over achieving their everyday work goals. Within these forms of work organization not only managers but also employees have to make decisions about how, when, and where to work and how to increase their own efficiency (Garhammer 2002; Höge 2011). Whereas the major aims of these new forms of management are to rationalize work and increase its productivity and flexibility, they also include key criteria of traditional human-oriented work design approaches; namely, provision of autonomy and latitude in decision-making (e.g. Hackman and Oldham 1976; Karasek 1979). Indeed, providing more responsibility and flexibility and encouraging workers' creativity and engagement offer opportunities for personal growth, learning, and coordination of work and life (Oldham and Hackman 2010). However, increasing autonomy and demands for self-organization can also be associated with additional effort that may overtax employees' capacities and lead to stress (Höge 2011; Voss and Pongratz 1998). In addition, such approaches allow employees discretion regarding both their work procedures and coping behaviors when confronted with high or overtaxing work demands. This may lead to new forms of coping reactions that can increase the risks of detrimental effects on employee health (Kaur et al. 2010). The specific choices of coping behaviors become more important.

Performance and goal-oriented work systems can be considered *indirect* leadership systems. Based on the theory of goal setting (Locke and Latham 2002), indirect leadership systems such as management by objectives aim to get employees to identify with organizational goals. Employees are encouraged to act as entrepreneurs ("entreployees", Voss and Pongratz 1998) and to orient their performance toward benchmarks and key performance indicators. Employees' engagement with attaining goals and avoiding failure is no longer controlled by external rewards or direct punishment but by personal interest in meeting these key performance indicators. Failing to meet work goals then translates into personal insufficiency (Peters 2011). As a consequence, employees will continue to strive for their work goals even when these goals cannot be reached with the given resources and capacities.

Against this background, employees are increasingly behaving in self-exploitative ways that are detrimental to their health (Docherty et al. 2002; Kieschke and Schaarschmidt 2008). In numerous case studies, Krause et al. (2012) have identified behavioral strategies used by employees to cope with increased requirements for flexibility and high work demands. On the one hand, these behaviors are functional in terms of dealing with stressful work situations and achieving work goals, thereby promoting self-esteem, motivation, and satisfaction. On the other, they have disadvantages regarding health and recovery. Accordingly, we have named these coping behaviors self-endangering work behavior (Krause et al. 2012). Self-endangering work behavior is characterized by actions that aim to deal with work-related demands but simultaneously increase the likelihood of health problems and impede necessary recovery from work-related stress. Examples of behaviors aimed at attaining internalized but potentially overtaxing work goals are extending work time, not taking time to recover, work intensification and sick presenteeism (Baeriswyl 2014; Semmer et al. 2010). Other behaviors involve taking risks to reach goals with less effort, for example by skipping security regulations and therefore risking adverse effects.

A range of case studies have revealed a variety of behaviors that can be considered self-endangering (Krause et al. 2012). Building on experiences gained in several health promotion projects and investigating employees working in flexible work systems, Krause et al. (2014) have systematically explored different forms of behavior that have the aim of coping with high work demands and the potential to harm individual health. They have identified the following eight types of behavior: (1) extension of work time and not taking time to recover; (2) work intensification; (3) sickness presenteeism; (4) abuse of stimulants in an attempt to optimize internal states; (5) abuse of sedative substance to facilitate relaxation; (6) reducing quality of work; (7) failure to comply with security regulations; and (8) faking.

Extension of work time and not taking time to recover means reducing private and family time and recovery periods in general in favor of work time. In Krause et al.'s 2012 study, employees reported different forms of extending work time when confronted with high work demands. These included not only long working hours and working overtime but also extending availability for work during off-job time (Dettmers and Bamberg 2013; Pangert and Schüpbach 2013). These findings closely match trends observed in national and international representative surveys (e.g., Eurofound 2012). One Swiss survey revealed that 20 % of Swiss employees work at least once a week during off-job time (Krieger et al. 2012) and 12 % work more than 10 h a day at least six times a month. These behaviors are reportedly associated with perceived stress, increased burnout levels, and health-related problems (Grebner et al. 2010; see also Wirtz 2010).

Another frequently reported behavior is *work intensification*. This means working at an increased pace and multitasking along with refraining from both work breaks and social interactions at work (Korunka and Kubicek 2013). There has been a great deal of research under the heading of *work intensification* (also referred to as intensification of work; e.g., Green 2004), particularly in the medical care context (e.g., Horner et al. 2012). Empirical findings suggest that work intensification can have positive consequences such as feelings of capability or delight (similar to consequences of challenge stressors, e.g., Widmer et al. 2012) or negative consequences such as impaired health (Korunka and Kubicek 2013). Another adverse side effect is that employees report less communication and social support. Rau (2012) warns that permanently intensifying work may lead to continuous activation even after leaving work, reducing sleep quality and increasing the risk of cardiovascular diseases.

Sick presenteeism means going to work when sick. Various studies have investigated this (e.g., Aronsson et al. 2000; Hägerbäumer 2011). The specific conditions found in flexible, goal-oriented leadership systems with high work demands promote this kind of behavior (Aronsson and Gustafsson 2005; Böckerman and Laukkanen 2009, 2010; Caverley et al. 2007; Claes 2011; Deery et al. 2014; Hägerbäumer 2011; Hansen and Andersen 2008; Heponiemi et al. 2010; Johns 2011). Employees report that their work will not get done while they are on sick leave and will therefore have accumulated when they return to work. Grebner et al. (2010) have shown that almost every second working person in Switzerland reports having gone to work when sick. The harmful effects of sickness presenteeism for employees and organizations are well documented. Sickness presenteeism is a predictor of different indicators of poor health (e.g., Bergström et al. 2009; Conway et al. 2014; Kivimäki et al. 2005) and future sickness absence (Bergström et al. 2009; Janssens et al. 2013). Sickness presenteeism has repeatedly been shown to be positively associated with emotional exhaustion (Demerouti et al. 2009; Hägerbäumer 2011; Lu et al. 2013a, b).

The above practices (also reported by Baeriswyl 2014) are not the only self-endangering behavior used by employees. For example, they actively try to modify internal states to cope better with high work demands (Krause et al. 2012). This may include *consuming substances to enhance their performance*. Employees report consuming caffeine and other stimulants. This so-called brain doping or neuro-enhancement may include taking pharmaceutical stimulants such as methylphenidate and even illegal amphetamines or cocaine. In one study in Germany, 1.5 % of employees and 5 % of university students admitted taking neuro-enhancers (Kowalski 2013). Given that such behavior is viewed negatively, it can be assumed that the actual number of employees taking drugs to cope with increasing demands is even higher. In addition, Krause et al. (2014) mention *substance abuse for recovery purposes*. Both types of substance-taking behavior aim to increase or restore the capacity to perform to cope with potentially overtaxing work demands.

Another type of self-endangering behavior designed to cope with high demands is *risky behavior and failure to comply with safety regulations* in reaction to time pressure (Leitner et al. 1987; Mearns and Hope 2005). Other potential reactions to excessive work demands include general *reduction of quality* (Welsh and Ordóñez 2014) and *faking* behavior (Krause et al. 2014). These may well have long-term negative effects on employees' work-related self-esteem and job security.

In summary, we see self-endangering work behavior as behaviors that employees deliberately implement when confronted with high work demands.

These behaviors aim at coping with the demands but are associated with detrimental effects. Thus, self-endangering work behaviors may increase negative outcomes of high work demands. We assume that autonomous and flexible forms of work in particular enforce this type of behavior. After adopting organizational work goals as their own personal goals, employees select behaviors for reaching these goals, even when they are unattainable with the given resources. In so doing, they intensify their work pace, do not take time to recover and manipulate their internal states to perform. Success in fulfilling their work goals may have the positive consequence of a boost in self-esteem (Widmer et al. 2012) or at least the satisfaction of not having failed to reach the goal (Peters 2011). However, this kind of behavior is not sustainable in the long run. Negative long-term effects on health and well-being are inevitable. Hence, self-endangering behavior can be considered a mediating mechanism between excessive work demands and negative effects on health and well-being (Baeriswyl 2014).

3 Self-Endangering Work Behavior and Coping

Self-endangering work-related behavior is not the only possible reaction to high or overtaxing work demands. Coping research (e.g., Carver and Scheier 1998; Lazarus and Folkman 1984) emphasizes the relevance of cognitive, emotional, and behavioral reactions to work demands. Transactional stress theory defines coping as "constantly changing cognitive and behavioral efforts to manage specific external and/or internal demands that are appraised as taxing or exceeding the resources of the person" (Lazarus and Folkman 1984, p. 141). The cognitive, emotional, or behavioral reaction to a stressor is a crucial determinant of the actual effect of that stressor on an individual. In this sense, coping reactions mediate the effect of stressors on well-being (Lazarus and Folkman 1984). Stress and coping theories such as Carver and Scheier's (1998) theory on the self-regulation of behavior demonstrate that both the stress caused by work demands (such as high workload) and the ways in which individuals deal with these demands contribute to the genesis of short- and long-term mental and physical illnesses (see also Bamberg et al. 2003; Greif et al. 1991; Lohmann-Haislah 2012; Mearns and Cain 2003; Zapf and Semmer 2004). The specific nature of individual coping reactions to stress determines whether those individuals stay healthy or become ill (Gutiérrez Doña 2002). There are many ways of categorizing coping (Carver and Connor-Smith 2010); Lazarus and Folkman (1984) identify two specific coping types: problem-focused coping and emotion-focused coping. A large proportion of the emotion-focused coping types are cognitive processes such as "distancing," "positive comparisons," or "avoidance", the main goal of which is to reduce emotional distress (Carver and Scheier 1998; Carver et al. 1989; Lazarus and Folkman 1984). In contrast, problem-focused coping comprises behaviors and ways of thinking that aim to solve a current problem or remove a given stressor. Strategies in this category include modifications of the environment and cognitive and motivational changes ("inward-directed strategies"). In addition, a relevant distinction is often made between active coping (engagement coping, approach coping) and avoidance coping (disengagement coping) (Billings and Moos 1981; Carver and Connor-Smith 2010; Carver and Scheier 1998; Carver et al. 1989; Nielsen and Knardahl 2014). The former refers to dynamic efforts to manage the situational stressor and emerging aversive emotions (Carver and Connor-Smith 2010). In their theory on the self-regulation of behavior, Carver and Scheier (1998) establish a link between active coping and goal achievement. Active coping can be seen as "continued engagement with goals that the stressor is threatening" (Carver and Scheier 1998, p. 214). This implies attempts to maintain a set goal. It includes both problem- and emotion-focused strategies, for instance seeking support, planning, problem solving, and cognitive restructuring. In contrast, avoidance or disengagement coping aims to prevent additional confrontations with the stressor and is associated negative feelings by abandoning the goal. Most of these types of coping are emotion-focused. Examples are denial, behavioral disengagement, and wishful thinking.

There is empirical evidence that avoidance coping has negative effects on health and well-being (Evans et al. 2004; Nielsen and Knardahl 2014; Wallace et al. 2010). This is mainly because this coping type "does nothing about the threat's existence and its eventual impact" (Carver and Connor-Smith 2010, p. 686). In contrast, active coping strategies can be positively associated with well-being (Nielsen and Knardahl 2014; Wallace et al. 2010). For example, Nielsen and Knardahl (2014) have shown that engagement coping correlates negatively with impaired well-being. Active coping also buffers the effect of workload on burnout (Wallace et al. 2010). Furthermore, problem solving, cognitive restructuring, and seeking advice are positively related to job satisfaction and work engagement (Rothmann et al. 2011; Welbourne et al. 2007).

If self-endangering work behavior as an active behavioral reaction to excessive work demands is integrated into the theoretical coping framework, self-endangering work behavior can be considered a form of active coping: When confronted with high work demands, employees try to achieve the given work goal and do not abandon it despite being faced with hindrances and having insufficient resources. This is characteristic of active coping (Carver et al. 1989; Semmer and Meier 2009). With respect to health-related outcomes, however, the focus on goal attainment is disadvantageous: whereas self-endangering work behavior is directed toward goal attainment and may even succeed in achieving that, this success is built on detrimental behaviors that may impair well-being and health. Furthermore, when self-endangering work behaviors are employed to achieve potentially overtaxing work goals, these overtaxing goals become stabilized and legitimized within the organization, creating a permanent stressor of overload. Thus, self-endangering work behaviors do not contribute to reducing stressors, as would be expected for problem-focused coping.

In summary, self-endangering work behavior can be seen as a coping reaction to high work demands. Because self-endangering work behavior includes retaining work goals, it can be considered a form of active coping. However, rather than buffering the effects of demands and reducing stressors, self-endangering work behavior has predictable detrimental effects on well-being. We believe that, because self-endangering behavior goes beyond the traditional distinction between active and avoidance coping, it has an incremental value for explaining the effects of stressors on health and well-being (Dettmers and Deci 2014).

4 Self-Endangering Work Behavior, Work Engagement, Workaholism, and Overcommitment

As far as the principle of excessive work behavior is concerned, the concept of self-endangering work behavior may resemble other established constructs about organizational behavior such as work engagement (Schaufeli et al. 2002), workaholism (Schaufeli et al. 2008), and overcommitment (Siegrist et al. 2004). However, there are significant differences between these phenomena.

Compared with *work engagement*, different psychological levels are in focus. Whereas self-endangering work behavior is a specific observable behavior, work engagement is conceptualized as a psychological state of mind. Schaufeli et al. (2002, p. 74) define work engagement as "a positive, fulfilling, work-related state of mind that is characterized by vigor, dedication, and absorption." Moreover, according to the job demands–resources model (Bakker and Demerouti 2007), work engagement is part of the motivational process. Engaged employees enjoy working and this leads to positive outcomes (Schaufeli et al. 2008). In contrast, self-endangering work behavior potentially impairs health (Baeriswyl et al. 2014). Baeriswyl et al. (2014) have demonstrated that the self-endangering work behavior of sick presenteeism mediates the relationship between job stressors and burnout. Thus, compared with work engagement, self-endangering behaviors concerns different phenomial levels (behavior vs. psychological state) and processes (health impairment vs. motivational process).

The core of the concept of *workaholism* is an inner drive to work (Schaufeli et al. 2008). Workaholics "work so hard out of an inner compulsion, need, or drive, and not because of external factors such as financial rewards, career perspectives, organizational culture, or poor marriage" (Schaufeli et al. 2008, p. 175). This inner drive or addiction component is not considered to be part of the concept of self-endangering work behavior presented here. Rather, we understand self-endangering work behavior as a reaction designed to cope with excessive work demands that occurs specifically under the condition of high requirements for self-organization and self-leadership (Baeriswyl et al. 2014; Krause et al. 2012, 2014). Nonetheless, despite these differences in the underlying causes, the behavioral consequences of workaholism may resemble self-endangering work behaviors in many ways.

Finally, *overcommitment* (Siegrist et al. 2004) may have similarities with the concept of self-endangering work behavior. This is particularly true for its theoretical conceptualization. Overcommitment is defined as an enduring cognitive-motivational pattern of maladaptive coping with demands characterized by excessive striving and an inability to withdraw from obligations (Siegrist et al. 2004). Overcommitted individuals are driven by their high need for control and approval. They tend to permanently overtax their own resources, which may eventually result in exhaustion and breakdown (Joksimovic et al. 1999); thus, in this respect, overcommitment is similar to the presented construct of self-endangering work behavior. However, when it comes to its operationalization, Siegrist et al.'s (2004) *overcommitment* scale does not focus primarily on specific behaviors but rather on psychological consequences ("strain") of an excessive engagement with work (e.g., being unable to switch off mentally). Thus, self-endangering work behavior is a distinct construct that focuses exclusively on specific behaviors without capturing the psychological consequences of its operationalization.

5 Effects of Self-Endangering Work Behavior

In contrast to both the above-mentioned work engagement and active and problem-focused coping (Nielsen and Knardahl 2014; Wallace et al. 2010), we suggest that self-endangering work behaviors may have harmful effects. As an attempt to cope with high work demands, self-endangering work behavior may mediate the detrimental effects of excessive work demands on well-being; however, if it results in successfully coping with challenging demands, it may also contribute to feelings of accomplishment, competence, or pride (Widmer et al. 2012). The challenge–hindrance framework states that positive effects of challenge stressors are always accompanied by parallel effects on strain and exhaustion. Correspondingly, we assume that when it comes to self-endangering work behaviors, psychological costs such as irritation, exhaustion, feelings of insufficiency, or psychosomatic complaints will—in the long term—prevail over potential positive effects (Hockey 1997; Schönpflug 1987; Semmer et al. 2010).

Most of the initial studies investigating the effects of self-endangering work behaviors have confirmed its assumed detrimental effects (Baeriswyl et al. 2014; Deci et al. [Coping in Flexible Working Conditions – Engagement, Disengagement and Self-Endangering Strategies, manuscript under review]; Dettmers and Deci 2014). Furthermore, self-endangering work behavior contributes to explaining psychosomatic complaints and emotional exhaustion after controlling for the direct effects of stressors and other coping behaviors such as engagement or disengagement coping as conceptualized in Carver et al.'s (1989) COPE inventory (see also Dettmers and Deci 2014). Baeriswyl et al. (2014) have tried to integrate the self-endangering work behavior of presenteeism into the job demands–resources model. Results of a study on 579 teachers reveal that presenteeism mediates the effect of work demands on burnout.

So far, there is no evidence for the assumed positive short-term effects of self-endangering work behavior. Studies trying to test this relationship (e.g., Dettmers 2014) have failed to relate self-endangering behavior to perceptions of challenge or challenging effects of stressors such as time pressure.

6 Consequences for Prevention

Flexible work arrangements that build on the capacity of employees for self-organization and self-leadership are an increasing phenomenon in the world of work. From the point of view of occupational health, there is a need to identify criteria for analyzing, evaluating, and designing flexible work arrangements because the associated growing requirements for flexibility and self-organization may overtax the capabilities and internal resources of employees. Furthermore, there is an increased risk of inadequate coping reactions when faced with high work demands. In this chapter, we have introduced the concept of self-endangering work behavior as a means of coping with high demands for flexible and autonomous forms of work organizations; it is a mechanism aimed at mediating the negative effects of stressors that has incremental value above the effect of the stressor and traditionally conceptualized coping behaviors. The few studies conducted on this topic have confirmed this view. Based on theoretical assumptions and empirical results, we propose that health-oriented prevention should take into account the need to reduce self-endangering work behaviors, particularly in organizations using indirect leadership techniques that aim at employees to identify with organizational work goals. These organizations have to be aware that the adoption of excessive work goals may have detrimental effects on employees via the use of self-endangering work behavior. A health-oriented strategy would have to systematically assess the occurrence of self-endangering work behavior and initiate processes that encourage employees to reflect on and change these behaviors. The identification of risks to health before problems actually occur could be achieved by including assessment of self-endangering work behavior in early warning systems within general work-related risk assessment and employee surveys. Finally, employees should be encouraged to take decisions and be flexible in meeting not only organizational requirements but also their personal needs and interests, and to openly discuss work goals that seem to be unattainable with the given resources and deadlines. These topics should be part of an open culture of negotiation between leaders and followers within an organization.

Generally, assessing self-endangering work behavior may be a promising way of identifying excessive work demands. For the analysis, evaluation, and design of flexible work, we propose that, alongside traditional criteria such as ergonomic and task design, managers also need to consider coping strategies and their effects on well-being. Human-oriented work design should minimize the use of self-endangering work behavior as a prerequisite for achieving work goals. If analysis detects self-endangering work behaviors, this may indicate that work organization and performance management are exceeding individual capacities (e.g., unrealistic work goals and fixed non-negotiable deadlines). Assessing self-endangering work behaviors may be a promising way of initiating reflective processes within an organization's staff aimed at finding new ways of coping actively that have less adverse effects, and of developing a problem focus that is sustainable in the long term. Well-validated instruments that focus on a variety of self-endangering behaviors (Deci et al. under Review; Krause et al. 2014) or single aspects such as presenteeism (Hägerbäumer 2011) are useful tools for promoting prevention within these new forms of work.

References

Allen TD, Johnson RC, Kiburz KM et al (2013) Work–family conflict and flexible work arrangements: deconstructing flexibility. Pers Psychol 66(2):345–376

Allvin M, Aronsson G, Hagström T et al (2011) Work without boundaries: psychological perspectives on the new working life. Wiley, Chichester

Aronsson G, Gustafsson K (2005) Sickness presenteeism: prevalence, attendance pressure factors, and an outline of a model for research. J Occup Environ Med 47(9):958–966

Aronsson G, Gustafsson K, Dallner M (2000) Sick but yet at work. An empirical study of sickness presenteeism. J Epidemiol Community Health 54(7):502–509

Badura B, Ducki A, Schröder H et al (2012) Fehlzeiten-Report 2012. Gesundheit in der flexiblen Arbeitswelt: Chancen nutzen—Risiken minimieren [Absenteeism report 2012. Health in the world of flexible work: making the most of chances, minimizing risks]. Springer, Berlin (in German)

Baeriswyl S (2014) Exploring the processes leading to emotional exhaustion—two extensions of the JD–R approach. Outline for Doctoral Thesis, University of Applied Sciences and Arts Northwestern Switzerland, Olten

Baeriswyl S, Krause A, Kunz Heim D (2014) Arbeitsbelastungen, Selbstgefährdung und Gesundheit bei Lehrpersonen—eine Erweiterung des Job Demands–Resources Modells [Job demands, self-endangering behavior and occupational health of teachers]. Emp Päd 28(2): 128–146 (in German)

Bakker AB, Demerouti E (2007) The job demands-resources model: state of the art. J Manage Psychol 22(3):309–328

Bamberg E, Busch C, Ducki A (2003) Betriebliches Stress- und Ressourcenmanagement. Strategien und Methoden für die neue Arbeitswelt [Managing organizational stress and resources: strategies and methods for the new world of work]. Huber, Bern (in German)

Bergström G, Bodin L, Hagberg J et al (2009) Does sickness presenteeism have an impact on future general health? Int Arch Occup Environ Health 82(10):1179–1190

Billings AG, Moos RH (1981) The role of coping responses and social resources in attenuating the stress of life events. J Behav Med 4(2):139–157

Böckerman P, Laukkanen E (2009) What makes you work while you are sick? Evidence from a survey of workers. Eur J Pub Health 20(1):43–46

Böckerman P, Laukkanen E (2010) Predictors of sickness absence and presenteeism: does the pattern differ by a respondent's health? J Occup Environ Med 52(3):332–335

Carver CS, Connor-Smith J (2010) Personality and coping. Ann Rev Psychol 61:679–704

Carver CS, Scheier MF (1998) On the self-regulation of behavior. Cambridge University Press, Cambridge

Carver CS, Scheier MF, Weintraub JK (1989) Assessing coping strategies: a theoretically based approach. J Pers Soc Psychol 56(2):267–283

Caverley N, Cunningham JB, MacGregor JN (2007) Sickness presenteeism, sickness absenteeism, and health following restructuring in a public organization. J Manage Stud 44(2):304–319

Claes R (2011) Employee correlates of sickness presence: a study across four European countries. Work Stress 25(3):224–242

Conway PM, Hogh A, Rugulies R et al (2014) Is sickness presenteeism a risk factor for depression? A Danish 2-year follow-up study. J Occup Environ Med 56(6):595–603

Deci N, Dettmers J, Krause A, Berset M (under review) Coping in flexible working conditions—engagement, disengagement and self-endangering strategies

Deery S, Walsh J, Zatzick CD (2014) A moderated mediation analysis of job demands, presenteeism, and absenteeism. J Occup Organ Psychol 87:352–369
de Menezes LM, Kelliher C (2011) Flexible working and performance: a systematic review of the evidence for a business case. Int J Manage Rev 13:452–474
Demerouti E, Le Blanc PM, Bakke AB et al (2009) Present but sick: a three-wave study on job demands, presenteeism and burnout. Career Dev Int 14:50–68
Dettmers J (2014) Selbstgefährdendes Bewältigungsverhalten und die Wirkung von Challenge-Stressoren [Self-endangering coping behavior and the effect of challenge stressors]. In: Abstracts of the 49th Kongress der Deutschen Gesellschaft für Psychologie, University of Ruhr, Bochum, 21–25 Sept 2014 (in German)
Dettmers J, Bamberg E (2013) Rufbereitschaft als erweiterte Verfügbarkeit für die Erwerbsarbeit [On-call as extended availability for gainful employment]. In: Dettmers J, Fietze S, Friedrich N, Keller M (eds) Rufbereitschaft. Hampp, Mering (in German)
Dettmers J, Deci N (2014) Coping with time pressure—The mediating role of different coping strategies. In: Abstracts of the international conference healthy at work, Leuphana University Lüneburg, 16–17 May 2014
Docherty P, Forslin J, Shani AB (2002) Creating sustainable work systems: emerging perspectives and practice. Psychology, New York
Ducki A (2009) Arbeitsbedingte Mobilität und Gesundheit—Überall dabei—nirgendwo daheim [Work-related mobility and health: always working; never at home]. In: Badura B, Schröder H, Klose J, Macco K (eds) Fehlzeiten-Report 2009. Arbeit und Psyche: Belastungen reduzieren—Wohlbefinden fördern. Springer, Berlin, pp 61–70 (in German)
Eurofound (2012) Fifth European working conditions survey. Publications Office of the European Union, Luxembourg
Evans GD, Bryant NE, Owens JS et al (2004) Ethnic differences in burnout, coping, and intervention acceptability among childcare professionals. Child Youth Care Forum 33(5):349–371
Felstead A, Jewson N (1999) Global trends in flexible labour. Macmillan, London
Gajendran RS, Harrison DA (2007) The good, the bad, and the unknown about telecommuting: meta-analysis of psychological mediators and individual consequences. J Appl Psychol 92: 1524–1541
Garhammer M (2002) Pace of life and enjoyment of life. J Happiness Stud 3(3):217–256
Graf A (2012) Selbstmanagement-Kompetenz in Unternehmen nachhaltig sichern. Leistung, Wohlbefinden und Balance als Herausforderung [Sustaining self-management competence in companies: the challenge of achievement, well-being, and balance]. Springer Gabler, Wiesbaden (in German)
Grebner S, Berlowitz I, Alvarado V et al (2010) Stress bei Schweizer Erwerbstätigen. Zusammenhänge zwischen Arbeitsbedingungen, Personenmerkmalen, Befinden und Gesundheit [Stress in Swiss workers: relations between working conditions, personality characteristics, well-being, and health]. Staatssekretariat für Wirtschaft SECO, Bern (in German)
Green F (2004) Why has work effort become more intense? Ind Relat 43:709–741
Greif S, Bamberg E, Semmer N (1991) Psychischer Streß am Arbeitsplatz. Hogrefe, Göttingen, pp 91–119 (in German)
Gutiérrez Doña EB (2002) Coping with stress at work: a longitudinal study on health outcomes and quality of life. Doctoral dissertation, Free University of Berlin. Available from http://edocs.fu-berlin.de/diss/servlets/MCRFileNodeServlet/FUDISS_derivate_000000000898/00_Gutierrez.pdf. Accessed 10 Mar 2015
Hackman J, Oldham G (1976) Motivation through the design of work: test of a theory. Organ Behav Hum Perform 16(2):250–279
Hägerbäumer M (2011) Ursachen und Folgen des Arbeitens trotz Krankheit—Implikationen des Präsentismus für das betriebliche Fehlzeiten- und Gesundheitsmanagement [Causes and consequences of working when sick: implications of presenteeism for absence and health

management in companies]. Doctoral dissertation, University of Osnabrück. Available from https://repositorium.uni-osnabrueck.de/bitstream/urn:nbn:de:gbv:700-201112158616/1/thesis_haegerbaeumer.pdf. Accessed 10 Mar 2015 (in German)
Hansen CD, Andersen JH (2008) Going ill to work—what personal circumstances, attitudes and work-related factors are associated with sickness presenteeism? Soc Sci Med 67:956–964
Heponiemi T, Elovainio M, Pentti J et al (2010) Association of contractual and subjective job insecurity with sickness presenteeism among public sector employees. J Occup Environ Med 52:830–835
Hockey GRJ (1997) Compensatory control in the regulation of human performance under stress and high workload: a cognitive-energetical framework. Biol Psychol 45:73–93
Höge T (2011) Perceived flexibility requirements at work and the entreployee-work-orientation: concept and measurement. Psychol Everyday Act 4:3–21
Höge T, Hornung S (2013) Perceived flexibility requirements: exploring mediating mechanisms in positive and negative effects on worker well-being. Econ Ind Democr 0:1–24
Horner RD, Matthews G, Yi MS (2012) A conceptual model of physician work intensity: guidance for evaluating policies and practices to improve health care delivery. Med Care 50:654–661
Janssen D, Nachreiner F (2004) Flexible Arbeitszeiten [Flexible working hours]. Schriftenreihe der Bundesanstalt für Arbeitsschutz und Arbeitsmedizin. Bundesanstalt für Arbeitsschutz und Arbeitsmedizin, Dortmund (in German)
Janssens H, Clays E, De Clerq B et al (2013) The relation between presenteeism and different types of future sickness absence. J Occup Health 55:132–141
Johns G (2011) Attendance dynamics at work: the antecedents and correlates of presenteeism, absenteeism, and productivity loss. J Occup Health Psychol 16:483–500
Joksimovic L, Siegrist J, Meyer-Hammer M et al (1999) Overcommitment predicts restenosis after coronary angioplasty in cardiac patients. Int J Behav Med 6:356–369
Joyce K, Pabayo R, Critchley JA et al (2010) Flexible working conditions and their effects on employee health and wellbeing. Cochrane Database Syst Rev. doi:10.1002/14651858.CD008009.pub2
Karasek RA (1979) Job demands, job decision latitude, and mental strain: implications for job redesign. Adm Sci Q 24(2):285–308
Kaur S, Kremer M, Mullainathan S (2010) Self-control and the development of work arrangements. Am Econ Rev 100:624–628
Kieschke U, Schaarschmidt U (2008) Professional commitment and health among teachers in Germany: a typological approach. Learn Instr 18:429–437
Kivimäki M, Head J, Ferrie JE et al (2005) Working while ill as a risk factor for serious coronary events: the Whitehall II Study. Am J Public Health 95:98–102
Korunka C, Kubicek B (2013) Beschleunigung im Arbeitsleben—neue Anforderungen und deren Folgen [Speeding up working life: new demands and their consequences]. In: Junghanns G, Morschhäuser M (eds) Bundesanstalt für Arbeitsschutz und Arbeitsmedizin. Immer schneller, immer mehr. Springer, Wiesbaden, pp 17–39 (in German)
Kowalski H (2013) Gehirndoping am Arbeitsplatz [Neuro-Enhancement]. In: Badura B, Ducki A, Schröder H, Klose J, Meyer M (eds) Fehlzeiten-Report 2013. Verdammt zum Erfolg—Die süchtige Arbeitsgesellschaft? Springer, Berlin, pp 27–34 (in German)
Krause A, Dorsemagen C, Stadlinger J et al (2012) Indirekte Steuerung und interessierte Selbstgefährdung: Ergebnisse aus Befragungen und Fallstudien. Konsequenzen für das Betriebliche Gesundheitsmanagement [Indirect control and self-endangering behavior]. In: Badura B, Ducki A, Schröder H, Klose J, Meyer M (eds) Fehlzeiten-Report 2012. Gesundheit in der flexiblen Arbeitswelt: Chancen nutzen—Risiken minimieren. Springer, Berlin, pp 191–202 (in German)
Krause A, Baeriswyl S, Berset M et al (2014) Selbstgefährdung als Indikator für Mängel bei der Gestaltung mobil-flexibler Arbeit: Zur Entwicklung eines Erhebungsinstruments [Self-endangering behavior as an indicator for shortcomings in the design of mobile and flexible work]. Wirt Psych 4(14):49–59 (in German)

Krieger R, Pekruh U, Lehmann M et al (2012) Fünfte Europäische Erhebung über die Arbeitsbedingungen 2010. Ausgewählte Ergebnisse aus Schweizer Perspektive. Staatssekretariat für Wirtschaft SECO, Bern (in German)

Lazarus RS, Folkman S (1984) Stress, appraisal and coping. Springer, New York

Leitner K, Greiner B, Oesterreich R et al (1987) Analyse psychischer Belastung in der Arbeit: das RHIA-Verfahren; Handbuch. Verlag TÜV Rheinland, Köln (in German)

Locke EA, Latham GP (2002) Building a practically useful theory of goal setting and task motivation: a 35-year odyssey. Am Psychol 57(9):705

Lohmann-Haislah A (2012) Stressreport Deutschland 2012. Psychische Anforderungen, Ressourcen und Befinden [Stress report Germany 2012: psychological demands, resources, and well-being]. Bundesanstalt für Arbeitsschutz und Arbeitsmedizin, Dortmund (in German)

Lu L, Cooper CL, Lin HY (2013a) A cross-cultural examination of presenteeism and supervisory support. Career Dev Int 18:440–456

Lu L, Lin HY, Cooper CL (2013b) Unhealthy and present: motives and consequences of the act of presenteeism among Taiwanese employees. J Occup Health Psychol 18:406–416

Mearns J, Cain JE (2003) Relationship between teachers' occupational stress and their burnout and distress: roles of coping and negative mood regulation expectancies. Anxiety Stress Coping 16(1):71–82

Mearns K, Hope L (2005) Health and well-being in the offshore environment: the management of personal health (No. 305). HSE Research Report 376

Moldaschl M, Voss GG (2003) Subjektivierung von Arbeit [Subjectivizing work]. Rainer Hampp, München (in German)

Näswall K, Hellgren J, Sverke M (2008) The individual in the changing working life. Cambridge University Press, Cambridge

Nielsen MB, Knardahl S (2014) Coping strategies: a prospective study of patterns, stability, and relationships with psychological distress. Scand J Psychol 55(2):142–150

Oldham GR, Hackman JR (2010) Not what it was and not what it will be: the future of job design research. JOB 479:463–479

Pangert B, Schüpbach H (2013) Die Auswirkungen arbeitsbezogener erweiterter Erreichbarkeit auf Life-Domain-Balance und Gesundheit [The effects of work-related extended availability on life-domain-balance and health]. Bundesanstalt für Arbeitsschutz und Arbeitsmedizin BAuA, Dortmund (in German)

Peters K (2011) Indirekte Steuerung und interessierte Selbstgefährdung: Eine 180-Grad-Wende bei der betrieblichen Gesundheitsförderung [Indirect control and an interest in self-threatening behavior: an about turn in company health promotion]. In: Kratzer N, Dunkel W, Becker K, Hinrichs S (eds) Arbeit und Gesundheit im Konflikt: Analysen und Ansätze für ein partizipatives Gesundheitsmanagement. Ed. Sigma, Berlin, pp 105–122 (in German)

Pfeiffer S (2012) Technologische Grundlagen der Entgrenzung: Chancen und Risiken [Technological foundations of exclusion: chances and risks]. In: Badura B, Ducki A, Schröder H, Klose J, Meyer M (eds) Fehlzeiten-Report 2012. Gesundheit in der flexiblen Arbeitswelt: Chancen nutzen—Risiken minimieren. Springer, Berlin, pp 15–21 (in German)

Posthuma RA, Campion MC, Masimova M et al (2013) A high performance work practices taxonomy. Integrating the literature and directing future research. J Manage 39:1184–1220

Rau R (2012) Erholung als Indikator für gesundheitsförderlich gestaltete Arbeit [Recuperation as an indicator for health-promoting work design]. In: Badura B, Ducki A, Schröder H, Klose J, Meyer M (eds) Fehlzeiten-Report 2012. Gesundheit in der flexiblen Arbeitswelt: Chancen nutzen—Risiken minimieren. Springer, Berlin, pp 181–190 (in German)

Rothmann S, Jorgensen LI, Hill C (2011) Coping and work engagement in selected South African organisations. SA J Ind Psychol 37(1):1–11

Rousseau DM (1997) Organizational behavior in the new organizational era. Annu Rev Psychol 48:515–546

Schaufeli WB, Salanova M, González-Romá V et al (2002) The measurement of engagement and burnout: a two sample confirmatory factor analytic approach. J Happiness Stud 3(1):71–92

Schaufeli WB, Taris TW, Van Rhenen W (2008) Workaholism, burnout, and work engagement: three of a kind or three different kinds of employee well-being? Appl Psychol 57(2):173–203
Schönpflug W (1987) Beanspruchung und Belastung bei der Arbeit—Konzepte und Theorien [Strain and stress at work: concepts and theories]. In: Kleinbeck U, Rutenfranz J (eds) Enzyklopädie der Psychologie: Themenbereich D Praxisgebiete, Serie III Wirt-schafts-, Organisations- und Arbeitspsychologie, Band I Arbeitspsychologie. Hogrefe, Göttingen, pp 130–184 (in German)
Semmer NK, Meier LL (2009) Individual differences, work stress and health. In: Cooper CL, Quick JC, Schabracq M (eds) International Handbook of Work and Health Psychology, 3rd edn. Wiley, Chichester, pp 99–121
Semmer NK, Grebner S, Elfering A (2010) "Psychische Kosten" von Arbeit: Beanspruchung und Erholung, Leistung und Gesundheit [Psychological costs of work]. In: Kleinbeck U, Schmidt K-H (eds) Enzyklopädie der Psychologie: Themenbereich D Praxisgebiete, Serie III Wirtschafts-, Organisations- und Arbeitspsychologie, Band 1 Arbeitspsychologie. Hogrefe, Göttingen, pp 325–370 (in German)
Siegrist J, Starke D, Chandola T, Godin I, Marmot M, Niedhammer I, Peter R (2004) The measurement of effort–reward imbalance at work: European comparisons. Soc Sci Med 58(8):1483–1499
Voss GG, Pongratz HJ (1998) Der Arbeitskraftunternehmer. Eine neue Grundform der 'Ware Arbeitskraft'? [The entreployee: a new form of labor power]. Kölner Z Soziol Soz 50:131–158 (in German)
Wallace SL, Lee J, Lee SM (2010) Job stress, coping strategies, and burnout among abuse-specific counselors. J Employ Couns 47:111–122
Welbourne JL, Eggerth D, Hartley TA et al (2007) Coping strategies in the workplace: relationships with attributional style and job satisfaction. J Vocat Behav 70(2):312–325
Welsh DT, Ordóñez L (2014) The dark side of consecutive high performance goals: linking goal setting, depletion, and unethical behavior. Organ Behav Hum Decis Process 123:79–89
Widmer PS, Semmer NK, Kälin W et al (2012) The ambivalence of challenge stressors: time pressure associated with both negative and positive well-being. J Vocat Behav 80:422–433
Wirtz A (2010) Gesundheitliche und soziale Auswirkungen langer Arbeitszeiten [Effects of long working hours on health and social life]. Bundesanstalt für Arbeitsschutz und Arbeitsmedizin BAuA, Dortmund (in German)
Zapf D, Semmer NK (2004) Stress und Gesundheit in Organisationen [Stress and health in organizations]. Enzyklopädie der Psychologie, Themenbereich D, Serie III 3:1007–1112 (in German)
Zok K, Dammasch H (2012) Flexible Arbeitswelt: Ergebnisse einer Beschäftigtenbefragung [Flexible working world: results of an employee survey]. In: Badura B, Ducki A, Schröder H, Klose J, Meyer M (eds) Fehlzeiten-Report 2012: Gesundheit in der flexiblen Arbeitswelt. Springer, Berlin, pp 39–52 (in German)

Work-Related Anxieties and Work-Place Phobia: A Topical Phenomenon at the Interface of Clinical and Organizational Research and Practice

Beate Muschalla

Abstract

Work-places inherently contain various anxiety-provoking characteristics, including rivalries between colleagues, control issues with supervisors, demands for achievement, dangerous environments, and uncertainty concerning future developments. Work-related anxieties can present in various guises and include situational anxieties, social anxieties, anxiety about being inadequate or underachieving at work, health-related anxieties, or—in the worst case—complex work-place phobias. Work-related anxieties lead to problems with participation in work resulting in (long-term-) sick leave or even disability pensions. In 10 years of research on work-related anxieties, we have found that 30–60 % of individuals undergoing medical/vocational rehabilitation have work-related anxieties that often pose barriers to returning to work. Even 5 % of mentally healthy employees admit that they sometimes ask for a sick leave certificate when experiencing work-related anxieties. Future research on work-related mental health problems should not only focus on rehabilitation and treatment of work-related anxieties, but also earlier stages. The concept of work-related anxieties gives rise to ideas that can be useful for mental health-oriented job analysis, employee–work-place fit, and job design for employees with different psychological constitutions and capacities.

This chapter is a translated and modified version of a recently published paper: Muschalla B (2014) Arbeitsbezogene Ängste in Forschung und Praxis—ein aktuelles Schnittstellenphänomen. Z Arb Organ 58:206–214.

B. Muschalla (✉)
Department of Work and Organizational Psychology, University of Potsdam, Karl-Liebknecht-Strasse 24-25, 14476 Potsdam-Golm, Germany
e-mail: beate.muschalla@gmx.de

M. Wiencke et al. (eds.), *Healthy at Work*, DOI 10.1007/978-3-319-32331-2_5

1 Introduction

Mental health is a problem of growing importance in the work-place. At over 60 days, the mean duration of sick leave for mental disorders is about twice as long as that of sick leave for somatic disorders (Dewa et al. 2014). Thirteen per cent of people with depression or anxiety disorders retire early as compared with 4 % of controls (Wedegaertner et al. 2013). Scientists and politicians need to develop ideas for preventing and addressing work-related mental health problems. In this chapter, we will introduce a clinical framework for considering work-related anxieties (Linden and Muschalla 2007; Muschalla 2014) with the aim of facilitating future discussions on mental health in the work-place. This framework provides those involved in work and organizational research and practice with a means of focusing on specific fundamental intra-psychic experiences of, and reactions to, work-place conditions. We will first present a short overview on present concepts and empirical findings concerning work and mental health. Next, we will introduce concepts about how work-places may trigger anxiety and present the differential diagnosis of work-related anxieties. We will also report the operationalization and empirical evidence on work-related anxieties that we have developed over 10 years of our own research. Finally, we will discuss how the concept of work-related anxieties may contribute to the field of mental health research and practice.

2 Mental Disorders and Work from a Clinical Perspective

Epidemiologic studies over some decades have shown that the 12-month prevalence of mental disorders in persons aged between 18 and 65 years is about 30 % (Wittchen et al. 2011). Mental disorders have a chronic course in 40 % of subjects (Linden et al. 2012) and are often accompanied by enduring disability because of attendant limitations in capacity and ability to participate (Muschalla et al. 2012). These problems manifest especially in the context of the work-place (Blank et al. 2008; Greenberg et al. 1999; Haslam et al. 2005; Linden and Weidner 2005; Muschalla et al. 2012; Stansfeld et al. 2012). With a mean of 40–60 or more days of sick leave absence for each subject (Dewa et al. 2014; Techniker Krankenkasse 2012), mental disorders are responsible on average for more sick leave absences than cardiovascular diseases and musculoskeletal disorders (Techniker Krankenkasse 2012, WidO Wissenschaftliches Institut der AOK 2011). Because work-related mental health problems have specific clinical implications, some specific work-related diagnoses such as work-related reactive depression have been suggested (Mezerai et al. 2006). In psychosomatic rehabilitation clinics, physicians and therapists are daily confronted with the question of vocational reintegration of about 80 % of their patients after (long-term) sick leave absence. Specific work-related anxieties complicate return to work in 60 % of cases (Deutsche Rentenversicherung Bund 2005; Linden and Muschalla 2007). Meanwhile, data concerning international developments and evaluations of work-related interventions for mental health issues have been published (e.g., Beutel et al. 2004; Hillert et al. 2009; Lagerveld et al. 2012; Noordik et al. 2013).

3 Work and Mental Health from a Work and Organizational Psychology Research Perspective

In work and organizational psychology, there is a long tradition of research on coping capacities in relation to work (e.g., Frese and Fay 2001; Kaluza 2011; Schaarschmidt and Fischer 2001), as well as on working conditions and their relationship with mental health. In this research, the concepts of burnout and stress are widely used (Berger et al. 2012; Burisch 2010; Limm et al. 2011; Maslach and Jackson 1981; Semmer 2006).

The term *burnout* describes work-related stress with the dimensions of emotional exhaustion, depersonalization, and reduced personal accomplishment. Burnout is not a diagnosis (Berger et al. 2012); however, this term is occasionally used as a diagnosis in clinical practice when the intention is to hint that there is a mental health problem related to the working situation. A mental health problem that is *caused by overloading at work* seems to gain more societal acceptance than does a "mental disorder." The term work-related anxieties is more specific than the term burnout in that it more accurately describes psychopathological qualities of anxiety and provides a better basis for interventions.

When originally introduced by Selye (1956), the term *stress* described an acute physiological reaction that was repeatedly activated by exposure to an ongoing stressor. In themselves, such physiological reactions are neutral and do not imply anything about the emotional states that the affected individual associates with the activating factor(s). Lazarus proposed a concept of transactional stress that introduced the idea that individuals cognitively evaluate potentially stressful situations (Lazarus 1993). This model explains why anxiety may develop as a result of an individual's perception of a situation as "dangerous." More recent research has concentrated on describing subclinical stress reactions such as irritation (Mohr et al. 2007). The differential diagnostic concept of work-related anxieties goes beyond describing physiological stress reactions and understanding anxiety as global phenomena; rather, it focuses on various facets of the experience of anxiety and different forms of behavioral reactions involving avoidance of the work-place (Muschalla and Linden 2009, 2013a). Assessing work-related anxieties in this more detailed way can provide hints about important work–health issues such as the following. How can an employee with a certain anxiety be advantageously deployed? How can provocation of anxiety in work-places be minimized? When and how should supervisors react to a coworker's problematic behavior, such as frequent absences or avoidance of certain tasks (Muschalla and Linden 2013a)?

4 How Does Work Provoke Anxiety?

Until now, there has been broad research on the relationship between work and mental health problems (e.g., Angerer et al. 2014; Rau and Henkel 2013; Stansfeld and Candy 2006). In the following, we will describe how various conditions at work can provoke anxiety. Work-places inherently contain the following potentially anxiety-provoking stimuli:

- The experience or even imagining failure is an unconditional anxiety-provoking stimulus. *Requirements for achievement* can be perceived as qualitatively or quantitatively overloading; for example, work duties exceeding the employee's capacities or a persistently excessive work load (Oppolzer 2010). High requirements are perceived as problematic when accompanied by little possibility of control and little social support at work (Sanne et al. 2005). In addition to external achievement requirements, personal expectations concerning work results—that is, motives for personal achievement or level of ambition—are also relevant (Heckhausen 1989; Hoppe 1930; McClelland et al. 1976). When there is an ongoing experience of not succeeding, anxiety about failing can develop.
- *Supervisors* are also potentially anxiety-provoking, solely because of their controlling and sanctioning functions (Gilbreath and Benson 2004; Panse and Stegmann 2007; Tepper 2007). Regular communication across hierarchic levels and improving transparency can help to minimize unnecessary development of anxiety (Perlow and Williams 2003).
- Employees are often grouped together for most of their time at work in small spaces such as offices. Thus, it is unsurprising that *social conflicts* may arise at work; these are often unavoidable. "Pack behavior and pecking order"—that is, the natural battle between members of a group for their own rank (Yalom 1970), competitive behavior between group members (Stroebe et al. 2007; Thomas and Hynes 2007), and used danger signals such as gaze and body language (Eibl-Eibesfeldt 2004)—can be anxiety-provoking. Additionally, perceived bullying or discrimination may trigger anxiety (Bowling and Beehr 2006; Escartín et al. 2011; Freudenberger 1974; McDonald 2012; Morgan et al. 2013).
- In addition to conflicts with colleagues, *endangerment by third parties*, such as verbal or physical assaults by students, patients, customers, or even raids, can occur. In particular, individuals working as fire fighters, police officers, psychiatric nurses, and bank employees are exposed to these dangers (Alexy and Hutchins 2006; Johnson 2009; Kuchenbecker and Amann 2010; Laposa et al. 2003; MacDonald et al. 2003; Mahan et al. 2010; Price et al. 2006).
- Accidents or other *dangers in the work environment*, such as allergens or chemical substances (Nakazawa et al. 2005; Nicholson and Vincenti 1994), may both contribute to health problems and trigger anxiety concerning danger to health.
- Experiencing *little control* is also an unconditional anxiety-provoking stimulus (Christianson et al. 2011), especially for persons with a low tolerance for

uncertainty (Becker and Margraf 2002; Linden and Hautzinger 2008). At work there may be little scope for action, little job control (Bundesanstalt für Arbeitsschutz und Arbeitsmedizin [BAuA, Federal Institute for Occupational Safety and Health] 2012; Rau and Henkel 2013) and a lack of transparency that can intensify the experience of a lack of control and provoke anxiety. Examples are uncertainty about who will next be transferred or downsized (Forcella et al. 2007), the introduction of new technologies (Beutel et al. 2004; Smith et al. 1999; Hallowell 1999), or even wide-reaching changes such as anticipated company closure or merger (Campbell and Pepper 2006). Additionally, daily minor sources of uncertainty, such as unreliable work equipment, lack of information, and changes in work duties without warning (Leach et al. 2013), can provoke worries and tension.

- The loss of a work-place can feel like a threat to existence and is a frequently mentioned burdensome life event (Holmes and Rahe 1967; Spera et al. 1994). Loss of a work-place also means a loss of social status and therefore a psychological burden (Strazdins et al. 2004). Even subjectively perceived work-place insecurity may feel threatening (Campbell and Pepper 2006; Elst et al. 2014; Forcella et al. 2007; Mohr and Otto 2011; Paul and Moser 2009).

5 How Can Anxiety at Work Be Conceptualized?

According to Lang's three components model (1968), anxiety can be described on three levels that are independent of its development: the levels of cognition, physiological reaction, and avoidance behavior. Anxiety is always a dimensional phenomenon: experiencing anxiety does not equate with having an anxiety disorder. The clinical significance of anxiety depends on the intensity and frequency of the anxiety reaction and the quality of the stimulus (American Psychiatric Association [APA] 1994; Gastpar et al. 2003). Anxiety at work can be described from different perspectives. First, it can be described as a *global construct*, as in the sense of a subclinical proneness to anxiety (e.g., Mohr et al. 2007; Spielberger et al. 1983). Second, it can be described according to the *anxiety-provoking stimulus* (e.g., Kittner 2003; Payne et al. 1982), or, third, according to the *psychopathological processes* involved, for example anxiety in the form of excessive worrying or in the form of panic symptoms with flight reaction (Gastpar et al. 2003). In the next section, the latter, a psychopathologically based differential diagnostic concept of work anxiety, will be introduced.

6 Psychopathological Conceptualization of Work-Related Anxieties

Both global measures of anxiety (Spielberger et al. 1983) and stimulus-oriented concepts (Payne et al. 1982) treat anxiety as a one-dimensional phenomenon. They involve no differentiation between various qualities of anxiety in the sense of psychopathological differential diagnoses. However, a more differentiated understanding of anxiety is a prerequisite for curing affected people, successful vocational reintegration and return to work, and effective work-place adjustments for employees with tendencies to anxiety.

The concept of work-related anxieties has been developed in psychosomatic rehabilitation, a setting in which people with work-place-related mental health problems are over-represented (Linden and Muschalla 2007; Muschalla and Linden 2009). According to the classifications of anxiety disorders in internationally accepted diagnostic manuals, the Diagnostic and Statistical Manual of Mental Disorders (DSM) (APA 1994) and International Classification of Diseases (World Health Organization 1992), as well as the Arbeitsgemeinschaft für Methodik und Dokumentation in der Psychiatrie (AMDP) manual for the assessment and documentation of psychopathology (AMDP, Guy et al. 1982), different types of anxiety can occur in work contexts. We have developed a typology of specific work-related anxieties and evaluated it based on several studies of psychosomatic rehabilitation patients who were enrolled in our studies because of work-place problems and long-term sick leave (Linden and Muschalla 2007; Muschalla 2008). These are the common work-related anxieties:

- *Stimulus-related phobic anxiety and avoidance behavior at work.* This category includes specific anxieties towards defined places, certain tasks, or specific situations at work. Affected persons react with tension, anxiety, or even panic when confronted with, or even thinking about, a specific place, work task or (non-social) situation at work. Anxiety decreases when the anxiety-provoking stimulus can be avoided. Anxiety-provoking stimuli include computer programs, scaffolds, specific work duties, working times, work-places, and work environments, such as night shift or a single duty with onerous responsibilities. These anxieties are often "learned," characteristically developing after an initial event (e.g., anxiety after having made a serious mistake with a new computer program or having had an accident on a scaffold).
- *Social anxiety at work.* Social anxieties are related to contacts with colleagues, supervisors, or third parties such as clients, students, or patients. These anxieties may reflect a constitutional social insecurity and gaze aversion or have been learned as a result of poor social competency and negative reactions from others. Social anxieties are characterized by one of two extremes: being excessively and awkwardly shy or exhibiting aggressive behavior towards others. Employees with social anxieties have difficulty participating adequately in social situations at work, maintaining their viewpoints in a working team, or exposing themselves to criticism from others, for example when required to give an oral presentation at a conference or in a team meeting. Problems may also occur in the tearoom or canteen when taking a break with colleagues.
- *Health- and body-related anxieties.* This category includes all types of hypochondriac anxieties, such as excessive worrying about one's own physical integrity. Within the context of work, this includes the belief that some aspect of the work environment may be harmful to health (e.g., the toner powder in the

office printers or background noise in an open-plan office) or exacerbate an existing illness (e.g., the belief that back pain is aggravated by work or that work stress may cause another myocardial infarction). Affected persons experience symptoms such as an accelerated heart rate or mild paresthesiae in various body parts. As a consequence, they may perform required functions incorrectly and avoid specific work duties or work-places they perceive as hazardous to their health or both.

- *Anxiety about inadequacy at work*. Anxieties about inadequacy include anxieties about being insufficiently qualified, easily over-challenged, not knowing enough, or not being sufficiently competent to perform required tasks and therefore being prone to making mistakes and failing. This category includes anxiety about changes, such as undertaking new tasks or learning new procedures that are necessary for incorporating structural, personnel, or technical changes in the firm. Persons with such anxieties often react tensely and nervously and may appear to be relatively helpless or to be seeking help.
- *Work-related general worrying*. Worrying is a type of anxiety that occurs mainly "in the head." It is characterized by relatively constant tendencies to generalized worrying about minor and everyday matters at work, anticipation of problems everywhere, and great difficulty in dealing with uncertainty. There is a constant worrisome thinking about work problems, even in free time. Individuals with severe work-related worrying see potential worries and problems that might be averted by particular actions everywhere. For example, affected individuals may worry about whether the computer has been turned off correctly, that certain tasks may not be finished correctly by colleagues, or even that they will be sacked from their jobs. People with generalized worrying are often over-conscientious about fulfilling their working duties and may even keep an eye on colleagues for whom they feel responsible. These people are usually very attentive to all official announcements from the firm. They get feedback that they should "not always think about problems" or colleagues may ask them to stop seeking information about, and attempting to control, other's work. Their family and friends express concern that the worrying person is unable to get away from work, ask for a vacation, or take sick leave because of their inability to tolerate the uncertainty associated with being absent from work.
- *Work-place phobia*. Work-place phobia is a specific form of work-related anxiety. According to the World Health Organization definitions of phobias (1992), an individual has a work-place phobia when approaching, being at, or simply imagining the work-place leads to acute physiological arousal that amounts to a panic reaction. As a consequence, these individuals often exhibit avoidance behavior in the form of taking sick leave (Smith 2009), which in turn operates as a negative reinforcement for the anxiety (Skinner 1969). There is often generalization of the avoidance behavior, for example avoidance of the street where the firm is located, avoidance of events at which one might meet colleagues or supervisors, or even arousal when speaking about work (Haines et al. 2002).

7 Psychopathological Differential Diagnostic Operationalization of Work-Related Anxieties and Empirical Evidence

The *Job Anxiety Scale* (JAS) is a self-rating instrument that describes the psychopathological dimensions of work-related anxieties presented in Sect. 5 (Linden et al. 2008; Muschalla et al. 2010). This scale is based on psychopathological concepts of cognitive, emotional, and physiological anxiety symptoms. The JAS is not designed to make diagnoses; rather, it quantifies the perceived degree of work-related anxieties. It covers the following main dimensions: stimulus-related anxieties and avoidance behavior, social anxiety and cognitions of persecution, health- and body-related anxieties, perception of insufficiency, and work-related general anxieties. The 13 items of the JAS provide a means of economic "Workplace phobia screening" (Muschalla and Linden 2008, 2009, 2011) that particularly assesses work-related anxiety symptoms according to the two factors of panic and avoidance.

The semi-structured *Work-Anxiety Interview* (Linden and Muschalla 2007; Muschalla and Linden 2013a) is available for categorical differential diagnosis of work-related anxieties. In this tool, the categories of anxiety disorders have been adapted for the work context from the DSM-IV criteria (APA 1994) for anxiety disorders. Studies using the Work-Anxiety Interview criteria have reported that over 60 % of persons undergoing psychosomatic rehabilitation have some type of work-related anxiety. Fourteen percent of these patients reportedly have a specific stand-alone type of work-related anxiety without a simultaneous general (not work-specific) anxiety disorder (Linden and Muschalla 2007). The partial independence of work-related anxieties from general mental disorders has been confirmed by a number of studies of subjects with psychosomatic and somatic disorders (Muschalla et al. 2010; Muschalla and Linden 2009, 2012, 2013a). A specific socio-medical problem associated with work-related anxieties is long-term sick leave. Long-term sick leave leads to both socioeconomic and existential problems for those affected. Work-related anxieties are related to sick leave in a specific way (Muschalla 2008; Smith 2009). Five percent of individuals undergoing somatic rehabilitation and 17 % of those with psychosomatic disorders have the most severe form of workplace anxiety, namely work-place phobia (Linden et al. 2009; Muschalla 2008). In one study of subjects with psychosomatic disorders with work-place phobia, 80 % were no longer at work: 57 % were on ongoing sick leave (average duration 24 weeks) and the other 23 % had become unemployed as a result of the illness (Muschalla 2008; Muschalla and Linden 2009). Work-place phobia is a specific mental health problem with its own clinical significance. Muschalla and Linden (2014) reported that 10 % of patients with chronic mental illness in a general practice had work-place phobia. Of a sample of presently employed persons who were explicitly not being treated for mental health problems, 5 % reported tendencies to avoid their work-places because of work-related anxiety (Muschalla et al. 2013).

Another characteristic that differentiates work-related anxieties from general anxiety disorders is that their incidence differs between occupations: specific social anxieties reportedly occur more frequently in office employees and anxiety about being inadequate is more frequent in health practice employees, whereas worrying has a similar incidence in most occupational settings (Muschalla and Linden 2013b). Different types of work-related anxiety also result in different impairments: whereas work-related worrying is accompanied by difficulty in making decisions, more overwork hours and ongoing tension, work-place phobia is characteristically associated with the longest sick leave durations and most frequent work-place losses (Muschalla 2008).

8 How May the Concept of Work-Related Anxieties Be Used in the Future?

For both research and practice concerning work, some key points are relevant:

- When *analyzing work-place-related mental health problems*, it is important to keep in mind that the different specific stimuli in different occupations result in different types of work-related anxieties and consequently types of impairment that must be recognized and dealt with.
- On the level of (therapeutic) *interventions for affected individuals*, the type of anxiety must first be identified to enable selection of an appropriate intervention (Linden and Hautzinger 2008). Additionally, for *work-place interventions*—that is, interventions from a work-place and organizational perspective—knowledge about the specific anxieties that may occur in particular types of work is important. When analyzing possible hazards, potentially anxiety-provoking factors must be identified and the work-place designed to minimize development of anxiety (e.g., installing alarm systems and arranging de-escalation trainings or both for service employees often confronted by angry customers).
- When *selecting personnel*—that is, choosing employees for different tasks—whether individuals are particularly vulnerable or predisposed to various specific anxieties should be assessed. In terms of work psychology, we are referring here to the person–environment/role fit (Kulik et al. 1987). The person–environment-fit perspective is useful when considering work-place adjustment, analyzing the demands of a job (Cramer and Davidhizar 2000; French 1973; Marcus 2011), and designing work-places for people with disabilities (Hillert et al. 2009). Considering this perspective facilitates practicing inclusion and integration (United Nations 2006; Bundesministerium für Arbeit und Sociales [Federal Ministry of Labor and Social Services] 2011) of employees with tendencies to specific (even subclinical) anxiety. For example, an employee who is extremely conscientious but has tendencies to social anxiety.

In the future, implementing concepts for sustainably integrating people with mental disorders into the workforce will be of interest. Early intervention for (Nash-Wright 2011) and prevention of work-place absence because of work-related anxieties can be accomplished if known current work-place demands (BAuA 2012; Organisation for Economic Co-operation and Development [OECD] 2012) and thus their potential anxiety-provoking aspects are identified. The concept of work-related anxieties can be helpful for conceptualizing and analyzing psychological hazards or work-place adjustments: Bringing together clinical (differential diagnosis of anxieties) and work-organizational-psychological aspects (triggers for anxiety in the work-place) makes it possible to assess (with the aim of prevention) specific aspects of the work-place that have been clinically established as having potentially unfavorable impacts that could lead to sick leave and work disability. Epidemiologic studies have shown that about 30 % of the general population has mental health disorders (Wittchen et al. 2011) and therefore reduced potential to achieve. For these workers, some aspects of work and required achievement can be overtaxing. Examples include clocked work, competition-oriented work, and work requirements that are too high for employees with anxiety about inadequacy. Other examples are the type and degree of social exposure for employees with social anxiety and the work-place environment for employees with hypochondriac anxiety (Muschalla and Linden 2013a).

Psychological demand and hazard analysis can be conceptualized such that they cover the potentially anxiety-provoking and anxiety-relevant aspects of work (see Sect. 4). As to *person–environment fit*, attempts can be made to match work-place demands and other work factors with an individual's vulnerability to anxiety. In the context of *hazard analysis*, whether the majority of employees with the same duties report high work-related anxiety under the specific work conditions in question should be checked.

Finally, from a *scientific* point of view further research is needed: it is still unclear which work-place factors will lead to employees with specific clinical problems requiring sick leave. Developing and evaluating concepts of clinical work demand and hazard analysis may be helpful (e.g., Metz and Rothe 2014; Molodynski et al. 2013; Zielke 2001).

Appendix: Differential Diagnosis of Work-Related Anxieties

Modified and Shortened Version with Permission from Muschalla and Linden 2013a; Muschalla 2008

1. **Stimulus-related phobic anxiety and avoidance behavior at work** This category includes specific anxieties towards a defined place, a certain task, or specific situations at work. Affected persons react with tension, anxiety, or even panic when confronted with, or even thinking about, specific places, work tasks, or (non-social) situations at work. Anxiety decreases when the anxiety-provoking stimulus can be avoided. Anxiety-provoking stimuli include computer programs, scaffolds, specific work duties, working times, work-places, and work environments, such as night shift or a single duty with onerous responsibilities. These anxieties are often "learned," characteristically developing after an initial event (e.g., anxiety

(continued)

after having made a serious mistake with a new computer program or having had an accident on a scaffold).
2. **Social anxiety at work** Social anxieties are related to contacts with colleagues, supervisors, or third parties such as clients, students, or patients. These anxieties may reflect a constitutional social insecurity and gaze aversion or have been learned as a result of poor social competency and negative reactions from others. Social anxieties are characterized by one of two extremes: being excessively and awkwardly shy or exhibiting aggressive behavior towards others. Employees with social anxieties have difficulty participating adequately in social situations at work, maintaining their viewpoints in a working team, or exposing themselves to criticism from others, for example when required to give an oral presentation at a conference or in a team meeting. Problems may also occur in the tearoom or canteen when taking a break with colleagues.
3. **Health- and body-related anxieties** This category includes all types of hypochondriac anxieties, such as excessive worrying about one's own physical integrity. Within the context of work, this includes the belief that some aspect of the work environment may be harmful to health (e.g., the toner powder in the office printers or background noise in an open-plan office) or exacerbate an existing illness (e.g., the belief that back pain is aggravated by work or that work stress may cause another myocardial infarction). Affected persons experience symptoms such as an accelerated heart rate or mild paresthesiae in various body parts. As a consequence, they may perform required functions incorrectly and avoid specific work duties or work-places they perceive as hazardous to their health or both.
4. **Anxiety about inadequacy at work** Anxieties about inadequacy include anxieties about being insufficiently qualified, easily over-challenged, not knowing enough, or not being sufficiently competent to perform required tasks and therefore being prone to making mistakes and failing. This category includes anxiety about changes, such as undertaking new tasks or learning new procedures that are necessary for incorporating structural, personnel, or technical changes in the firm. Persons with such anxieties often react tensely and nervously and may appear to be relatively helpless or seeking help.
5. **Work-related general worrying** Worrying is a type of anxiety that occurs mainly "in the head." It is characterized by relatively constant tendencies to generalized worrying about minor and everyday matters at work, anticipation of problems everywhere, and great difficulty in dealing with uncertainty. There is a constant worrisome thinking about work problems, even in free time. Individuals with severe work-related worrying see potential worries and problems that might be averted by particular actions everywhere. For example, affected individuals may worry about whether the computer has been turned off correctly, that certain tasks may not be finished correctly by colleagues, or even that they will be sacked from their jobs. People with generalized worrying are often over-conscientious about fulfilling their working duties and may even keep an eye on colleagues for whom they feel responsible. These people are usually very attentive to all official announcements from the firm. They get feedback that they should "not always think about problems" or colleagues may ask them to stop seeking information about and attempting to control other's work. Their family and friends express concern that the worrying person is unable to get away from work, ask for a vacation, or take sick leave because of their inability to tolerate the uncertainty associated with being absent from work.

(continued)

6. **Work-place phobia**
Work-place phobia is a specific form of work-related anxiety. According to the World Health Organization definitions of phobias (1992), an individual has a work-place phobia when approaching, being at, or simply imagining the work-place leads to acute physiological arousal that amounts to a panic reaction. As a consequence, these individuals often exhibit avoidance behavior in the form of taking sick leave (Smith 2009), which in turn operates as a negative reinforcement for the anxiety (Skinner 1969). There is often generalization of the avoidance behavior, for example avoidance of the street where the firm is located, avoidance of events at which one might meet colleagues or supervisors, or even arousal when speaking about work (Haines et al. 2002).

References

Alexy EM, Hutchins JA (2006) Workplace violence: a primer for critical care nurses. Crit Care Nurs Clin North Am 18:305–312

Angerer P, Glaser J, Gündel H et al (eds) (2014) Psychische und psychosomatische Gesundheit in der arbeit. Wissenschaft, Erfahrungen, Lösungen aus Arbeitsmedizin, Arbeitspsychologie und Psychosomatischer Medizin. Ecomed Medizin, Heidelberg (in German)

APA (1994) Diagnostic and statistical manual of mental disorders (DSM-IV). American Psychiatric Association, Washington

BAuA (2012) Stressreport Deutschland 2012. Psychische Anforderungen, Ressourcen und Befinden. BauA, Federal Institute for Occupational Safety and Health, Berlin (in German)

Becker ES, Margraf J (2002) Generalisierte Angststörungen. Ein Therapieprogramm. Beltz, Weinheim (in German)

Berger M, Linden M, Schramm E Wagner S et al (2012) Positionspapier der Deutschen Gesellschaft für Psychiatrie, Psychotherapie und Nervenheilkunde (DGPPN) zum Thema Burnout. Available via: http://www.dgppn.de/fileadmin/user_upload/_medien/download/pdf/stellungnahmen/2012/stn-2012-03-07-burnout.pdf. Accessed 16 Feb 2015 (in German)

Beutel ME, Gerhard C, Wagner S et al (2004) Reduction of technology fears in psychosomatic rehabilitation—concepts and results based on a computer training for older employees. Z Gerontol Geriatr 37:221–230 (in German)

Blank I, Peters J, Pickvance S et al (2008) A systematic review of the factors which predict return to work for people suffering episodes of poor mental health. J Occup Rehabil 18:27–34

Bowling N, Beehr TA (2006) Workplace harassment from the victim's perspective: a theoretical model and meta-analysis. J Appl Psychol 91:998–1012

Bundesministerium für Arbeit und Sociales (ed) (2011) Unser Weg in eine inklusive Gesellschaft. Der Nationale Aktionsplan der Bundesregierung zur Umsetzung der UN-Behindertenrechtskonvention. Bundesministerium für Arbeit und Soziales, Berlin (in German)

Burisch M (2010) Das Burnout-Syndrom. Theorie der inneren Erschöpfung. Springer, Berlin (in German)

Campbell R, Pepper L (2006) Downsizing and social cohesion: the case of downsizing survivors. New Solut 16(4):373–393

Christianson JP, Jennings JH, Ragole T et al (2011) Safety signals mitigate the consequences of uncontrollable stress via circuit involving the sensory insular cortex and bed nucleus of the stria terminalis. Biol Psychiatry 70:458–464

Cramer C, Davidhizar R (2000) The health care employee with an "attitude". Hosp Mater Manage Q 22:27–33

Deutsche Rentenversicherung Bund (2005) Qualitätsbericht 2005 Rehazentrum Seehof. Deutsche Rentenversicherung Bund, Berlin (in German)

Dewa CS, Loong D, Bonato S (2014) Work outcomes of sickness absence related to mental disorders: a systematic literature review. BMJ Open 4:e005533. doi:10.1136/bmjopen-2014-005533
Eibl-Eibesfeldt I (2004) Die Biologie des menschlichen Verhaltens. Grundriss der Humanethologie. Blank, Vierkirchen-Pasenbach (in German)
Elst TV, de Witte H, de Cuyper N (2014) The Job Insecurity Scale: a psychometric evaluation across five European countries. Eur J Work Organ Psychol 23:364–380
Escartín J, Zapf D, Arrieta C et al (2011) Worker's perception of workplace bullying: a cross-cultural study. Eur J Work Organ Psychol 20:178–205
Forcella L, Di Donato A, Coccia U et al (2007) Anxiety, job stress and job insecurity among teachers with indefinite or definite time contract. G Ital Med Lav Ergon 29:683–686 (in Italian)
French JRP (1973) Person role fit. Occup Mental Health 3:15–20
Frese M, Fay D (2001) Personal initiative: an active performance concept for work in the 21st century. Res Organ Behav 23:133–187
Freudenberger HJ (1974) Staff burnout. J Soc Issues 30:159–165
Gastpar MT, Caspar S, Linden M (eds) (2003) Psychiatrie und Psychotherapie. Springer, Berlin (in German)
Gilbreath B, Benson PG (2004) The contribution of supervisor behaviour to employee psychological well-being. Work Stress 18:255–266
Greenberg PE, Sisitsky T, Kessler RC et al (1999) The economic burden of anxiety disorders in the 1990s. J Clin Psychiatry 60:427–435
Guy W, Ban TA, Bobon D, AMDP et al (eds) (1982) The AMDP system. Manual for the assessment and documentation of psychopathology. Springer, Berlin
Haines J, Williams CL, Carson JM (2002) Workplace phobia: psychological and psychophysiological mechanisms. Int J Stress Manage 9(3):129–145
Hallowell EM (1999) The human moment at work. Harv Bus Rev 77(58–64):66
Haslam C, Atkinson S, Brown SS et al (2005) Anxiety and depression in the workplace: effects on the individual and organisation (a focus group investigation). J Affect Disord 88:209–215
Heckhausen H (1989) Motivation und Handeln. Springer, Berlin (in German)
Hillert A, Müller-Fahrnow W, Radoschewski FM (eds) (2009) Medizinisch-beruflich orientierte Rehabilitation. Grundlagen und klinische Praxis. Deutscher Ärzte, Köln (in German)
Holmes TH, Rahe RH (1967) The social readjustment rating scale. J Psychosom Res 11:213–218
Hoppe F (1930) Erfolg und Misserfolg. Dissertation, Verlagsbuchhandlung Julius Springer, Berlin (in German)
Johnson SL (2009) Improving the school environment to reduce school violence: a review of the literature. J Sch Health 79:451–465
Kaluza G (2011) Stressbewältigung. Trainingsmanual zur psychologischen Gesundheitsförderung. Springer, Berlin (in German)
Kittner C (2003) Angst im Job. Hampp, München (in German)
Kuchenbecker M, Amann S (2010) Seminar Prävention von Raubüberfällen und Gewaltereignissen. RKW Rationalisierungs- und Innovationszentrum der deutschen Wirtschaft e.V, Eschborn (in German)
Kulik CT, Oldham GR, Hackman JR (1987) Work design as an approach to person-environment fit. J Vocat Behav 31:278–296
Lagerveld SE, Blonk RW, Brenninkmeijer V (2012) Work-focused treatment of common mental disorders and return to work: a comparative outcome study. J Occup Health Psychol 17:220–234
Lang PJ (1968) Fear reduction and fear behavior: problems in treating a construct. In: Shlien JM (ed) Research in psychotherapy, vol 3. American Psychological Association, Washington, pp 90–103
Laposa JM, Alden LE, Fullerton LM (2003) Work stress and posttraumatic stress disorder in ED nurses/personnel. J Emerg Nurs 29:23–28

Lazarus RS (1993) From psychological stress to the emotions: a history of changing outlooks. Annu Rev Psychol 44:1–22
Leach D, Hagger-Johnson GH, Doerner N et al (2013) Developing a measure of work uncertainty. J Occup Organ Psychol 86:85–99
Limm H, Gündel H, Heinmüller M et al (2011) Stress management interventions in the workplace improve stress reactivity: a randomised controlled trial. Occup Environ Med 68:126–133
Linden M, Hautzinger M (2008) Verhaltenstherapie. Theoretische und empirische Grundlagen und klinische Anwendungsprinzipien. In: Möller HJ, Laux G, Kapfhammer HP (eds) Psychiatrie und Psychotherapie (Band 1, Allgemeine Psychiatrie, S.). Springer, Berlin, pp 743–774 (in German)
Linden M, Muschalla B (2007) Anxiety disorders and workplace-related anxieties. J Anxiety Disord 21:467–474
Linden M, Muschalla B, Glatz J et al (2009) Abschlussbericht zum Forschungsprojekt Arbeitsplatzbezogene Ängste und Arbeitsplatzphobie bei Patienten in der psychosomatischen und kardiologischen Rehabilitation. Berlin: Deutsche Rentenversicherung Bund (in German)
Linden M, Muschalla B, Kessler U et al (2012) Reha in der Hausarztpraxis. Rehabedarfsfeststellung, -zugang, -steuerung, -koordinierung, -optimierung und -verstetigung bei psychischen Erkrankungen unter Mitwirkung niedergelassener Ärzte. Eine Untersuchung zum Stand und zu den Optimierungsmöglichkeiten durch ein Rehakonsil. Abschlussbericht zum Modellprojekt. Deutsche Rentenversicherung Bund, Berlin (in German)
Linden M, Muschalla B, Olbrich D (2008) Die Job-Angst-Skala (JAS). Ein Fragebogen zur Erfassung arbeitsplatzbezogener Ängste. Z Arb Orga 52:126–134
Linden M, Weidner C (2005) Arbeitsunfähigkeit bei psychischen Störungen. Nervenarzt 76: 1421–1431 (in German)
MacDonald HA, Colota V, Flamer S et al (2003) Posttraumatic stress disorder (PTSD) in the workplace: a descriptive study of workers experiencing PTSD resulting from work injury. J Occup Rehabil 13:63–77
Mahan PL, Mahan MP, Park NJ et al (2010) Work environment stressors, social support, anxiety, and depression among secondary school teachers. Am Assoc Occup Health Nurs 58:197–205
Marcus B (2011) Anforderungsanalyse. In: Marcus B (ed) Personalpsychologie. Springer, Wiesbaden, pp 11–28 (in German)
Maslach C, Jackson SE (1981) The measurement of experienced burnout. J Occup Behav 2: 99–113
McClelland DC, Atkinson JW, Clark RA et al (1976) The achievement motive. Appleton-Century-Crofts, New York
McDonald PK (2012) Workplace sexual harassment 30 years on: a review of the literature. Int J Manage Rev 14:1–17
Metz A-M, Rothe H-J (2014) Screening psychischer Arbeitsbelastungen. In: Bundesanstalt für Arbeitsschutz und Arbeitsmedizin (ed) Gefährdungsbeurteilung psychischer Belastung. Erfahrungen und Empfehlungen. Erich Schmidt, Berlin, pp 212–217 (in German)
Mezerai M, Dahane A, Tachon JP et al (2006) Dépression dans le milieu du travail. Presse Med 35: 823–830 (in French)
Mohr G, Otto K (2011) Health effects of unemployment and job insecurity. In: Antoniou AS, Cooper C (eds) New directions in organizational psychology and behavioral medicine. Gower, Surrey, pp 289–311
Mohr G, Rigotti T, Müller A (2007) Irritations-Skala zur Erfassung arbeitsbezogener Beanspruchungsfolgen. Hogrefe, Göttingen (in German)
Molodynski A, Linden M, Juckel G et al (2013) The reliability, validity, and applicability of an English language version of the Mini-ICF-APP. Soc Psychiatry Psychiatr Epidemiol 48: 1347–1354
Morgan WB, Walker SS, Hebl M et al (2013) A field experiment: reducing interpersonal discrimination toward pregnant job applicants. J Appl Psychol 98:799–809

Muschalla B (2008) Work-anxieties and workplace phobia. A concept of domain-specific mental disorders. Dissertation, Universität Potsdam
Muschalla B (2014) Arbeitsbezogene Ängste in Forschung und Praxis—Ein aktuelles Schnittstellenphänomen. Z Arb Organ 58:206–214 (in German)
Muschalla B, Linden M (2008) Die Arbeitsplatzphobieskala. Ein Screening-Instrument für die medizinische Rehabilitation. Ärztl Psychotherapie 3:258–262
Muschalla B, Linden M (2009) Workplace phobia—A first explorative study on its relation to established anxiety disorders, sick leave, and work-directed treatment. Psychol Health Med 14: 591–605
Muschalla B, Linden M (2011) Die Arbeitsplatzphobieskala—Ein Screeninginstrument in der psychosomatischen Rehabilitation. DRV-Schriften 93:80–81 (in German)
Muschalla B, Linden M (2012) Specific job-anxiety in comparison to general psychosomatic symptoms at admission, discharge and six months after psychosomatic inpatient treatment. Psychopathology 45:167–173
Muschalla B, Linden M (2013a) Arbeitsplatzbezogene Ängste und Arbeitsplatzphobie. Phänomenologie, Diagnostik, Behandlung, Sozialmedizin. Kohlhammer, Stuttgart (in German)
Muschalla B, Linden M (2013b) Different workplace-related strains and different workplace-related anxieties in different professions. J Occup Environ Med 55:978–982
Muschalla B, Linden M (2014) Workplace phobia, workplace problems, and work ability in primary care patients with chronic mental disorders. J Am Board Fam Med 27:486–494
Muschalla B, Linden M, Olbrich D (2010) The relationship between job-anxiety and trait-anxiety—A differential diagnostic investigation with the Job-Anxiety-Scale and the State-Trait-Anxiety-Inventory. J Anxiety Disord 24:366–371
Muschalla B, Vilain M, Lawall C (2012) Participation restrictions at work indicate participation restrictions in other domains of life. Psychol Health Med 17:95–104
Muschalla B, Heldmann M, Fay D (2013) The significance of job-anxiety in a working population. Occup Med 63:415–421
Nakazawa H, Ikeda H, Yamashita T et al (2005) A case of sick building syndrome in a Japanese office worker. Ind Health 43:341–345
Nash-Wright J (2011) Dealing with anxiety disorders in the workplace: importance of early intervention when anxiety leads to absence from work. Prof Case Manage 16:55–59
Nicholson PJ, Vincenti GE (1994) A case of phobic anxiety related to the inability to smell cyanide. Occup Med 44:107–108
Noordik E, van der Klink JJ, Geskus RB et al (2013) Effectiveness of an exposure-based return-to-work program for workers on sick leave due to common mental disorders: a cluster-randomized controlled trial. Scand J Work Environ Health 39:144–154
OECD (ed) (2012) Sick on the job? Myths and realities about mental health and work. OECD, Paris
Oppolzer A (2010) Psychische Belastungsrisiken aus Sicht der Arbeitswissenschaft und Ansätze für die Prävention. In: Badura B, Schröder H, Klose J, Macco K (eds) Fehlzeiten-Report 2009. Arbeit und Psyche: Belastungen reduzieren—Wohlbefinden fördern. Springer, Berlin, pp 13–22 (in German)
Panse W, Stegmann W (2007) Angst–Macht–Erfolg. Erkennen Sie die macht der konstruktiven angst. Volk, München
Paul KI, Moser K (2009) Unemployment impairs mental health. Meta-analyses. J Vocat Behav 74: 264–282
Payne RL, Fineman S, Jackson PR (1982) An interactionist approach to measuring anxiety at work. J Occup Psychol 55:13–25
Perlow L, Williams S (2003) Is silence killing your company? Harv Bus Rev 81:52–58, 128
Price JL, Monson CM, Callahan K et al (2006) The role of emotional functioning in military-related PTSD and its treatment. J Anxiety Disord 20:661–674
Rau R, Henkel D (2013) Zusammenhang von Arbeitsbelastungen und psychischen Erkrankungen. Review der Datenlage Nervenarzt 84:791–798 (in German)

Sanne B, Mykletun A, Dahl AA et al (2005) Testing the Job Demand-Control-Support model with anxiety and depression as outcomes: the Hordaland Health Study. Occup Med 55:463–473
Schaarschmidt U, Fischer AW (2001) Bewältigungsmuster im Beruf. Vandenhoeck & Ruprecht, Göttingen (in German)
Selye H (1956) The stress of life. McGraw-Hill, New York
Semmer NK (2006) Job stress interventions and the organization of work. Scand J Work Environ Health 32:515–527
Skinner BF (1969) Contingencies of reinforcement. Prentice Hall, Englewood Cliffs
Smith ME (2009) Work phobia and sickness leave certificates. Afr J Psychiatry 12:249–252
Smith MJ, Conway FT, Karsh BT (1999) Occupational stress in human computer interaction. Ind Health 37:157–173
Spera SP, Buhrfeind ED, Pennebaker JW (1994) Expressive writing and coping with job loss. Acad Manage J 37:722–733
Spielberger CD, Gorsuch RL, Lushene R et al (1983) Manual for the state-trait anxiety inventory. Consulting Psychologists Press, Palo Alto
Stansfeld S, Candy B (2006) Psychosocial work environment and mental health—a meta-analytic review. Scand J Work Environ Health 32:443–462
Stansfeld S, Shipley MJ, Head J et al (2012) Repeated job strain and the risk of depression: longitudinal analyses from the Whitehall TT study. Am J Public Health 102:2360–2366
Strazdins L, D'Souza RM, Lim LL et al (2004) Job strain, job insecurity, and health: rethinking the relationship. J Occup Health Psychol 9:296–305
Stroebe W, Jonas K, Hewstone M (2007) Sozialpsychologie. Eine Einführung. Springer, Berlin (in German)
Techniker Krankenkasse (2012) Gesundheitsreport 2012. Band 27 Teil 1: Arbeitsunfähigkeiten. Techniker Krankenkasse, Hamburg (in German)
Tepper BJ (2007) Abusive supervision in work organizations: review, synthesis and research agenda. J Manage 33:261–289
Thomas M, Hynes C (2007) The darker side of groups. J Nurs Manage 15:375–385
United Nations (2006) Convention on the rights of persons with disabilities. Available from: http://www.un.org/disabilities/documents/convention/convoptprot-e.pdf. Accessed 16 Feb 2015
Wedegaertner F, Arnhold-Kerri S, Sittaro NA et al (2013) Depression- and anxiety-related sick leave and the risk of permanent disability and mortality in the working population in Germany: a cohort study. BMC Public Health 13:145. doi:10.1186/1471-2458-13-145
Wissenschaftliches Institut der AOK (2011) Pressemitteilung Fehlzeitenreport 2011. Available from http://www.aok-bv.de/imperia/md/aokbv/presse/pressemitteilungen/archiv/2011/pm_wido_2011-16-08_fzr2011_v02_final.pdf. Accessed 16 Feb 2015 (in German)
Wittchen HU, Jacobi F, Rehm J et al (2011) The size and burden of mental disorders and other disorders of the brain in Europe 2010. Eur Neuropsychopharmacol 21:655–679
World Health Organization (1992) International statistical classification of diseases and related health problems (10th revision). World Health Organization, Geneva
Yalom ID (1970) The theory and practice of group psychotherapy. Basic, New York
Zielke M (2001) Entwicklung und Begründung eines Analysemodells des Arbeits- und Leistungsprozesses (AMALPROZESS). In: Zielke M, von Keyserlingk H, Hackhausen W (eds) Angewandte Verhaltensmedizin in der Rehabilitation. Pabst Science, Berlin, pp 629–667 (in German)

Healthy Aging at Work

Jan Oltmanns, Götz Richter, Ben Godde, and Ursula M. Staudinger

Abstract

In Germany as well as in most Western industrialized countries, the work-force is rapidly "aging". Modern companies therefore increasingly need to abandon established approaches to occupational safety and pursue preventive strategies to protect and restore well-being, work ability, and good health across the entire lifespan. Hence, effective occupational health management needs to encompass strategic work and career designs. In response to these challenges, an interdisciplinary group of researchers at the Jacobs Center on Lifelong Learning and Institutional Development of the Jacobs University Bremen has zoomed in on two concepts that have so far received little attention in research on occupational health management; namely, person–environment fit and work–task mobility. *Person–environment fit* refers to the fit between individuals (e.g., abilities, behaviors, goals, attitudes) and their work environments (e.g., job profile, demands, support structures, culture). *Work–task mobility* refers to careers involving repeated intra-organizational changes of work tasks at the same

---

Since July 2015, Jan Oltmanns works as an HR specialist for Daimler AG at Mercedes-Benz, Bremen.

J. Oltmanns (✉)
Daimler AG, Mercedes-Benz, Bremen, Germany
e-mail: jan.oltmanns@daimler.com

G. Richter
Bundesanstalt für Arbeitsschutz und Arbeitsmedizin, Dortmund, Germany
e-mail: richter.goetz@baua.bund.de

B. Godde
Jacobs University Bremen, Bremen, Germany
e-mail: b.godde@jacobs-university.de

U.M. Staudinger
Columbia Aging Center, Columbia University, New York, NY, USA
e-mail: umstaudinger@columbia.edu

M. Wiencke et al. (eds.), *Healthy at Work*, DOI 10.1007/978-3-319-32331-2_6

level of job complexity (no promotion or demotion). This chapter reports on two research projects, *Demopass* and *Mobilis*, that aim to investigate *person–environment fit* and *work–task mobility*, respectively, as two important tools for systemic and dynamic occupational health management in times of demographic change.

1 Healthy Aging at Work

Modern occupational health psychology, whether in the field of research or professional practice, aims to create "healthy workplaces in which people may produce, serve, grow, and be valued" (Quick et al. 1997, p 3). Rather than targeting only selected domains of health, such as, for instance, focusing on the prevention of coronary heart diseases or various chronic conditions, effective health promotion in the work-place means targeting working individuals as whole persons, including their: (1) work–life balance; (2) growth and development; (3) health and safety; (4) recognition; and (5) involvement. Because health results from the interplay of multiple interconnected factors (Adkins et al. 2000), it requires ongoing constant attention and regeneration, including after the achievement of optimal health (Grawitch et al. 2006).

Against this background, it is surprising to note that the terms "age" or "aging", which have strong associations with physical and psychological health (see Baltes et al. 2006; Dekkers-Sánchez et al. 2008; Lidwall et al. 2009; Schaie 2005), have thus far received only scant attention in organizational health research. More often than not, chronological age has merely been used as a covariate or confound (De Lange et al. 2006; Schalk et al. 2010). Given the aging of the work-force, this is an oversight with serious consequences that therefore warrants remedy. In 2013, the employment rate in Germany was at its highest level since the reunification of Germany in 1989 (Statistisches Bundesamt [Federal Statistics Office] 2014). This positive trend has been largely based on a steady increase in the employment rate of older age groups. Within the last decades, the employment rate of the work population between 50 and 64 years of age has stepped up by approximately 20 % whereas the share of working individuals between 30 and 49 years of age decreased by approximately 15–20 % (Fuchs et al. 2011). In 2012, 77 % of the 55–59-year-olds and 48 % of the 60–64-year-olds were economically active (Statistisches Bundesamt 2014). Additionally, Fuchs et al. (2011) have projected that labor participation of older men and women is still growing and will continue to do so.

The "aging" of the German work-force is a positive development that has important consequences (Leber et al. 2013) and is mainly attributable to changes in employment behavior, which in turn are attributable to a number of factors (Mümken and Brussig 2012). One such factor is an increasing shortage of qualified younger adults seeking work. Another is policy changes. Over the past few years, politicians have closed existing paths to early retirement or made these paths unattractive by imposing financial penalties on employees who retire early. Therefore, employees stay with their companies and on the labor market for longer. Thus, even more measures are required to promote the employability of older workers and to assure them a prolonged and healthy working life (Buchholz et al. 2013).

In Germany, the proportion of older employees in mentally and physically challenging work environments such as industrial assembly-lines and rotating shift work is also increasing (European Commission 2014; Statistisches Bundesamt 2013). In many cases, such work environments involve a combination of repetitive tasks with little autonomy and high time pressure and performance demands. At the same time, industrial environments offer high job security and attractive salary arrangements. Therefore, many employees remain in these jobs and expose themselves to the challenging conditions of highly productive industrial work for most of their lives, very often for 20, 30 years or more.

Production companies of the twenty-first century face the challenge of preparing their aging work-forces for the ever-increasing pressures of economization and innovation. Industrial workers are increasingly confronted with repeated changes in their operational sequences because of advances in digitization (Kagermann et al. 2012) as well as new tools, machines, and materials. Additionally, modern organizations are very often subject to change themselves, for instance as a consequence of repeated reorganizations and constant change management (Bundesanstalt für Arbeitsschutz und Arbeitsmedizin [BauA; Federal Institute for Occupational Safety and Health] 2013). More than ever, modern companies (and in particular their human resources [HR] departments and production planners) need to abandon established approaches to occupational safety and instead pursue preventive strategies to protect and restore the "work ability" (Tuomi et al. 2006) of their aging work-forces (Buss and Kuhlmann 2013). Sustainable HR policies and modern production work must allow for longer career paths, in some cases for the whole of an individual's working life (Staudinger and Kocka 2010). To maximize the potential of experienced work-forces, preventive measures and career designs must be core elements in setting up production lines, when developing new products, and discussing individual career paths (Sonntag 2014; Staudinger and Bowen 2011). To maintain and protect health, proficiency and motivation across a person's entire working life is a responsibility for all managerial decision makers and asks for a holistic approach to prevention (BAuA 2013).

In response to these challenges, an interdisciplinary group of researchers at the Jacobs Center on Lifelong Learning and Institutional Development of the Jacobs University Bremen (Germany) has focused on two concepts that so far have received little attention in research on occupational health management; namely, person–environment fit and work–task mobility.

2 Person–Environment Fit and Work–Task Mobility

Person–environment fit describes the fit between individuals—that is, their abilities, behaviors, goals, or attitudes—and their work environments—that is, job profile, demands, support structures, and culture (Staudinger et al. 2011). Health promotion in the twenty-first century needs to focus on both individuals and their health behaviors, as well as on work settings, for example the characteristics of work tasks or supervisor(s). Achieving and maintaining fits between these three levels (employee, work group/supervisor, work characteristics) seem to be crucial elements in effective occupational health management. This was one of the central tenets to be tested by the *Demopass* project (funded by the Bundesministerium für Bildung und Forschung [Federal Ministry of Education and Research] and European Social Fund), which was carried out between 2007 and 2010 at the Jacobs Center on Lifelong Learning and Institutional Development.

A second concept relates to work–task mobility; that is, intra-organizational changes in work tasks at the same level of job complexity (no promotion or demotion). Each work-task change implies learning new skills or material. Large-scale longitudinal studies have provided sound evidence that highly complex work (work that requires thought and independent judgment) positively affects brain and cognitive aging (e.g., Schooler et al. 1999; Suo et al. 2012). Conversely, low job complexity is reportedly associated with a negative impact on cognition and brain aging (Gajewski et al. 2010; Marquie et al. 2014). However, because highly complex work is by definition linked with high levels of education, not everyone has access to such jobs. Extant research has not yet studied ways of designing low-complexity jobs to counteract their detrimental effects on the aging of the brain and cognition. This was the central aim of the *Mobilis* project (funded by the Volkswagen Foundation). In particular, *Mobilis* investigated work–task mobility as a dynamic characteristic of the work setting that may buffer the long-term negative effects of low-complexity work.

The results of both research projects, *Demopass* and *Mobilis*, indicated that *person–environment fit* and *work–task mobility* are both important tools for modern occupational health efforts in these times of demographic change. Therefore, we will devote the remainder of this chapter to describing these two projects and their findings in greater detail.

3 Demopass: Healthy Aging at Work: A Matter of Fit

In its most general sense, person–environment fit relates to fits between attributes of individuals and the characteristics of their (work) settings. However, industrial and organizational psychological research has elaborated the concept of fit in recent years. For example, fits between the skills and abilities of employees and their job-specific tasks and duties are referred to as person–job fits (Edwards 1991; O'Reilly 1977). Fits between the attitudes of employees and those of their colleagues are referred to as person–group fits (Adkins et al. 1996; Kristof-Brown and Stevens 2001). Person–organization fits refer to when employees share the value system and goals of their companies (Chatman 1989; Kristof 1996).

To investigate fits between employees, supervisor/work groups, and work characteristics, the *Demopass* project conducted a survey of more than 1000 employees from five German companies from various industries. A special feature of the *Demopass* project was that although all three levels (employee, supervisor/

work group, work characteristics) were considered individually, their effects on each other were also examined (Staudinger and Kocka 2010). Work group membership was recorded anonymously and the corresponding supervisors were also interviewed to enable comparisons between statements of managers and those of their employees. The sample included 1032 employees with 119 work groups and their 119 supervisors. The ages of the respondents ranged from 21 to 62 years; 177 (17 %) were younger than 29 years old, 547 (53 %) were between 30 and 45 years, and 282 (27 %) were older than 45 years (3 % did not indicate their age). The sample covered a great variety of educational levels, industrial sectors and business divisions (e.g., production, human resources, financial services, and research and development).

Numerous studies have provided indications that the organizational climate—the psychological environment or atmosphere of an organization—can have a major impact on work-related attitudes and productivity (e.g., Bowen et al. 2010; Nerdinger et al. 2008; Staudinger 2015a). Older workers often have to contend with negative stereotypes. For instance, older employees are often assumed to be less flexible, less willing to learn, less open to change and more expensive than their younger colleagues (Bellmann 2002; Posthuma and Campion 2009; Staudinger and Noack 2009). Use and perception (conscious or unconscious) of such stereotypes can contribute to the emergence of a negative psychological old age climate. Therefore, the goals of the *Demopass* project were to: (1) explore how people who differ in characteristics such as age, gender, and education perceive the age climate of their organization; and (2) assess the associations between age climate, job satisfaction, and work motivation. As to the first question, the *Demopass* study identified no differences in perception of the age climate according to age, gender, level of education, length of employment, or type of job-task. However, regardless of age, employees reportedly stated they were less likely to change their jobs when they perceived their organizations' age climates as positive (irrespective of their current job satisfaction, educational level, length of employment or gender; Bowen and Staudinger 2011). Similarly, turnover intentions seemed to be linked with fits between employees' and supervisors' self-perceptions of their own aging processes. That is, the employees who expressed the strongest turnover intentions were those whose supervisors had negative self-perceptions of aging while the employees had more positive self-perceptions. Employees who themselves reported negative self-perceptions (and whose supervisors also did) expressed significantly fewer turnover intentions. However, this association was only true when supervisors exhibited negative self-perceptions of aging. In other words, turnover intentions are unaffected by self-perceptions of aging when supervisors have positive self-perceptions of their own aging processes.

Similarly, in the *Demopass* project we found that an employee's "promotion orientation" may depend on characteristics of the age climate (Bowen and Staudinger 2013). In this study, there was an age-related difference in promotion orientation only when employees' work groups perceived the age climate as negative. In such cases, promotion orientation was negatively related to age. However, age was not related to promotion orientation in employees whose work groups had predominantly positive age climates. That is, employees who perceived the psychological age climate in their organizations as positive strove for improvements as much as their younger colleagues. In sum, these findings suggest that a positive psychological age climate can support employees' job satisfaction as well as their positive development (and with that their mental health).

A special form of person–job fit relates to the fit between the demands of an organization and the skills and abilities of an individual (Edwards 1991; O'Reilly 1977; also see Muchinsky and Monahan 1987). Demands–abilities fit plays an important role in personnel selection and has accordingly received attention in occupational stress research (e.g., McGrath 1976). A high demands–abilities fit can be crucial in preventing age-related cognitive deficits (Feuerhahn et al. 2011). However, it is important to note that both demands that are too high and those that are too low are negatively associated with good health. Conversely, early detection and prevention of misfits can help to maintain and improve work performance, especially in older employees. *Demopass* investigated how concepts such as work ability (Tuomi et al. 2006), self-efficacy (Stajkovic and Luthans 1998), and job satisfaction are affected by high versus low demands–abilities fits (Warr et al. 1979). Analyses showed that, for instance, a good fit between the self-rated motor control skills of older industrial workers and their actual job demands was associated with greater perceived self-efficacy. That is, self-efficacy was high when self-rated abilities (high/low) corresponded with the actual work demands (high/low). When motor skills did not match actual work demands, there was a negative correlation between self-efficacy and self-rated motor control skills (also see Trautmann et al. 2011).

Additionally, high versus low demands–abilities fits between self- and supervisor ratings of abilities may be able to predict ability to work and job satisfaction (Trautmann et al. 2011). In the *Demopass* project, we found that and employee's work ability is positively influenced by a fit between self- and supervisor ratings of, for instance, the ability to learn, regardless of whether the conclusion is that the ability to learn is high or low. That is, with regard to work ability, the rating itself is not crucial, but rather the fit with the supervisor rating. This was especially true for workers in the middle-aged (30–45 years) and older (46–66 years) age groups, and less so for younger workers. In terms of the ability to handle complex tasks, the fit between self- and supervisor ratings affected job satisfaction (although predominantly in older employees and clerical staff). However, unlike in the previous example, the fit between self- and supervisor rating of the ability to handle complex tasks was only important when the supervisor rating was high. Rating of ability as high by both supervisor and employee was associated with greater job satisfaction. In contrast, high supervisor ratings were associated with less job satisfaction if the employee did not allocate the same rating. When supervisors rated their employees' abilities to handle complex tasks as low, the fit between self- and supervisor rating seemed to be irrelevant.

The *Demopass* project also tapped special training opportunities for older employees. Job-related trainings and further education are important factors in maintaining performance and productivity at all ages (Noethen and Voelpel

2011). Nevertheless, the results of the *Demopass* study clearly showed that employees aged over 55 years participated in vocational training programs less often than their younger colleagues (also see Ng and Feldman 2008). There may be several reasons for this. First, analyses of the *Demopass* data revealed that employees assign lower ratings to the importance of training programs when they expect few benefits from those programs. This relationship appeared to become stronger with age. Additionally, we found that older workers were more afraid of failure than their younger colleagues, probably because their last learning experiences had often occurred several years previously and they underestimated their learning ability. However, self-confidence in learning ability can be significantly influenced by supervisors. The *Demopass* data suggest that, especially for older and less-qualified employees, self-confidence concerning learning can benefit from supervisor support (e.g., through identification of training needs, motivating their team members, or acting as mentors). Unfortunately, the data also indicate that older and less-qualified employees were much less likely to be classified by their supervisors as "willing" to take part in trainings and further education. On average, younger employees were much more often rated as willing to take part in trainings and further education. Additionally, the probability of a positive fit between self- and supervisor ratings of training willingness decreased monotonously with each additional year.

Finally, key concepts of psychological stress research can also be interpreted as matters of fit. Perceptions by individuals that the relationship between their efforts (psychological, physical, and temporal) and their rewards (in the form of salary, career opportunities, job security, and personal recognition) is out of balance may indicate poor "effort–reward fits". Such poor fits are associated with psychological stress, burnout, and job dissatisfaction (Siegrist 2002; Tsutsumi and Kawakami 2004). In contrast, numerous studies have consistently shown that a strong effort–reward fit is associated with greater job satisfaction, better performance, and reduced perceptions of stress (e.g., Kristof-Brown et al. 2005; Verquer et al. 2003). In line with these findings, the results of the *Demopass* project indicate very clearly that effort–reward imbalances characterized by high levels of effort and low levels of reward are negatively associated with health and health-related variables. In particular, we found that a strong fit between effort and reward correlates with positive mood, greater job satisfaction, and better overall health. In contrast, a poor fit between effort and reward is correlated with negative mood and increased intention to change jobs, as well as with burnout and concentration deficits (Feuerhahn et al. 2011).

4 Mobilis: Healthy Aging at Work: A Matter of Mobility

In the *Mobilis* project, the research focus was on cognitive health (i.e., the preservation and improvement of good cognitive functioning). In particular, we were interested in finding a feasible way of promoting cognitive functioning in middle-aged production workers, an occupational group that has been associated with poorer cognitive performance as well as with an increased risk of cognitive impairment and dementia.

"Cognitive aging" describes the fact that the cognitive system changes with age. On a behavioral level, many cognitive abilities, including memory, executive functions, and processing speed, demonstrate a monotonic and linear decline after a peak in early adulthood (Baltes et al. 2006). On a neurophysiological level, cognitive aging denotes fundamental transformations in the structure of the brain; these include gray matter volume loss through cell shrinkage, synapse loss, and dendritic regression (e.g., Grady 2012; Raz et al. 2005). Notwithstanding these age-related changes, the human brain has enormous potential for plasticity; that is, for structural and functional modifications in response to cognitive challenges and environmental demands (e.g., Draganski and May 2008; Greenwood 2007). There is ample research evidence that, for example, a cognitively stimulating lifestyle is positively associated with cognitive performance (for recent reviews see Hertzog et al. 2009; Jopp and Hertzog 2007; Voelcker-Rehage et al. 2010, 2011) as well as with greater maintenance of gray matter volume in the frontal, parietal, and temporal regions (Bartrés-Faz et al. 2009; Valenzuela et al. 2008; Voelcker-Rehage and Niemann 2013).

Because work is an important part of everyday life, working conditions and cognitive work demands may have strong influences on brain and cognition. Schooler and colleagues (Kohn and Schooler 1978) were among the first to focus on working conditions and their cumulative effect on cognitive performance. These authors established a positive relationship between job complexity (work that requires thought and independent judgment) and intellectual flexibility (Schooler et al. 1999). Irrespective of age and level of education, exposure to greater complexity at work for 30 years was associated with better cognitive functioning than in those with exposure to less complexity. Subsequent studies (mainly epidemiological) have corroborated Schooler et al.'s findings. For instance, greater job complexity is reportedly associated with a lower risk of cognitive impairment (Bosma et al. 2003) as well as of dementia and Alzheimer disease (Andel et al. 2005; Potter et al. 2007). High job complexity has similar positive effects on a neurophysiological level; however, published reports concerning this are scarce. Preliminary evidence suggests that high job complexity (supervisory experience) is positively associated with less loss of brain volume in older persons (Suo et al. 2012).

Conversely, low job complexity seems to be associated with poorer cognitive functioning and reduced brain activity (as shown, for example, in task-switching tests). Gajewski et al. (2010) compared the performances of various age groups of assembly-line workers with those of industry workers in more flexible and self-determined working conditions (e.g., quality control) on a task-switching task. They found that assembly-line work had a weak detrimental effect on performance in the two younger groups of workers (18–30 years of age) but a more severe effect in older subjects (48–66 years of age). Older assembly-line workers performed more poorly and had less working memory capacity and worse error monitoring than the same age group in the control group. Another study found that prolonged exposure to rotating shift work (e.g., for 5 or 10 years) negatively affected cognitive performance (Marquie et al. 2014).

However, it is noteworthy that high job complexity is necessarily linked with high levels of education; thus, not everyone has access to high-complexity jobs. It has not yet been determined how to design low-complexity jobs to counteract their detrimental effects on brain and cognition. Laboratory studies suggest that positive changes in the brain and cognition are triggered specifically by novel experiences and learning new skills, such as learning to juggle or to decipher Morse code (Boyke et al. 2008; Hultsch et al. 1999; Park et al. 2014; Schmidt-Wilcke et al. 2010; also see Bowen et al. 2010). Therefore, the *Mobilis* project investigated the effects of repeated novel experiences on workers in long-term low-complexity work. The research question was whether repeated confrontation with novel work situations (versus prolonged routine work tasks) helps to maintain brain matter and cognition, even in middle-aged subjects with less complex jobs.

With this aim, we performed a case-control quasi-experimental field study in which we compared the performance on a number of psychological and neurophysiological performance tests of middle-aged male production workers in a globally acting industrial company in north Germany. All participants had worked on the assembly line of this company or in similar monotonous work-places without interruption for at least 17 years. To control for selection biases, we diligently applied a matching procedure to a large number of baseline variables thought to affect the outcome measure (Holland and Rubin 1988; Rubin 1974). Baseline variables we considered important matching variables included cognitive functioning, leisure time activities (current and at baseline), and openness to experience. With the consent of the works council, the HR department of the company pre-selected more than 3000 production workers who had been continuously employed full-time in jobs of low complexity (no promotion or demotion) during the phase of 17 years under investigation. We matched participants on all relevant covariates on the basis of a screening questionnaire. In subsequent semi-structured telephone interviews, we constructed individual work biographies to determine the number of work-task changes (WTCs). We classified all participants with zero or one WTC in 17 years as relatively non-mobile and all those with two or more WTCs in 17 years as highly mobile. On the basis of this matching procedure we identified 19 pairs of relatively non- and highly mobile participants ($N = 38$). These 38 participants were then invited to take part in the *Mobilis* study.

To assess the effect of repeated versus rare WTCs on cognitive functioning, we tested processing speed and working memory. Additionally, participants underwent brain magnetic resonance imaging to enable comparison of gray matter volumes between participants with high and low WTCs. All analyses (behavioral and neurophysiological) were controlled for cognitively stimulating leisure time activities.

We found that repeated lateral and intra-organizational WTCs had beneficial effects on the brains and cognition of middle-aged production workers. On a behavioral level, highly mobile individuals performed better on processing speed and working memory than relatively non-mobile individuals. On a neurophysiological level, these differences were accompanied by greater gray matter volumes in important brain regions. In particular, WTCs may positively affect gray matter volume in the striatal, frontal, and insular regions. According to longitudinal neurophysiological research, these regions are particularly prone to the processes of cognitive decline (Raz et al. 2005) and have been shown to be critically involved in skill acquisition, learning processes, and cognitive functioning (Doyon and Benali 2005; Seidler 2010; Shohamy and Wimmer 2013). Additionally, we found that brain volumes in these regions were positively correlated with processing speed performance.

Mobilis is not only one of the first projects to relate working conditions to brain structure and cognitive performance, but is also the first study to find a feasible way (independent of cognitively stimulating leisure time behavior) of diminishing the cumulative long-term effects of low job complexity on cognition. By providing cognitive stimulation through repeated lateral and intra-organizational changes of work tasks, WTC offers a way of fostering both cognitive functioning and gray matter volume at work not only for workers in highly complex occupations but also for those in relatively non-complex occupations. Therefore, WTC offers a means of counteracting the detrimental effects of low-complexity occupations.

5 Discussion

Production companies increasingly need to pursue preventive strategies for protecting and restoring the work ability of their aging work-force. Sustainable occupational health management must encompass preventive work and career designs. The maintenance and protection of health and motivation across a person's entire working life should be core elements of all HR managerial decisions. The evidence provided concerning the importance of person–environment fit and changing work tasks suggests that occupational health approaches in the twenty-first century need to be systemic and dynamic (see also Staudinger 2015b; Staudinger and Bowen 2011).

Systemic refers to the practice of considering employees' health in the context of work tasks, colleagues, supervisors, and organizational characteristics. In this framework, the *Demopass* study showed that several types of fit or misfit (e.g., effort–reward imbalance, demands–abilities, perceptions of aging–training willingness) between individuals and their work groups or supervisors can severely affect health and health-related variables such as job satisfaction, mood, intentions to change jobs, promotion orientation, self-efficacy, participation in training, work ability, burnout risk, and concentration deficits.

Dynamic occupational health management refers to work careers unfolding across time and having a cumulative impact on employees' health. Rather than waiting until constraints in work capacity occur or the number of sick leave days increases, it seems worthwhile in terms of both productivity and health to compose careers that are characterized by in-time work-task changes. The *Mobilis* project's findings indicate that repeated lateral and intra-organizational WTCs have cumulative beneficial effects on the cognitive systems of middle-aged production workers and offer a means (that is independent of cognitively stimulating leisure time behavior) of counteracting the detrimental effects of relatively non-complex occupations. Effective prevention must pursue long-term strategies and address both job and career design as well as further training and qualification (BAuA/ Initiative Neue Qualität der Arbeit [New Initiative on Quality of Work] 2013).

Preserving well-being, work ability, and good productivity are on many managers' agendas. Although it is generally accepted that the aging of the work-force requires systematic management of health, skills, and motivation (Naegele and Walker 2006), in many cases the measures taken are poorly integrated and coordinated. The foundations for employees to stay fit and healthy until retirement age have yet to be laid, particularly in production industries (manufacturing, assembling) (Sonntag 2014).

Preventive occupational health management faces the challenge that many detrimental changes, especially age-related processes, develop slowly over time. For instance, the risk of chronic diseases increases with age. Working conditions that impose heavy physical or mental demands or both, for instance heavy lifting or carrying, loud noise, excessive heat and cold, monotony, strong time pressure, and shift work, can aggravate these processes. Similarly, many older employees are subject to unintended but prolonged deskilling and downgrading processes (e.g., through adapted work-places and fewer training opportunities). As a consequence, older employees often suffer from profound qualification and competence deficits.

Against the background of these challenges, the two concepts *fit* and work-task *mobility* provide valuable instruments for helping strategic occupational health managers and HR departments to set up preventive strategies designed to protect and restore the work ability of their aging work-forces. With the concept of fit, modern companies gain a crucial *systemic* variable for describing, measuring, and designing the interplay of individual and organizational factors that affect the long-term maintenance of work ability and may therefore help to protect and develop individual resources. To enable practitioners to establish feasible in-house solutions on the basis of the fit concept, we developed a comprehensive toolbox within the *Demopass* project that aims to provide both precise diagnostic instruments and concrete policy recommendations for mastering the challenges of an "aging" work-force and demographic changes (see Staudinger et al. 2011). In a similar vein, work-task mobility can serve as a strategy for *dynamically* preserving cognitive health and employability. The *Mobilis* project has provided a useful avenue (independent of cognitively stimulating leisure time behavior) for counteracting the cumulative long-term effects of relatively non-complex jobs on cognitive aging. This study therefore provides an important indication of how future personnel and career development could and should look like in industrial work-places.

In this chapter we aimed to emphasize that effective occupational health management cannot merely focus on older workers who have already experienced detrimental age-related changes in cognitive functioning and mental flexibility. Rather, it is crucial to promote lifelong learning and employability right from the start of a person's working life (Staudinger and Bowen 2011; Staudinger and Kocka

2010). This requires a comprehensive approach that also includes younger workers (Grabbe and Richter 2014). Because working life is like a long-distance run (Höpflinger 2008), success is not achieved just before the finish line, but results from the initial strategy. Demographic change requires a paradigm shift. The primacy of short-term success must be replaced by a more differentiated and sustainable culture of learning in and through work.

References

Adkins CL, Ravlin EC, Meglino BM (1996) Value congruence between co-workers and its relationship to work outcomes. Group Organ Manage 21(4):439–460

Adkins JA, Quick JC, Moe KO (2000) Building world class performance in changing times. In: Murphy LR, Cooper CL (eds) Healthy and productive work: an international perspective. Taylor & Francis, Philadelphia, pp 107–132

Andel R, Crowe M, Pedersen NL et al (2005) Complexity of work and risk of Alzheimer's disease: a population-based study of Swedish twins. J Gerontol B Psychol Sci Soc Sci 60(5):251–258

Baltes PB, Lindenberger U, Staudinger UM (2006) Lifespan theory in developmental psychology. In: Lerner RM (ed) Handbook of child psychology, 6th edn. Wiley, New York, pp 569–664

Bartrés-Faz D, Solé-Padullés C, Junqué C et al (2009) Interactions of cognitive reserve with regional brain anatomy and brain function during a working memory task in healthy elders. Biol Psychol 80(2):256–259

Bellmann L (2002) Das IAB Betriebspanel: Konzeption und Anwendungsbereiche [Institute for employment research panel: conception and application]. Allg Stat Arch 86:177–188 (in German)

Bosma H, van Boxtel MPJ, Ponds R et al (2003) Mental work demands protect against cognitive impairment: MAAS prospective cohort study. Exp Aging Res 29(1):33–45

Bowen CE, Staudinger UM (2011) Die Bedeutung des Arbeitsklimas und seiner Facetten für das Personalmanagement im Demografischen Wandel [The relevance of the work climate and its facets for HR management in demographic change]. In: Staudinger UM, Godde B, Heidemeier H, Kudielka BM, Schömann K, Stamov-Rossnagel C, Voelpel S (eds) Den demografischen Wandel meistern: Eine Frage der Passung. W Bertelsmann, Bielefeld, pp 59–77 (in German)

Bowen CE, Staudinger UM (2013) Relationship between age and promotion orientation depends on perceived older worker stereotypes. J Gerontol B Psychol Sci Soc Sci 68(1):59–63

Bowen CE, Noack MG, Staudinger UM (2010) Aging in the work context. In: Schaie KW, Willis SL (eds) Handbook of the psychology of aging, 7th edn. Elsevier Academic, San Diego, pp 263–277

Boyke J, Driemeyer J, Gaser C et al (2008) Training-induced brain structure changes in the elderly. J Neurosci 28(28):7031–7035

Buchholz S, Rinklake A, Blossfeld H-P (2013) Umkehr von Frühverrentung in Deutschland. Eine Längsschnittanalyse der Auswirkungen der jüngsten Rentenreformen auf den Zeitpunkt des Erwerbsausstiegs und die Rentenhöhe [Revising early retirement in Germany. A longitudinal analysis of the consequences of the latest pension reform on the time of retirement and annuity rate]. Z für Bevölk [Comp Popul Stud] 38(4):907–936 (in German)

Bundesanstalt für Arbeitsschutz und Arbeitsmedizin (BAuA) (2013) Arbeitnehmer in Restrukturierungen. Gesundheit und Kompetenz erhalten [Employees in reorganization processes. Maintaining health and competence]. W Bertelsmann, Bielefeld (in German)

Bundesanstalt für Arbeitsschutz und Arbeitsmedizin (BAuA)/Initiative Neue Qualität der Arbeit (INQA) (2013) Securing the future with prevention. Strategies for a world of work aligned to demographic change [PDF]. Available from http://www.inqa.de/SharedDocs/PDFs/EN/second-memorandum-english.pdf?__blob=publicationFile. Accessed 16 Mar 2015

Buss K-P, Kuhlmann M (2013) Akteure und Akteurskonstellationen alter(n)sgerechter Arbeitspolitik [Protagonists and their constellation in age-based labor policy]. WSI-Mitteilungen 66(5):350–359 (in German)

Chatman JA (1989) Improving interactional organizational research: a model of person-organization fit. Acad Manage Rev 14(3):333–349

Dekkers-Sánchez PM, Hoving JL, Sluiter JK et al (2008) Factors associated with long-term sick leave in sick-listed employees: a systematic review. Occup Environ Med 65:153–157

De Lange A, Taris T, Jansen P, Smulders P, Houtman I, Kompier M (2006) Age as a factor in the relation between work and mental health: results from the longitudinal TAS survey. Occup Health Psychol Eur Perspect Res Educ Pract 1:21–45

Doyon J, Benali H (2005) Reorganization and plasticity in the adult brain during learning of motor skills. Curr Opin Neurobiol 15(2):161–167

Draganski B, May A (2008) Training-induced structural changes in the adult human brain. Behav Brain Res 192(1):137–142

Edwards JR (1991) Person-job fit: a conceptual integration, literature review, and methodological critique. In: Cooper CL, Robertson IT (eds) International review of industrial and organizational psychology. Wiley, Oxford, pp 283–357

European Commission (2014) For a European industrial renaissance. Communication from the Commission to the European Parliament, the Council, the European economic and social committee, and the committee of the regions [PDF]. Available from http://eur-lex.europa.eu/legal-content/EN/TXT/PDF/?uri=CELEX:52014DC0014&from=EN. Accessed 23 Mar 2015

Feuerhahn N, Stamov-Rossnagel C, Kudielka BM (2011) Passung in organisationen [Fit in organizations]. In: Staudinger UM, Godde B, Heidemeier H, Kudielka BM, Schömann K, Stamov-Rossnagel C, Voelpel S (eds) Den demografischen Wandel meistern: Eine Frage der Passung. W. Bertelsmann, Bielefeld, pp 37–58 (in German)

Fuchs J, Söhnlein D, Weber B (2011) Projektion des Arbeitskräfteangebots bis 2050: Rückgang und Alterung sind nicht mehr aufzuhalten [Projection of the German labor supply until 2050: recession and aging cannot be stopped]. IAB-Kurzbericht, Nürnberg (in German)

Gajewski PD, Wild-Wall N, Schapkin SA et al (2010) Effects of aging and job demands on cognitive flexibility assessed by task switching. Biol Psychol 85(2):187–199

Grabbe J, Richter G (2014) Arbeits-und Beschäftigungsfähigkeit–Grundlage von Innovations-und Wettbewerbsfähigkeit [Work ability and employability—A basis for innovation and competitiveness]. In: Klaffke M (ed) Generationen-Management. Springer Fachmedien, Wiesbaden, pp 83–106 (in German)

Grady C (2012) The cognitive neuroscience of ageing. Nat Rev Neurosci 13(7):491–505

Grawitch MJ, Gottschalk M, Munz DC (2006) The path to a healthy workplace: a critical review linking healthy workplace practices, employee well-being, and organizational improvements. Consulting Psychol J Pract Res 58(3):129–147. doi:10.1037/1065-9293.58.3.129

Greenwood PM (2007) Functional plasticity in cognitive aging: review and hypothesis. Neuropsychology 21(6):657–673

Hertzog C, Kramer AF, Wilson RS et al (2009) Enrichment effects on adult cognitive development: can the functional capacity of older adults be preserved and enhanced? Psychol Sci Pub Int 9(1):1–65

Holland P, Rubin D (1988) Causal inference in retrospective studies. Eval Rev 12(3):203–231

Höpflinger F (2008) Berufskarriere ist Langstreckenlauf, kein Sprint [Careers are long-distance runs, not sprints]. Personalführung 9:26–31

Hultsch DF, Hertzog C, Small BJ et al (1999) Use it or lose it: engaged lifestyle as a buffer of cognitive decline in aging? Psychol Aging 14(2):245–263

Jopp D, Hertzog C (2007) Activities, self-referent memory beliefs, and cognitive performance: evidence for direct and mediated relations. Psychol Aging 22(4):811–825

Kagermann H, Wahlster W, Helbig J (2012) Umsetzungsempfehlungen für das Zukunftsprojekt Industrie 4.0—Abschlussbericht des Arbeitskreises Industrie 4.0 [Recommended action plans for the future project industry 4.0—Final report of the work group industry 4.0]. Forschungsunion im Stifterverband für die Deutsche Wissenschaft, Berlin (in German)

Kohn ML, Schooler C (1978) The reciprocal effects of the substantive complexity of work and intellectual flexibility: a longitudinal assessment. Am J Sociol 84(1):24–52
Kristof AL (1996) Person-organization fit: an integrative review of its conceptualizations, measurement, and implications. Pers Psychol 49(1):1–49
Kristof-Brown AL, Stevens CK (2001) Goal congruence in project teams: does the fit between members' personal mastery and performance goals matter? J Appl Psychol 86(6):1083–1095
Kristof-Brown AL, Zimmerman RD, Johnson EC (2005) Consequences of individual's fit at work: a meta-analysis of person-job, person-organization, person-group, and person-supervisor fit. Pers Psychol 58(2):281–342
Leber U, Stegmaier J, Tisch A (2013) Altersspezifische Personalpolitik: Wie Betriebe auf die Alterung ihrer Belegschaften reagieren [Age-specific personnel policy: how industrial companies react to the aging of their employees]. IAB-Kurzbericht, Nürnberg (in German)
Lidwall U, Bergendorff S, Voss M et al (2009) Long-term sickness absence: changes in risk factors and the population at risk. Int J Occup Med Environ Health 22:157–168
Marquie JC, Tucker P, Folkard S et al (2014) Chronic effects of shift work on cognition: findings from the VISAT longitudinal study. Occup Environ Med. doi:10.1136/oemed-2013-101993
McGrath JE (1976) Stress and behavior in organizations. In: Dunnette M (ed) Handbook of industrial and organizational psychology. Rand McNally, Chicago, pp 1351–1395
Muchinsky PM, Monahan CJ (1987) What is person-environment congruence? Supplementary versus complementary models of fit. J Vocat Behav 31(3):268–277
Mümken S, Brussig M (2012) Alterserwerbsbeteiligung in Europa. Deutschland im internationalen Vergleich [Labor participation among older employees in Europe. Germany in an international comparison]. Altersübergangs Report Number 2012-01 (in German)
Naegele G, Walker A (2006) A guide to good practice in age management. Office for Official Publications of the European Communities, Luxembourg. Available from http://eurofound.europa.eu/sites/default/files/ef_publication/field_ef_document/ef05137en_0.pdf. Accessed 16 Mar 2015
Nerdinger FW, Blickle G, Schaper N (2008) Arbeits-und Organisationspsychologie [Occupational and organizational psychology]. Springer, Heidelberg (in German)
Ng TWH, Feldman DC (2008) The relationship of age to ten dimensions of job performance. J Appl Psychol 93(2):392–423
Noethen D, Voelpel S (2011) Der Einfluss von Führungskräften auf Einstellungen und Verhaltensweisen ihrer Mitarbeiter und Konsequenzen für relevante Handlungsfelder im demografischen Wandel [The impact of leaders on their members' attitudes and behavior and consequences for relevant fields of action in the context of demographic change]. In: Staudinger UM, Godde B, Heidemeier H, Kudielka BM, Schömann K, Stamov-Rossnagel C, Voelpel S (eds) Den demografischen Wandel meistern: Eine Frage der Passung. W. Bertelsmann, Bielefeld, pp 79–96 (in German)
O'Reilly CA III (1977) Personality—job fit: Implications for individual attitudes and performance. Organ Behav Hum Perform 18(1):36–46
Park DC, Lodi-Smith J, Drew L et al (2014) The impact of sustained engagement on cognitive function in older adults: the synapse project. Psychol Sci 25(1):103–112
Posthuma RA, Campion MA (2009) Age stereotypes in the workplace: common stereotypes, moderators, and future research directions. J Manage 35(1):158–188
Potter GG, Helms MJ, Burke JR et al (2007) Job demands and dementia risk among male twin pairs. Alzheimers Dement 3(3):192–199
Quick JC, Camara WJ, Hurrell JJJ, Johnson JV, Piotrowski CS, Sauter SL, Spielberger CD (1997) Introduction and historical overview. J Occup Health Psychol 2:3–6. doi:10.1037/1076-8998.2.1.3
Raz N, Lindenberger U, Rodrigue KM et al (2005) Regional brain changes in aging healthy adults: general trends, individual differences and modifiers. Cereb Cortex 15(11):1676–1689
Rubin D (1974) Estimating causal effects of treatments in randomized and nonrandomized studies. J Educ Psychol 66(5):688–701

Schaie KW (2005) Developmental influences on intelligence: the Seattle Longitudinal Study. Oxford University Press, New York

Schalk R, van Veldhoven M, de Lange AH, De Witte H, Kraus K, Stamov-Roßnagel C et al (2010) Moving European research on work and ageing forward: overview and agenda. Eur J Work Organ Psychol 19(1):76–101. doi:10.1080/13594320802674629

Schmidt-Wilcke T, Rosengarth K, Luerding R et al (2010) Distinct patterns of functional and structural neuroplasticity associated with learning Morse code. Neuroimage 51(3):1234–1241

Schooler C, Mulatu MS, Oates G (1999) The continuing effects of substantively complex work on the intellectual functioning of older workers. Psychol Aging 14(3):483–506

Seidler RD (2010) Neural correlates of motor learning, transfer of learning, and learning to learn. Exerc Sport Sci Rev 38(1):3–9

Shohamy D, Wimmer GE (2013) Dopamine and the cost of aging. Nat Neurosci 16(5):519–521

Siegrist J (2002) Effort-reward imbalance at work and health. In: Perrewe PL, Ganster DC (eds) Historical and current perspectives on stress and health. JAI Elsevier, Amsterdam, pp 261–291

Sonntag K (2014) Potenziale Erwerbstätiger bei verlängerter Lebensarbeitszeit [Employee potentials in prolonged working lives]. Available from http://www.gesamtmetall.de/gesamtmetall/meonline.nsf/id/PagePotenziale-Erwerbstaetiger-bei-verlaengerter-Lebensarbeitszeit--Chancen-und-Herausforderungen-f/$file/Potenziale-Erwerbstaetiger.pdf. Accessed 16 Mar 2015 (in German)

Stajkovic AD, Luthans F (1998) Self-efficacy and work-related performance: a meta-analysis. Psychol Bull 124(2):240–261

Statistisches Bundesamt (2013) Bevölkerung und Erwerbstätigkeit—Beruf, Ausbildung und Arbeitsbedingungen der Erwerbstätigen in Deutschland [Population and employment—Occupation, training, and working conditions of employees in Germany]. Available from https://www.destatis.de/DE/Publikationen/Thematisch/Arbeitsmarkt/Erwerbstaetige/BerufArbeitsbedingungErwerbstaetigen2010412127004.pdf;jsessionid=20E1959E06AF7E7CB91740F3E0BD0E34.cae1?__blob=publicationFile. Accessed 23 Mar 2015

Statistisches Bundesamt (2014) Erstmals 43 Millionen Erwerbstätige im Oktober 2014. [October 2014: For the first time more than 43 million in economic activity]. Available from https://www.destatis.de/DE/PresseService/Presse/Pressemitteilungen/2014/11/PD14_420_132pdf.pdf?__blob=publicationFile. Accessed 16 Mar 2015

Staudinger UM (2015a) Images of aging: outside and inside perspectives. Ann Rev Gerontol Geriatr 35(1):187–210

Staudinger UM (2015b) The future of aging: how I will live. In: Barysch K (ed) Our world and us: how our environment and our societies will change. Allianz SE, Munich, pp 137–149

Staudinger UM, Bowen CE (2011) A systemic approach to aging in the work context. Z Arbeitsmarktforschung 44(4):295–306

Staudinger U, Kocka J (2010) More years, more life. Recommendations of the joint academy initiative on aging. (Translation of "Gewonnene Jahre"; Aging in Germany Bd. 9). Nova Acta Leopoldina NF Bd, 108.

Staudinger UM, Noack CMG (2009) Die Wirkung von Altersbildern in Unternehmen [The effect of images of aging in organizations]. In: Ehmer J, Höffe O (eds) Bilder des Alters im Wandel. Historische, interkulturelle, theoretische und aktuelle Perspektiven (Altern in Deutschland Band 1), vol 99. Wissenschaftliche Verlagsgesellschaft mbH, Stuttgart, pp 197–205 (in German)

Staudinger UM, Godde B, Heidemeier H et al (2011) Den demografischen Wandel meistern: Eine Frage der Passung. W Bertelsmann, Bielefeld

Suo C, León I, Brodaty H et al (2012) Supervisory experience at work is linked to low rate of hippocampal atrophy in late life. Neuroimage 63(3):1542–1551

Trautmann M, Voelcker-Rehage C, Godde B (2011) Fit between workers' competencies and job demands as predictor for job performance over the work career [Passung zwischen Kompetenzen der Mitarbeiter und Anforderungen des Arbeitsplatzes als Prädiktor für Leistung über das Arbeitsleben]. J Lab Market Res 44(4):339–347. doi:10.1007/s12651-011-0078-2

Tsutsumi A, Kawakami N (2004) A review of empirical studies on the model of effort–reward imbalance at work: reducing occupational stress by implementing a new theory. Soc Sci Med 59(11):2335–2359

Tuomi K, Ilmarinen J, Jahkola A et al (2006) Arbeitsbewältigungsindex [Work ability index]. Schriftenreihe der Bundesanstalt für Arbeitsschutz und Arbeitsmedizin: Übersetzung Ü 14 (in German)

Valenzuela MJ, Sachdev P, Wen W et al (2008) Lifespan mental activity predicts diminished rate of hippocampal atrophy. PLoS One 3(7):e2598

Verquer ML, Beehr TA, Wagner SH (2003) A meta-analysis of relations between person–organization fit and work attitudes. J Vocat Behav 63(3):473–489

Voelcker-Rehage C, Niemann C (2013) Structural and functional brain changes related to different types of physical activity across the life span. Neurosci Biobehav Rev 37(9):2268–2295

Voelcker-Rehage C, Godde B, Staudinger UM (2010) Physical and motor fitness are both related to cognition in old age. Eur J Neurosci 31(1):167–176

Voelcker-Rehage C, Godde B, Staudinger UM (2011) Cardiovascular and coordination training differentially improve cognitive performance and neural processing in older adults. Front Hum Neurosci 5:26

Warr P, Cook J, Wall T (1979) Scales for the measurement of some work attitudes and aspects of psychological well-being. J Occup Psychol 52(2):129–148

Part II

The Importance of Leadership for Occupational Health

The Relationship Between Leadership and Health: A Comparison of General and Health-Focused Leadership Approaches

Stephan A. Boehm, Miriam K. Baumgärtner, and Lars M. Kreissner

Abstract

Keeping employees physically and psychologically healthy and thriving has increasingly become a challenge for organizations. Because poor health is associated with various negative consequences for the individual as well as at organizational level, researchers have started to investigate pathways for fostering, retaining, and restoring health and well-being in the work-place. Various studies underline the important role of leaders in creating work-place conditions that either positively or negatively influence employee health and well-being. In this chapter, we therefore differentiate between relationship-oriented and task-oriented leadership behaviors while providing an overview of the association between established leadership theories and indicators of employee health. Moreover, we focus on health-specific leadership styles, which are assumed to have a more direct effect on followers' health than more general styles. We have assumed that health-focused leadership skills complement general leadership skills. Avenues for future research are discussed and practical conclusions provided.

1 Employees' Health as a Challenge and Corporate Goal

Employees' physical and psychological health has recently become a prominent topic for both organizational research and practice. There are various possible reasons for this development. First, ample research has demonstrated that poor health and well-being are associated with several negative outcomes on both the company and individual level. For instance, employees with health-related

S.A. Boehm (✉) • M.K. Baumgärtner • L.M. Kreissner
Center for Disability and Integration, University of St. Gallen, Rosenbergstrasse 51, 9000 St. Gallen, Switzerland
e-mail: stephan.boehm@unisg.ch; miriam.baumgaertner@unisg.ch; lars.kreissner@unisg.ch

M. Wiencke et al. (eds.), *Healthy at Work*, DOI 10.1007/978-3-319-32331-2_7

problems have been found to be less productive, to be absent from work more often, and to make poorer-quality decisions and fewer overall contributions to their organizations than their healthy colleagues (Danna and Griffin 1999). In addition, they may personally experience various physiological, psychological, and emotional costs (Bourbeau et al. 1996; Cartwright and Cooper 1993).

Second, recent data from Germany indicate that sick days increased by 28 % between 2006 and 2013. On average, persons in the German workforce miss 14.7 days per year because of poor health, generating considerable costs for both companies and social security systems (TK Gesundheitsreport 2014). Hence, organizations are becoming increasingly interested in exploring how to optimize working conditions to minimize costly sick days.

Third, health statistics from many developed economies indicate that employees' psychological health is particularly jeopardized. The Organisation for Economic Co-operation and Development (OECD 2014) reported that in Switzerland mental health problems annually cost the economy 19 billion Swiss francs, which corresponds to 3.2 % of the gross domestic product. In addition, people with mental health problems account for almost 40 % of all new disability benefit claims. The situation is similar in Germany, where the number of psychologically induced sick days grew by a factor of 19 between 2004 and 2011, causing the German economy to lose 6.3 billion euros annually (Betriebskrankenkasse [BKK] 2012). While researchers debate whether an apparently more complex working life (e.g., intensified cost pressure, fierce competition, globalization, virtualization, and ubiquity) explains this ongoing increase, it can be concluded that, at minimum, there is a strong interrelationship between work and psychological health. In addition, persons with psychological illnesses are at particular risk of being marginalized in the labor market (Baldridge et al. in press) because of the widespread stigma associated with mental disability (Corrigan et al. 2005).

Fourth, employees' health can be expected to become increasingly important because of the demographic changes in most industrialized countries, which are characterized by low birth rates, an increasing life expectancy, and the aging of the large baby-boom generation born between 1946 and 1964 (Craig and Paganelli 2000; Truxillo and Fraccaroli 2013). To take pressure off the pension system, there has been an increased rise in the mandatory retirement age in many countries (Börsch-Supan 2003; Flynn 2010), accompanied by the abolishment of early retirement programs in most firms. Because of the substantial correlation between age and disability (Ilmarinen 1994, 2001; World Health Organization [WHO] 2011), it can be expected that many older employees will develop health constraints during their working lives. In support of this contention, studies show that older employees take more sick days. While persons take sick leave less frequently with increasing age, the duration of that leave increases (Badura et al. 2014). A recent report on workers' health in German companies revealed that despite employees aged over 50 years comprising less than one third of the overall workforce in 2013, their health issues led to absenteeism corresponding to 46 % of the costs of the missed gross value added (Badura et al. 2014). This equals a loss of productivity of 42.6 billion euros. In addition, it has become much harder to opt out of working life and request early retirement because of health constraints (Muller-Camen et al. 2011). Therefore, now that organizations cannot "rejuvenate" their workforce by routinely dismissing older employees, they have to invest in the work ability of their personnel. Thus, health promotion has become a core component of managing an aging workforce (Boehm et al. 2013).

Taken together, companies face at least two major challenges regarding their employees' health. To reach organizational goals, they have to invest in health promotion and ensure that their fit and healthy personnel retain a high capacity for work (i.e., they have to engage in *prevention behaviors*). Meanwhile, to secure the jobs and foster employees' recovery with existing health constraints, organizations need to engage in activities to restore those employees' health and well-being (i.e., they have to invest in *intervention behaviors*). Given both the importance of work for the self-concept of most individuals (Blustein 2008) and the negative economic implications of losing a job because of poor health, both organizations and society as a whole should have a key interest in fostering, retaining, and restoring health in the work-place.

As already stated, organizations seem to have recognized this challenge. Most companies currently invest in corporate health programs and worksite health promotion, including health education, screening, intervention programs, fitness and nutrition programs, and stress management seminars. The positive impact of such health promotion programs upon key financial indicators has been studied and proven extensively (Aldana 2001; Goetzel and Ozminkowski 2008; Heaney and Goetzel 1997; Pelletier 2001), providing further motivation for companies to engage in such activities. Recently, however, both research and practice seem increasingly interested in shedding light on additional organizational factors that might contribute to employees' long-term health and well-being. It has been proposed that supervisors, particularly the leadership behavior they display to their followers, are one such decisive organizational factor (Gilbreath and Benson 2004).

2 Definition of Employees' Health and Well-Being

In the preamble to the constitution of the WHO, health is defined as "a state of complete physical, mental and social well-being and not merely the absence of disease or infirmity" (WHO 1946). Thus, health promotion has to go beyond sickness prevention to also focus on positive states of health. Regarding the relationship between leadership and health, a wide range of different outcomes have been investigated. In accordance with the research tradition of organizational behavior, soft health-related factors have mostly been studied. These have often been subsumed under the umbrella terms of positive and negative employee well-being, which conceptualizes well-being as having two distinct dimensions rather than being a continuum with two poles (Gregersen et al. 2014). Positive health-related outcomes of leadership include job satisfaction, perceived psychological and physical health, safety, and positive moods. Some studies even consider that more distal outcomes such as job performance are health-related (e.g., Kuoppala et al. 2008). Regarding negative health indicators, stress, strain, anxiety, exhaustion, burnout, and depression have been examined (Kelloway and Barling 2010; Kuoppala et al. 2008). However, few studies, and these are mainly in the field of medicine and rehabilitation, have investigated relatively hard outcomes of leadership, such as blood pressure (Karlin et al. 2003), cardiovascular diseases (Wager et al. 2003), work-place injuries and accidents (Mullen and Kelloway 2009), and sick leave and disability pensions (Kuoppala et al. 2008). Broadly speaking, the more objective the health indicator, the fewer empirical studies have been performed.

3 The Relationship Between Leadership and Followers' Health and Well-Being

Research has clearly indicated that "good" leadership plays a key role in keeping employees healthy and happy (for recent reviews, see Kelloway and Barling 2010; Kuoppala et al. 2008). However, what constitutes "good" leadership in terms of fostering employees' physical and psychological health is much less clear (Eriksson et al. 2010). Many of the existing studies on the leadership–health relationship have investigated the potential impact of established leadership styles such as leader–member exchange (LMX; e.g., McGee et al. 1987; Rousseau et al. 2008) and transformational leadership (TFL; e.g., Arnold et al. 2007; Sosik and Godshalk 2000).

Another line of research has taken a more focused approach and proposed that health-focused leadership (HFL) styles have a more direct effect on followers' health than more general leadership styles (e.g., Gurt et al. 2011; Boehm and Baumgärtner 2014). In this chapter, we will present and discuss both of these approaches with the aim of providing readers with a holistic impression of the leadership–health relationship. In so doing, we will also discuss existing gaps in the literature and propose directions for future research.

First, however, we want to provide our definition of leadership. The subject of leadership is possibly one of the best researched topics in organizational behavior. Yukl (2012, p. 66) summarized the essence of leadership in organizations as "influencing and facilitating individual and collective efforts to accomplish shared objectives". In general, leaders' behavior can be classified into task-oriented (e.g., developing plans, determining schedules, clarifying goals, setting priorities, monitoring progress); relationship-oriented (e.g., supporting, developing, and empowering employees); change-oriented (e.g., encouraging innovation, facilitating learning, and communicating a vision); and externally oriented activities (e.g., networking, representing, boundary spanning) (Yukl 2006, 2012). In particular, task- and relationship-oriented behaviors seem likely to influence employees' health and well-being, particularly given that supervisors have power that they can use to reward or penalize health-relevant employee behavior (Kelloway and Barling 2010). Therefore, we will now focus on the potential health impact of relationship-oriented leadership (i.e., consideration behaviors, LMX, and TFL) and task-oriented leadership behaviors (i.e., initiating structure and transactional leadership [TAL]).

4 The Impact of Established Leadership Concepts on Followers' Health and Well-Being

4.1 Consideration and Initiating Structure

Consideration and initiating structure are two dimensions of a classical leadership concept developed in the 1950s at Ohio State University (Fleishman 1953). The consideration dimension is people-oriented and comprises leader behaviors that target employees' welfare and foster interpersonal relationships, mutual trust, and friendship (Judge et al. 2004). Consideration-oriented leaders are perceived as being approachable, friendly, and treating followers equally. In contrast, initiating structure is task-oriented. Task-oriented leaders focus on accomplishing goals; they define roles, initiate actions, and organize group activities. They let followers know what is expected of them and request compliance with rules, standards, and procedures (Judge et al. 2004).

Both consideration and initiating structure are reportedly associated with indicators of health. Gregersen et al. (2014) found both consideration and initiating structure to be positively correlated with job satisfaction and general health and negatively with emotional exhaustion, depersonalization, and perceived strain. In a systematic review, Skakon et al. (2010) concluded that consideration is positively associated with the affective well-being of employees and negatively with low stress levels among employees. In their meta-analysis, Kuoppala et al. (2008) found weak evidence for an association between consideration and job satisfaction, very weak evidence for a relationship between consideration and job well-being, weak evidence for a link between consideration and job performance, and very weak evidence for an association between consideration and sick leave. They failed to identify any studies that investigated potential relationships with early retirement.

4.2 Transactional Leadership

Transactional leadership refers to the bargain and exchange processes that typically occur between leaders and their followers (Burns 1978; Bass 1985). Transactional leaders have two dominant behavioral patterns (Howell and Avolio 1993). First, contingent reward refers to the establishment of constructive transactions or exchanges with their followers (Bass 1991). By clarifying the behaviors they expect from their employees and establishing rewards for meeting these demands, they provide motivation for followers to act according to agreed-upon objectives. Rewards may involve the leader's recognition of work accomplished, bonuses, or merit increases (Judge and Piccolo 2004). Second, "management by exception" describes behavior that includes corrective criticism, negative feedback, and negative reinforcement when employees do not meet organizational demands. These interventions can either be executed actively, by continuously monitoring for and reacting to potential mistakes, or passively, by intervening only when problems arise.

Reported results of research on the health-related implications of TAL are inconsistent. Judge and Piccolo (2004) performed a meta-analysis comparing the effects of transformational, transactional, and laissez-faire leadership on various leadership criteria. Compared with the other leadership styles, they found that contingent reward leadership was the most highly correlated with employee job satisfaction. However, there were too few studies to subject the relationship between management by exception and job satisfaction to meta-analysis. The passive form of management by exception was moderately negatively correlated with group/organization performance; this was not true for the active dimension. In a more recent meta-analysis, Skakon et al. (2010) investigated the effects of different leadership styles on more directly health-related outcomes, such as affective well-being, employee stress, and burnout and reported inconsistent results. While some studies found a significant relationship between TAL and employee stress or well-being, others did not. However, in a current empirical study, Gregersen et al. (2014) provided evidence that contingent reward behavior is positively associated with job satisfaction ($r = 0.45$) and general health ($r = 0.17$) and negatively with emotional exhaustion ($r = -0.25$), depersonalization ($r = -0.12$), and perceived strain ($r = -0.18$).

4.3 Transformational Leadership

Transformational leadership has been developed as a strategy for complementing and eventually replacing TAL (Burns 1978). TFL is often defined in terms of its potential effects on individuals, teams, and whole organizations (Van Knippenberg and Sitkin 2013); these include instilling pride, trust, and respect, inspiring innovation and overcoming of the status quo, and shifting motivation from self-interest to collective interest, including higher-order needs; it thus enables followers to perform beyond expectations (Bass 1985; Conger and Kanungo 1987; Shamir et al. 1993). According to Bass and colleagues (Avolio and Bass 2004; Bass 1985; Bass and Riggio 2006), transformational leaders engage in four related, yet distinct leadership behaviors that together form the TFL construct. First, they engage in idealized influence behaviors by providing appropriate role models and showing concern for the ethical and moral dimensions of their activities. Second, they provide inspirational motivation by developing and communicating a compelling vision for their sphere of influence. Third, they ensure intellectual stimulation and foster employees' thinking, thereby facilitating challenges to the status quo. Finally, they show individualized consideration by providing coaching and developmental activities for each of their followers. Hundreds of studies, including various meta-analyses (De Rue et al. 2011; Judge and Piccolo 2004; Lowe et al. 1996), have provided evidence that there is a positive relationship between TFL and various desirable outcome criteria at all levels of analysis.

As to effects on health, Gregersen et al. (2014) summarized the existing literature as showing a positive relationship between TFL and psychological affective well-being, job satisfaction, and other indicators of well-being and a negative relationship with burnout, stress, and strain. However, they also identified some studies that did not find support a negative relationship between TFL and negative well-being (i.e., stress and burnout). In their own empirical study, the authors confirmed the overall tendency described above and provided support for positive relationships between TFL and job satisfaction and general health, as well as negative relationships between TFL and emotional exhaustion, depersonalization, and perceived strain. This is in accordance with the findings of Judge and Piccolo's (2004) meta-analysis, which found that TFL is positively associated with followers' job satisfaction. Furthermore, in their overview of 12 published papers, Skakon et al. (2010) concluded that TFL is positively associated with job satisfaction and affective well-being and negatively with stress and burnout. However, Kuoppala et al. (2008) found only very weak evidence for relationships between TFL and job satisfaction and TFL and job well-being. These authors were unable to find any published studies regarding job performance, sick leave, and early retirement, which underlines the lack of studies investigating relationships between leadership behavior and more objective health-related outcome variables.

4.4 Leader–Member Exchange

Leader–member exchange (Graen 1976; Graen and Uhl-Bien 1995) is potentially the most well-known approach to operationalizing and assessing the quality of relationships between supervisors and their followers. LMX theory proposes that leaders develop different levels of exchange relationships with their followers (Sparrowe and Liden 1997), ranging from close, trusting, and reciprocal high-quality relationships to more exchange-based and formalized low-quality relationships (Gerstner and Day 1997; Graen and Uhl-Bien 1995; Wayne et al. 1997). Ample empirical research, including three meta-analyses, has demonstrated the positive effects of being in a high-quality relationship with one's supervisor (Dulebohn et al. 2012; Gerstner and Day 1997; Ilies et al. 2007). These positive effects include more job-related information, increased interaction, greater personal concern, and greater job direction, as well as more feedback, support, training, and developmental opportunities from supervisors (Dunegan et al. 1992, 2002; Gerstner and Day 1997).

LMX is reportedly a powerful resource regarding physical and psychological health outcomes. Studies have demonstrated that good-quality relationships are positively associated with employee job satisfaction and well-being and negatively with stress (Skakon et al. 2010). Gregersen et al. (2014) compared different leadership styles with regard to the following indicators of employee health:

consideration, initiating structure, TFL, contingent reward, and LMX, and assessed the relationships between these indicators and job satisfaction, general health, emotional exhaustion, depersonalization, and perceived strain. The authors concluded that the quality of the relationship between supervisor and employee is the best predictor for well-being, particularly for job satisfaction, and that the other leadership constructs examined did not add substantial additional variance.

5 The Impact of Health-Focused Leadership Concepts on Followers' Health and Well-Being

To date, research on the leadership–health relationship has mainly investigated the impact of rather broad leadership theories. However, there is agreement on the usefulness of specific predictors that predict specific behaviors (Ajzen and Fishbein 1977): congruence between predictor and outcome is associated with greater predictive validity (Schneider et al. 2011). Consequently, investigating a leader's explicit engagement in health-supporting behaviors may be a promising avenue for predicting health-specific outcomes. Despite the promising potential of such a domain-specific leadership style (Barling et al. 2002) in the field of health promotion, theoretical and empirical research on this topic is "almost non-existent" (Gurt et al. 2011, p. 111). We will now describe some new approaches in this domain.

5.1 Health-Specific Leadership

Gurt et al. (2011) introduced the construct of "health-specific leadership (HSL)" as "the leaders' explicit and therefore visible consideration of and engagement in employee health." They argued that HSL influences employees' health via two mechanisms: First, through a leader directly targeting promoting health; and second, through role modeling that positively affects employees' health-related behaviors. Gurt et al. (2011) also included task- and relationship-related aspects of leadership in the HSL scale. The items in this scale mainly focus on health-promoting communication, more precisely on discussion with and information from the supervisor concerning health-related issues. In their study, Gurt et al. (2011) examined associations between HSL, general leadership behavior, and employee strain (measured by irritation), including three mediating variables (namely, climate for health, role ambiguity, and job satisfaction). They demonstrated that HSL can be clearly differentiated from general leadership practices and that neither HSL nor general leadership is directly associated with irritation. General leadership was significantly related to all three mediators: better climate for health, less role ambiguity, and greater job satisfaction. HSL showed a significantly positive relationship with both role ambiguity and climate for health. In turn, the mediating variables of role ambiguity and job satisfaction were significantly associated with employee irritation.

5.2 Health-Oriented Leadership

Franke et al. (2014) introduced the construct of health-oriented leadership (HoL), which combines follower-directed health-promoting leadership (StaffCare) and self-directed health-promoting leadership (SelfCare). This approach distinguishes StaffCare as an external resource from SelfCare as an internal resource. Whereas leaders more strongly influence the external conditions, such as the task and working context, followers add individual motivational, cognitive, and behavioral factors related to health. Both StaffCare and SelfCare consist of three distinct components; namely, (1) health behavior; (2) value of health; and (3) health awareness. Health behavior comprises actions relevant to the promotion of health. In terms of StaffCare, health behavior refers to creating health-promoting working conditions, reinforcing employees' health-directed working behavior, and informing employees about issues that are relevant to health. "Value of health" captures the importance attached to health. In StaffCare, value of health is associated with leaders' concern for the health of their followers and the degree to which they feel responsible for it. Health awareness comprises the degree to which leaders are sensitive and attentive to and reflective about the health of employees and the factors influencing it. The concept is summarized in the "house of HoL" (Franke et al. 2014, p. 143), in which the SelfCare of leaders is the foundation for StaffCare and for role modeling, which, in turn, is related to the SelfCare of followers.

These authors found that StaffCare is associated with followers' health and well-being, namely with state of health, irritation, health complaints, and work-family conflict. StaffCare explains unique variance in health outcomes beyond TFL. As hypothesized, the relation between StaffCare and the outcome variables was mediated by followers' health-related attributes (SelfCare). More precisely, the followers' ratings of their supervisors' StaffCare were positively related to their SelfCare, which, in turn, was positively linked to their health.

5.3 Health-Focused Leadership

A third approach to health-related leadership behaviors stems from the work of Boehm and Baumgärtner (2014; see also Boehm and Dwertmann 2015 as well as Böhm and Baumgärtner 2016). Drawing from various disciplines such as organizational psychology, medicine, public health, and disability management, Boehm and Baumgärtner (2014) developed a conceptual and empirical model of HFL and postulated that prevention and intervention behaviors are significant predictors of employees' health. Regarding prevention behaviors, leaders should aim at providing positive resources while avoiding harmful influences on their employees' health such as constant work overload, overtime, and a paucity of opportunities to recover. Based on Tuomi et al.'s findings (2001) that a non-restful work environment, work overload, and dissatisfaction with work time arrangements negatively affect employees' ability to work, Boehm and Baumgärtner (2014) suggested that preventive behaviors specifically targeted at reducing such negative factors are likely positively associated with followers' ability to work. Consequently, supervisors should assign appropriate work-loads that match the personal and organizational resources of their followers (Demerouti et al. 2001). In addition, leaders should ensure that employees are able to achieve their goals within acceptable work hours and without regular overtime or forgoing holidays. Finally, supervisors should try to balance their followers' demands over the long term, making sure that after periods of intensive engagement there are opportunities for employees to recover and recharge their energy levels (Bruch and Menges 2010; Dwertmann and Kunz 2012).

Second, at the first sign that an employee is ill (including physical and mental illness), leaders should clearly intervene, ensuring that employees understand that their supervisors care about them and their health. Ample research has demonstrated that early interventions are an effective tool for fostering the return-to-work and long-term ability to work of employees who have been absent because of sickness (e.g., Hoefsmit et al. 2012). Moreover, poor leader support correlates positively with the duration of work disability (Janssen et al. 2003; Krause et al. 2001). Therefore, leaders must have a key interest in recognizing employees' health problems early and in reacting appropriately and in a timely manner. They must clearly communicate to followers that they do not regard illness as a "sign of weakness" and that they understand their employees' difficulties. Further, leaders should demonstrate that they regard employees' health as an important resource and that recovery has priority over short-term work goals. For followers with serious health issues, they should try to find joint solutions to handling and improving work conditions such as to support recovery. These solutions could include accommodating the work-place to the employee's disabilities or restructuring work tasks or work time.

Boehm and Baumgärtner (2014) have developed and empirically tested a model of the two dimensions of HFL. They based their scale development and model testing on a qualitative study involving 153 employees of a German car manufacturer and two quantitative studies, one involving 96 members of the general German working population and the other 1277 employees of a German public service organization. They found that prevention and intervention behaviors are significant predictors of employees' physical and psychological health states and that these, in turn, have significant relationships with both job performance and turnover intention. In a further study with 2858 employees of a German manufacturing company, they found a negative relationship of HFL with objectively measured sick days (measured in a 6-month period following the survey) over and above a potential effect of LMX.

6 Theoretical Considerations: Leadership and Employees' Health: Lessons Learnt and Ways Forward

As we have described in this chapter, there is both a growing practical need for and scholarly interest in studying the effects of leadership on employees' health and well-being. In light of the findings discussed in the previous sections, various avenues for further research seem promising.

First, from a theoretical point of view, it seems worth discussing and analyzing these new health-focused leadership approaches in an integrated manner. More specifically, scholars may wish to investigate how Gurt et al.'s (2011), Franke et al.'s (2014), and Boehm and Baumgärtner's (2014) approaches relate to each other, what ideas they have in common, and whether they could eventually be combined to develop clear recommendations for the most health-friendly leadership behaviors. On a related aspect, both researchers and practitioners should be keen to ascertain whether these health-focused approaches could extend or replace existing leadership approaches. For instance, can successful leadership behavior be composed of various elements such as goal-oriented TAL as a base, vision- and identity-oriented TFL as a supplement, and a health-focused style as an additional component? Wegge et al. (2014) have recently introduced an integrative model of "five pathways between leadership behavior and employee health" in which they differentiate between interpersonal or dyadic, team or organizational, and environmental or work-system levels. Building on this model and explaining in more depth how leadership behavior at different levels of analysis fosters or hampers health, while simultaneously taking into account relevant moderators and mediators, seems to be strongly indicated.

Second, from an empirical point of view, it seems important to investigate the amount of unique variance in followers' health and well-being that can be explained by each of these leadership approaches. Gregersen et al.'s study (2014), which compares several leadership approaches within one sample, is an example of such an approach. Such a study should include both established leadership approaches (e.g., LMX, TAL, TFL, authentic leadership, servant leadership) and health-focused approaches (e.g., HSL, HoL, HFL).

Third, and this is related to the previous suggestion, scholars should strive to investigate a variety of health-related outcome variables in such studies. In addition to more well-being-related constructs like job satisfaction, engagement, and commitment, it may be crucial to investigate more health-focused outcomes such as work ability and perceived health. Finally, the inclusion of objective, measurable health indicators such as blood pressure, cortisol concentration, sickness absence, early retirement, and disability pension may be the most promising strategy.

Finally, empirically strong designs should be used to investigate the potential benefits of health-focused leadership approaches. The statement "due to our cross-sectional data, we unfortunately cannot show causality and call for longitudinal and experimental future research" has unfortunately become a commonplace in organizational behavior research; this limitation is particular disappointing in studies on the leadership–health relationship. When investigating the health effects of leadership, researchers should try their utmost to identify effects rather than relationships. One approach to achieving this would be to borrow from medical research and perform randomized controlled trial designs. For example, a randomly selected group of supervisors could be trained and the health of their followers compared with the health of employees of untrained supervisors.

7 Practical Considerations: What Companies Should Do to Foster Employees' Health and Well-Being

Well-developed health-focused leadership skills complement good general leadership skills and protect employees' psychological and physical health. Leadership development is a proven effective intervention in occupational health psychology (Kelloway and Barling 2010). Thus, organizations should train their leaders in health-focused leadership behaviors. For instance, they could organize staff training workshops that inform leaders of their responsibility for and effect on their followers' health. Training content could be based on the job demands-resources model (Demerouti et al. 2001). Informing leaders about both work-related demands and resources and health-focused leadership approaches would provide them with starting points for promoting their employees' health. Furthermore, leaders who successfully engage in health-focused leadership behaviors could be held up as examples and serve as role models for others. Finally, leaders could be asked to suggest ways of integrating health-focused leadership behaviors into their daily routines.

To create an organization-wide climate of health-focused leadership, these trainings should address different hierarchical levels within the organization. We recommend starting at the top management level and cascading health-focused leadership behaviors down the organization to all supervisors. This is in line with the current trend toward focusing on leadership climates within firms in published reports about leadership (e.g., Boehm et al. 2015).

Moreover, organizations would be well advised to complement trainings in health-focused leadership with other health promotion activities (Heaney and Goetzel 1997) with the aim of fostering a general health-promoting climate (Wilson et al. 2004). Such activities might include health-education programs, medical check-ups and screenings, prevention-oriented changes to the work-place, or individual stress management programs.

References

Ajzen I, Fishbein M (1977) Attitude-behavior relations: a theoretical analysis and review of empirical research. Psychol Bull 84(5):888–918

Aldana SG (2001) Financial impact of health promotion programs: a comprehensive review of the literature. Am J Health Promot 15(5):296–320

Arnold KA, Turner N, Barling J, Kelloway EK et al (2007) Transformational leadership and psychological well-being: the mediating role of meaningful work. J Occup Health Psychol 12(3):193–203
Avolio BJ, Bass BM (2004) Multifactor leadership questionnaire manual and sampler set, 3rd edn. Mind Garden, Redwood City
Badura B, Ducki A, Schröder H, Klose J, Meyer M (eds) (2014) Fehlzeiten-Report 2014: Erfolgreiche Unternehmen von morgen—gesunde Zukunft heute gestalten. Springer, Berlin (in German)
Baldridge D, Beatty J, Boehm SA, Kulkarni M, Moore M (in press) People with (dis)abilities. In Colella AJ, King EB (eds) The Oxford handbook of workplace discrimination. Oxford University Press, New York
Barling J, Loughlin C, Kelloway EK (2002) Development and test of a model linking safety-specific transformational leadership and occupational safety. J Appl Psychol 87(3):488–496
Bass BM (1985) Leadership and performance beyond expectations. Free, New York
Bass BM (1991) From transactional to transformational leadership: learning to share the vision. Organ Dyn 18(3):19–31
Bass BM, Riggio RE (2006) Transformational leadership, 2nd edn. Erlbaum, Mahwah
BKK (Betriebskrankenkassen); Federal Association of Company Health Insurance Funds) (2012) BKK Faktenspiegel—BKK Gesundheitsreport 2012. BKK Factsheet—BKK Health report. BKK Bundesverband, Essen
Blustein DL (2008) The role of work in psychological health and well-being: a conceptual, historical, and public policy perspective. Am Psychol 63(4):228–240
Boehm SA, Baumgärtner MK (2014) Health-focused leadership: prevention and intervention as enablers of followers' health and performance. Paper presented at the 74th annual meeting of the Academy of Management, Philadelphia, 1–5 Aug 2014
Boehm SA, Dwertmann DJG (2015) Forging a single-edged sword: facilitating positive age and disability diversity effects in the workplace through leadership, positive climates, and HR practices. Work Aging Retire 1(1):41–63
Boehm SA, Schröder H, Kunze F (2013) Comparative age management: theoretical perspectives and practical implications. In: Field J, Burke R, Cooper C (eds) Sage handbook of aging, work and society. Sage, London, pp 211–237
Boehm SA, Dwertmann DJ, Bruch H et al (2015) The missing link? Investigating organizational identity strength and transformational leadership climate as mechanisms that connect CEO charisma with firm performance. Leader Q 26:156–171
Böhm SA, Baumgärtner MK (2016) Gesünder führen. Harv Bus Manag 38(2):6–9
Börsch-Supan A (2003) Labor market effects of population aging. Labour 17:5–44
Bourbeau J, Brisson C, Allaire S (1996) Prevalence of the sick building syndrome symptoms in office workers before and after being exposed to a building with an improved ventilation system. Occup Environ Med 53(3):204–210
Bruch H, Menges JI (2010) The acceleration trap. Harv Bus Rev 88(3):80–86
Burns JM (1978) Leadership. Harper and Row, New York
Cartwright S, Cooper C (1993) The psychological impact of mergers and acquisitions on the individual: a study of building society managers. Hum Relat 46:327–347
Conger JA, Kanungo RN (1987) Toward a behavioral theory of charismatic leadership in organizational settings. Acad Manage Rev 12(4):637–647
Corrigan PW, Kerr A, Knudsen L (2005) The stigma of mental illness: explanatory models and methods for change. Appl Prevent Psychol 11:179–190
Craig LA, Paganelli V (2000) Phased retirement: reshaping the end of work. Compensat Benefit Manage 16:1–9
Danna K, Griffin RW (1999) Health and well-being in the workplace: a review and synthesis of the literature. J Manage 25(3):357–384
Demerouti E, Bakker AB, Nachreiner F et al (2001) The job demands-resources model of burnout. J Appl Psychol 86(3):499–512

De Rue DS, Nahrgang JD, Wellman NED et al (2011) Trait and behavioral theories of leadership: an integration and meta-analytic test of their relative validity. Pers Psychol 64(1):7–52
Dulebohn JH, Bommer WH, Liden RC et al (2012) A meta-analysis of antecedents and consequences of leader-member exchange integrating the past with an eye toward the future. J Manage 38(6):1715–1759
Dunegan KJ, Duchon D, Uhl-Bien M (1992) Examining the link between leader member exchange and subordinate performance: the role of task analyzability and variety as moderators. J Manage 18(1):59–76
Dunegan KJ, Uhl-Bien M, Duchon D (2002) LMX and subordinate performance: the moderating effects of task characteristics. J Bus Psychol 17(2):275–285
Dwertmann DJG, Kunz JJ (2012) HR strategies for balanced growth. In: Mennillo G, Schlenzig T, Friedrich E (eds) Balanced growth: finding strategies for sustainable development. Springer, Berlin, pp 137–161
Eriksson A, Axelsson R, Axelsson SB (2010) Development of health promoting leadership—experiences of a training programme. Health Educ 110(2):109–124
Fleishman EA (1953) The description of supervisory behavior. J Appl Psychol 37(1):1–6
Flynn M (2010) The United Kingdom government's 'business case' approach to the regulation of retirement. Ageing Soc 30:421–443
Franke F, Felfe J, Pundt A (2014) The impact of health-oriented leadership on follower health: development and test of a new instrument measuring health-promoting leadership. Z Personalforschung 28(1–2):139–161
Gerstner CR, Day DV (1997) Meta-analytic review of leader–member exchange theory: correlates and construct issues. J Appl Psychol 82(6):827–844
Gilbreath B, Benson PG (2004) The contribution of supervisor behaviour to employee psychological well-being. Work Stress 18(3):255–266
Goetzel RZ, Ozminkowski RJ (2008) The health and cost benefits of work site health-promotion programs. Annu Rev Publ Health 29:303–323
Graen GB (1976) Role making processes within complex organizations. In: Dunnette MD (ed) Handbook of industrial and organizational psychology. Rand-McNally, Chicago, pp 1201–1245
Graen GB, Uhl-Bien M (1995) Relationship-based approach to leadership: development of leader-member exchange (LMX) theory of leadership over 25 years: applying a multi-level multi-domain perspective. Leader Q 6(2):219–247
Gregersen S, Vincent-Höper S, Nienhaus A (2014) Health–relevant leadership behaviour: a comparison of leadership constructs. Z Personalforschung 28(1–2):117–138
Gurt J, Schwennen C, Elke G (2011) Health-specific leadership: is there an association between leader consideration for the health of employees and their strain and well-being? Work Stress 25(2):108–127
Heaney CA, Goetzel RZ (1997) A review of health-related outcomes of multi-component worksite health promotion programs. Am J Health Promot 11(4):290–307
Hoefsmit N, Houkes I, Nijhuis FJ (2012) Intervention characteristics that facilitate return to work after sickness absence: a systematic literature review. J Occup Rehab 22(4):462–477
Howell JM, Avolio BJ (1993) Transformational leadership, transactional leadership, locus of control, and support for innovation: key predictors of consolidated-business-unit performance. J Appl Psychol 78(6):891–902
Ilies R, Nahrgang JD, Morgeson FP (2007) Leader-member exchange and citizenship behaviors: a meta-analysis. J Appl Psychol 92(1):269–277
Ilmarinen J (1994) Aging, work and health. In: Snel J, Cremer R (eds) Work and aging: a European perspective. Taylor and Francis, London, pp 47–63
Ilmarinen J (2001) Aging workers. Occup Environ Med 58:546–552
Janssen N, Van den Heuvel WPM, Beurskens AJHM et al (2003) The demand-control-support model as a predictor of return to work. Int J Rehabil Res 26(1):1–9

Judge TA, Piccolo RF (2004) Transformational and transactional leadership: a meta-analytic test of their relative validity. J Appl Psychol 89(5):755–768
Judge TA, Piccolo RF, Ilies R (2004) The forgotten ones? The validity of consideration and initiating structure in leadership research. J Appl Psychol 89(1):36–51
Karlin WA, Brondolo E, Schwartz J (2003) Workplace social support and ambulatory cardiovascular activity in New York City traffic agents. Psychosom Med 65(2):167–176
Kelloway EK, Barling J (2010) Leadership development as an intervention in occupational health psychology. Work Stress 24(3):260–279
Krause N, Dasinger LK, Deegan LJ et al (2001) Psychosocial job factors and return-to-work after compensated low back injury: a disability phase-specific analysis. Am J Ind Med 40(4): 374–392
Kuoppala J, Lamminpää A, Liira J et al (2008) Leadership, job well-being, and health effects—a systematic review and a meta-analysis. Occup Environ Med 50(8):904–915
Lowe KB, Kroeck KG, Sivasubramaniam N (1996) Effectiveness correlates of transformational and transactional leadership: a meta-analytic review of the MLQ literature. Leader Q 7(3): 385–425
McGee GW, Goodson JR, Cashman JF (1987) Job stress and job dissatisfaction: influence of contextual factors. Psychol Rep 61:367–375
Mullen JE, Kelloway EK (2009) Safety leadership: a longitudinal study of the effects of transformational leadership on safety outcomes. J Occup Organ Psychol 82(2):253–272
Muller-Camen M, Croucher R, Flynn M et al (2011) National institutions and employers' age management practices in Britain and Germany: 'path dependence' and option exploration. Hum Relat 64:507–530
OECD (2014) Mental health and work: Switzerland, doi:10.1787/9789264204973-en. Available from http://dx.doi.org/10.1787/9789264204973-en. Accessed 27 Jan 2014
Pelletier KR (2001) A review and analysis of the clinical-and cost-effectiveness studies of comprehensive health promotion and disease management programs at the worksite: 1998–2000 update. Am J Health Promot 16(2):107–116
Rousseau V, Aube C, Chiocchio F et al (2008) Social interactions at work and psychological health: the role of leader-member exchange and work group integration. J Appl Soc Psychol 38(7):1755–1777
Schneider B, Erhart MG, Macey WH (2011) Perspectives on organizational climate and culture. In: Zedeck S (ed) Handbook of industrial and organizational psychology. American Psychological Association Press, Washington, pp 373–414
Shamir B, House RJ, Arthur MB (1993) The motivational effects of charismatic leadership: a self-concept based theory. Organ Sci 4(4):577–594
Skakon J, Nielsen K, Borg V et al (2010) Are leaders' well-being, behaviours and style associated with the affective well-being of their employees? A systematic review of three decades of research. Work Stress 24(2):107–139
Sosik JJ, Godshalk VM (2000) Leadership styles, mentoring functions received, and job-related stress: a conceptual model and preliminary study. J Organ Behav 21(4):365–390
Sparrowe RT, Liden RC (1997) Process and structure in leader-member exchange. Acad Manage Rev 22(2):522–552
TK Gesundheitsreport (2014) Gesundheitsreport 2014—Veröffentlichungen zum Betrieblichen Gesundheitsmanagement der TK, Band 29. Techniker Krankenkasse, Hamburg (in German)
Truxillo DM, Fraccaroli F (2013) Research themes on age and work: introduction to the special issue. Eur J Work Organ Psychol 22(3):249–252
Tuomi K, Huuhtanen P, Nykyri E et al (2001) Promotion of work ability, the quality of work and retirement. Occup Med 51(5):318–324
Van Knippenberg D, Sitkin SB (2013) A critical assessment of charismatic–transformational leadership research: back to the drawing board? Acad Manage Ann 7(1):1–60
Wager N, Fieldman G, Hussey T (2003) The effect on ambulatory blood pressure of working under favourably and unfavourably perceived supervisors. Occup Environ Med 60(7):468–474

Wayne SJ, Shore LM, Liden RC (1997) Perceived organizational support and leader-member exchange: a social exchange perspective. Acad Manage J 40(1):82–111
Wegge J, Shemla M, Haslam SA (2014) Leader behavior as a determinant of health at work: specification and evidence of five key pathways. Z Personalforschung 28(1–2):6–23
WHO (1946) Preamble to the constitution of the World Health Organization as adopted by the international health conference, New York, 19—22 June 1946
WHO (2011) World report on disability 2011. World Health Organization, Geneva
Wilson MG, Dejoy DM, Vandenberg RJ et al (2004) Work characteristics and employee health and well-being: test of a model of healthy work organization. J Occup Organ Psychol 77(4): 565–588
Yukl G (2006) Leadership in organizations, 6th edn. Pearson, Upper Saddle River
Yukl G (2012) Effective leadership behaviors: what we know and what questions need more attention. Acad Manage Perspect 26(4):66–85

Four-Level Model of Health-Promoting Leadership

Erika Spiess and Peter Stadler

Abstract

We here present a management model developed by Spiess and Stadler. It is employers' responsibility to protect the health of their employees, and employees' well-being is a core resource for organizational success. Indeed, an organization cannot achieve long-term success without healthy employees. High-quality results can be achieved most effectively by healthy, motivated employees. Spiess' and Stadler's management model has four levels. The first, the heart of the model, is described as goal- and task-oriented management, in which leadership is defined as goal-oriented exertion of influence. The second level comprises staff-oriented management and support. The third level relates to the design of work and organizational processes, another important indicator of a good leadership culture. The fourth level refers to the creation of a health-promoting management and organizational culture. The consequences of using this model will be discussed.

1 Introduction

Health has become a highly pertinent and important topic in organizations. When suitable measures are implemented, good health leads to higher levels of employee happiness and stronger commitment to the organization (Avanzia et al. 2012; Gilbreath and Benson 2004; Gurt et al. 2011; Landy and Conte 2010; Skakon

E. Spiess (✉)
Department of Psychology, Ludwig-Maximilians-Universität, Leopoldstr. 13, 80802 München, Germany
e-mail: erika.spiess@psy.lmu.de

P. Stadler
Bavarian Health and Food Safety Authority, Pfarrstr. 3, 80538 München, Germany
e-mail: peter.stadler@lgl.bayern.de

M. Wiencke et al. (eds.), *Healthy at Work*, DOI 10.1007/978-3-319-32331-2_8

et al. 2010). As a consequence, positive effects may also be seen in terms of efficiency and organizational goals (Sosik and Godshalk 2000). In this context, supervisors play a major role by facilitating promotion of the concept of health in their function as role models. Quick et al. (2007, p. 189) refer to this thus: "The healthy leader is a touchstone for organizational health."

As a number of empirical studies have shown (Stadler and Spiess 2003), supervisors exercise considerable influence on the degree of stress in the workplace and thus on employees' well-being and health. Supervisors are responsible for the design and organization of work processes and procedures and therefore set the conditions for stress- and health-optimized work. Employee motivation and well-being depend substantially on social support measures, supervisors' leadership behavior and the targeted application of leadership instruments.

We will here discuss research results relating to the influence of negative leadership behavior on stress situations, employee well-being and health. A significant indicator of the presence of psychological stress and related health impairment is employee absenteeism.

Schmidt (1996) was able to prove that the behavior of supervisors significantly influences employee absenteeism. As to the supervisor-employee relationship, according to this study, employee absenteeism is particularly influenced by the willingness of supervisors to allow employees to have a voice and participate in decision-making. There is less absenteeism when supervisors are open to suggestions and include staff in setting tasks and making important decisions that concern staff.

According to a study by Cooper and Roden (1985), the two biggest sources of stress for tax officers in the United Kingdom are an autocratic style of leadership and not being consulted or included.

Laschinger et al. (1999) investigated the impact of leadership behavior on work effectiveness in a Canadian hospital. If a supervisor's behavior is encouraging, this has a positive influence on the employees, resulting in decreased tension in the workplace and increased work effectiveness.

Social support from co-workers and supervisors is also decisive for employees' well-being and their successful coping with burdensome tasks. Satisfactory social relationships act like a buffer against stress and reduce the risk of becoming ill. Various meta-analyses have determined that the following leadership behavior variants are demotivating, burdensome and inclined to promote absenteeism (Heckhausen 2000; Stadler and Spiess 2003):

- Concentration on work-related tasks while neglecting people-related tasks
- Authoritarian-type leadership
- Too little appreciation of employees' performance
- Too frequent and too subjective criticism
- Withholding information
- Conveying the meaningfulness of work poorly, unfair allocation of work and unequal treatment of employees
- Excessive control and supervision

- Unclear and constantly changing targets and guidelines
- Insufficient training of new staff or pre-existing staff for new tasks
- Too frequent transfers to different workplaces and changing work content on too short notice
- Not keeping promises concerning advancement possibilities
- Too few opportunities for further training
- Too little consideration of employees' personal career goals
- Unrealistic performance goals, leading to time pressure and working overtime
- Frequent interference in delegated areas (management action).

Organizations can use a variety of instruments to promote organizational health, such as forming committees, health groups, surveys, organizational health reports, risk assessment and medical examinations as preventive measures, staff talks, aptitude-based staff selection, preventive organization of work and technology, company-based doctoral care, company cafeterias that offer a variety of healthy food options, and health-promoting schemes and programs (e.g., check-ups) (Stadler and Spiess 2003).

What characterizes a "healthy" organization? Important aspects are freedom of action, autonomy, coherence (of the job), enabling learning and development, and facilitating possibilities for social interaction, social support and participation. These topics correspond to the classic criteria for the humane organization of work formulated by Cooper and Cartwright (1994).

2 A Model of Organizational Health

Hart and Cooper (2001) developed a heuristic model of organizational health in which the employee's well-being is central, influencing job performance. The employee's well-being is also determined by individual and organizational characteristics. Other influences include government, stakeholders, customers and business partners (see Fig. 1).

We have extended this model by incorporating organizational principles and structures (e.g., participation possibilities, incentive systems, advancement possibilities, transparency) together with management behavior and social support in the workplace. Client needs and job insecurity are influences from outside the workplace.

In our research, we have emphasized management behavior and social support in the workplace. In recent years, a number of research projects have focused on social support in various contexts (family, friends and work) (Glazer 2006). The concept of social support has its foundations in various research traditions that also deal with the inter-relationship of social support and mental health.

There are different forms of social support (Dücker 1995): material support (e.g., financial support), support in the form of helpful behavior (e.g., caring in the case of an illness), emotional support (e.g., affection, trust or sympathy), feedback (e.g.,

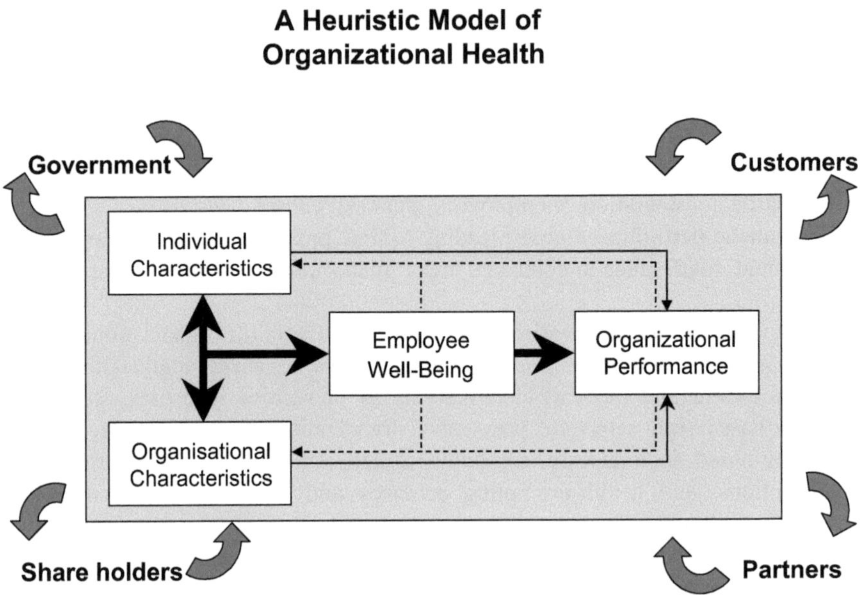

Fig. 1 Heuristic model of organizational health (adapted from Hart and Cooper 2001)

social confirmation), informative support, orientation assistance (e.g., advice), and positive social activities (e.g., fun and recreation) as part of a network.

Many studies (e.g., in the overview of Cohen and Wills 1985) have reported a positive relationship between social support at work and the well-being of those who receive the support. Frese and Semmer (1991) have named additional social support mechanisms. First, social support is a primary need: humans have a phylogenetic need to work in social groups. A lack of social support automatically leads to diminished well-being. Second, the positive feedback connected with social support directly affects self-confidence and thus other components of mental - well-being (Fig. 2).

Conceptually, social support can be divided into two major domains: emotional/ affective and informational/material/instrumental support (Madjar 2008; Podsiadlowski et al. 2013). Social support is reportedly a significant coping factor in reducing life-change stress and promoting psychological well-being (e.g., Adelman 1988). It has also proven to be particularly relevant in work-related contexts (e.g., Ng and Sorensen 2008).

The model of health-promoting leadership developed by Spiess and Stadler (2007) will now be presented. Employees' well-being is the core resource for organizational success: high-quality work outcomes can be achieved most effectively by healthy and motivated employees (Fig. 3).

The model has four levels. The *first* level is described as goal- and task-oriented management, which refers to the classic definition of leadership as a goal-oriented exertion of influence. This type of management presents itself in a communicative

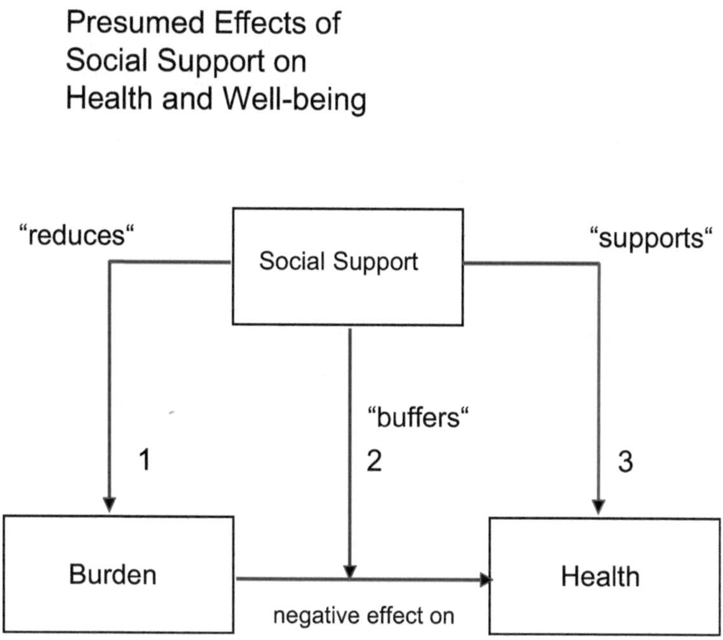

Fig. 2 Presumed effects of social support on health and well-being (adapted from Cohen and Wills 1985 and Stadler and Spiess 2003)

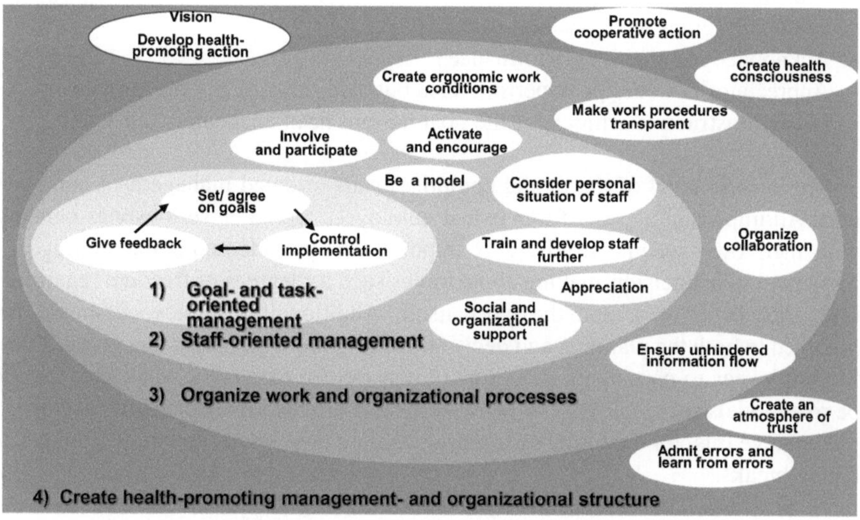

Fig. 3 Model of health-promoting management (adapted from Spiess and Stadler 2007)

way and in interaction with organizational structures, reflecting both the personality traits of the people involved and situational aspects (House et al. 2004).

Thus, the supervisor's first responsibility is to set goals, or better still to agree upon them. Staff should always be consulted, because this generates greater acceptance. Such consultation is likely to result in greater commitment to goals, because mutual agreement on goals can be perceived as a trust-building measure (the employee not as the recipient of orders, but rather as a co-actor in setting goals). Achieving goals, however, also has to be controlled and accompanied by feedback.

Numerous mistakes can be made in the cycle of setting goals—controlling achievement—feedback. Empirical studies have revealed that in organizational reality many mistakes are made (Spiess and Stadler 2007). Among other things, the set, mutually agreed-upon goals may be too challenging or not challenging enough, impossible to check, not concrete enough, or not realistic enough and therefore intermittently demotivating. Employees perceive too long or too short control (i.e., too much control and supervision) intervals or even forgoing controls (a laissez-faire management style) as burdensome and demotivating. If, when goals are agreed upon, various tasks and decision-making are delegated to employees and then withdrawn from them because of management action, the result is disappointment and demotivation.

There are certain rules for motivating feedback. It is important that the supervisor:

- Takes time for employees
- Sees them in the workplace
- Personally gives feedback on results
- Acknowledges positive performance
- Appreciates not only peak performance but also good steady performance
- Gives constructive criticism if the goal has not been adequately achieved.

The *second* level of the model comprises staff-oriented management and support. An important aspect of this is that employees are taken seriously as persons with their own specific needs. It is important to provide room for employees' participation in decision-making about topics such as degrees of freedom regarding time aspects at work and the work itself. Additionally, employees should be involved in both the planning and decision-making processes. Creating possibilities for employees to participate should be of concern for managers. Managers should be role models for employees. In terms of health-oriented management, executives can achieve this, for example, by insisting on both themselves and their employees taking breaks.

Employees often react negatively to work being organized without their involvement. In contrast, enabling staff to influence their working conditions by taking into account their experience, knowledge and needs increases their feelings of responsibility and acceptance. They generally know best what burdens them and how to deal with it.

The important thing is that supervisors activate and encourage their staff by expressing appreciation, appreciating the meaningfulness of their work and motivating them by implementing incentive systems.

Respect and appreciation are central to good collaboration: effective collaboration is distinguished by mutual appreciation. Such teams and organizations are characterized by employees being highly motivated and identifying with their work and with the organization and therefore having high creative potential (Brodbeck 2007). It is important that appreciation is not perceived merely as a technique, but rather as a lived fundamental principle.

Over- or under-challenging processes are equally frustrating, tiring and trust-eroding. To prevent this, employees must be assigned work that is consonant with their capabilities and qualifications. Because of their constant direct contact with staff, immediate supervisors in particular are in the best position to develop suitable targets. If there is a danger of the allotted work being over-challenging, supervisors should introduce appropriate additional qualification steps (or reduce the difficulty or amount of work). Supervisors who ensure that their employees are appropriately qualified by providing the necessary time and financial means demonstrate their appreciation and recognition that employees' opinions represent important resources for the success of a firm.

Employees' personal situations must also be taken into consideration. In particular, critical life phases (for instance, divorce and death of a family member) can easily lead to employees feeling over-challenged at work. It is important that supervisors are aware of such crises early and compensate accordingly (e.g., by work reduction or co-worker support in managing the workload). A prerequisite for this is that supervisors have built up relationships of trust with their staff, thus maximizing the chance that they will be aware of their employees' personal problems. Clearly this requires that the size of supervisors' sections, that is the number of staff directly responsible to each supervisor, be small enough to enable them to build up personal contact with each employee.

Moreover, co-worker support in managing the work load is an equally important determinant of staff health (Bono et al. 2007). On some occasions it is only possible to achieve a target or solve a problem relating to one's own work with the support of others—co-workers or supervisors. Difficulties are usually resolved more easily by working together. Support from others, particularly emotional support, makes it easier to bear burdens.

Inclusion, activation, appreciation and support facilitate building a good relationship and can occur via communication between supervisors and staff. Core criteria for these communication processes are:

- Clear, fear-free communication: staff will voice their ideas and views only if there is no fear of repercussions
- Active listening; that is, two-directional communication in which an effort is made to facilitate the clearest and most understandable possible expression of views and standpoints of both sides

- Genuineness and authenticity: employees sense play-acting, something that is especially crucial when there is no room for insincerity, leading to a sense that the communication lacks credibility, which can result in (partial) loss of trust
- No false promises or arousing false expectations.

From stress research (Lazarus and Folkman 1984), we know that the characteristics of the work situations or persons that relate positively to employees and help them cope with burdens better are useful resources. To enable employees to prevent and manage stress better, it is vital to promote in-house and external employee training. Participation in seminars aimed at augmenting skills (e.g., information technology) and increasing social competency such as time management, conducting discussions, dealing with conflict, and easing/defusing techniques can help expand resources.

What can be done to promote staff-oriented management? Some useful techniques are: recruiting the right person, guiding new employees through the socialization process (mentoring), pointing out development possibilities, agreeing on concrete goals, acknowledging performance, and assigning meaningful work.

The *third* level of this model comprises the design of work and organizational processes, another important indicator of a good leadership culture. This includes optimizing work conditions and organizational structures. Additionally, it is important to provide transparency and an unrestricted flow of information, for instance by thoroughly informing employees about organizational issues and changes and thus making organizational processes clear to everyone. Moreover, teamwork should be promoted; the development of team competence is an important aspect of this. In this context, immediately recognizing the potential for conflict is also important; for example, distributing work fairly to avoid conflict. Appropriate reward systems are also important.

It is eminently important to ensure transparency and information flow; for example, by informing staff in detail about organizational concerns and changes, thus rendering organizational processes transparent. Moreover, comprehensive information and transparency can increase employee acceptance of organizational changes.

The *fourth* level involves creating a health-promoting management and organizational culture: the presented measures can only be effective when the entire management and organizational culture is aligned with health-promoting measures. This requires working within an established framework of an overall cooperative organizational culture. Effective cooperation should not be based on announcements in mission statements, but should form an active part of the organization, for example by facilitating mutual support and both offering and accepting help. In this regard, a 'vision' in which health is given a high priority and is of value for employees is helpful. If the employees have co-created such a vision, it will generally be supported by the majority.

All the described measures can only achieve their complete potential if an organization's entire management and organizational structure are geared toward this. The goal is to create a health-promoting culture. Especially important in this context is how collaboration is designed within the framework of a cooperative organizational culture. Effective cooperation should not only be proclamations set forth in organizational guidelines, but rather must be actively lived in the organization. A vital component of this is admitting mistakes, not seeking to blame someone, learning from mistakes and developing technically "error-tolerant" systems.

An important component of differentiated personnel management is paying attention to the particular needs of various target groups. For instance, different ergonomic and social health-related measures are indicated for older than for younger employees. Similarly, special consideration should be given to the needs of women with children. Of utmost importance for the health and well-being of employees and their families is achieving a balance between work, family and leisure (work–life balance).

Management responsibility for promoting health should be underpinned by the development of leadership principles in which health consciousness and principles of cooperation are fundamental. Recommended leadership styles are participatory, change-oriented and cooperative. Avolio and Gardner (2005) have pointed out that it is important for supervisors to be authentic; that is, to appear credible to staff. In other words, supervisors should not expound the principles of health and then act to the contrary.

Results of studies relating to the impact of transformational leadership on health in the workplace are also available (Bass 1990, 1999). "Transformational leadership behaviors have been associated with both high levels of performance and follower health and well-being" (Nielsen and Cleal 2011, p. 344). Franke and Felfe (2011, p. 295) "found specific relations for the different transformational leadership dimensions with perceived strain. It was revealed that the relation between leaders' individual consideration and idealized influence (attribute) and perceived strain is moderated by affective organizational commitment." It is essential to propagate health-promoting leadership in organizations.

3 Conclusions

Employees are the core resource for achieving the goals of an organization. Promoting the well-being of employees is an important responsibility of management that maximizes achievement of long-term, high-quality performance. However, the reality is that health promotion has not yet been fully recognized as management's responsibility; consequently it has not been widely implemented. Research has shown that poor leadership style and behavior have a negative effect on employee motivation, health and attendance, clearly indicating the significance of employee-oriented and health-conscious leadership. Supervisors should implement the following four important health-promoting instruments and approaches:

- Goal and task-oriented management (agreement on goals, control and feedback)

- Staff-oriented management and support (function as role model, inclusion, activation, appreciation, promotion and support)
- Design of work and organizational processes (design of organizational processes, work conditions and collaboration according to ergonomic and stress-optimized criteria, ensuring transparency) health-promoting management and organizational culture (develop a vision of health-promoting action, promote cooperative action, create health consciousness and develop error-tolerant systems).

All these points must be anchored in the organization's guiding principles or visions as well as in the guidelines for organizational personnel development. Moreover, ongoing checking and cross-checking with the management principles is essential for ascertaining whether these guidelines are actually followed by leadership in real work situations. We therefore recommend they should be included in supervisor evaluations. Appropriate measures to implement include establishing health circles and medical check-ups for all staff, and using staff- and health-oriented leadership in leadership seminars, workshops and control trainings.

Because this model has not yet been empirically investigated, it has not been determined whether the four levels are independent of each other. In conclusion, it can be said that supervisors' responsibility for the health of their employees is important. Our model shows that it depends on various dimensions: behavior-oriented and organizational as well as organization-wide dimensions, such as organizational culture.

References

Adelman MB (1988) Cross-cultural adjustment: a theoretical perspective on social support. Int J Intercult Relat 12:183–204

Avanzia L, van Dick R, Fraccarolia F et al (2012) The downside of organizational identification: relations between identification, workaholism and well-being. Work Stress 26(3):289–307

Avolio BJ, Gardner WL (2005) Authentic leadership development: getting to the root of positive forms of leadership. Leader Q 16:315–338

Bass BM (1990) From transactional to transformational leadership: learning to share the vision. Organ Dyn 18(5):19–36

Bass BM (1999) Two decades of research and development in transformational leadership. Eur J Work Organ Psychol 8:9–32

Bono JE, Foldes HJ, Vinson G et al (2007) Workplace emotions: the role of supervision and leadership. J Appl Psychol 92(5):1357–1367

Brodbeck KH (2007) Entscheidung zur Kreativität (Decision to be creative). Wissenschaftliche Buchgesellschaft, Darmstadt

Cohen S, Wills TA (1985) Stress, social support, and the buffering hypothesis. Psychol Bull 98: 310–357

Cooper CL, Cartwright S (1994) Healthy mind; healthy organization—A proactive approach to occupational stress. Hum Relat 4:455–471

Cooper CL, Roden J (1985) Mental health and satisfaction amongst tax officers. Soc Sci Med 21(7):474–751

Dücker B (1995) Stress, Kontrolle und soziale Unterstützung im industriellen Bereich (Stress, control and social support in the industrial sector). Holos, Bonn
Franke F, Felfe J (2011) How does transformational leadership impact employees' psychological strain? Examining differentiated effects and the moderating role of commitment. Leadership 7: 295–316
Frese M, Semmer N (1991) Stressfolgen in Abhängigkeit von Moderatorvariablen: Der Einfluß von Kontrolle und sozialer Unterstützung (Stress consequences in dependence of moderator variables: the influence of control and social support). In: Greif S, Bamberg E, Semmer N (eds) Psychischer Stress am Arbeitsplatz (Psychological stress at work). Hogrefe, Göttingen, pp 135–153
Gilbreath B, Benson PG (2004) The contribution of supervisor behaviour to employee psychological well-being. Work Stress 18(3):255–266
Glazer S (2006) Social support across cultures. Int J Intercult Relat 30:605–622
Gurt J, Schwennen C, Elke G (2011) Health-specific leadership: is there an association between leader consideration for the health of employees and their strain and well-being? Work Stress 25(2):108–127
Hart PM, Cooper CL (2001) Occupational stress: toward a more integrated framework. In: Anderson N, Ones DS, Sinangil HK, Viswesvaran C (eds) Handbook of industrial, work and organizational psychology. Sage, London, pp 93–114
Heckhausen D (2000) Einflussfaktoren auf Fehlzeiten und Maßnahmen dagegen. Organisationsberatung Supervision Clin Manage 2:109–120
House RH, Hanges PJ, Javidan M (eds) (2004) Culture, leadership and organizations: the Globe study of 62 societies. Sage, Thousand Oaks
Landy FJ, Conte J (2010) Work in the 21th century, 3rd edn. Wiley, New York
Laschinger HKS, Wong C, McMahon L et al (1999) Leader behavior impact on staff nurse empowerment, job tension, and work effectiveness. J Nurs Admin 5:28–39
Lazarus RS, Folkman S (1984) Stress, appraisal and coping. Springer, New York
Madjar N (2008) Emotional and informational support from different sources and employee creativity. J Occup Organ Psychol 81(1):83–100
Ng TWH, Sorensen KL (2008) Toward a further understanding of the relationships between perceptions of support and work attitudes: a meta-analysis. Group Organ Manage 33(3): 243–268
Nielsen K, Cleal B (2011) Under which conditions do middle managers exhibit transformational leadership behaviors?—an experience sampling method study on the predictors of transformational leadership behaviors. Leader Q 22:344–352
Podsiadlowski A, Vauclair M, Spiess E et al (2013) Social support on international assignments: the relevance of socio-emotional support from locals. Int J Psychol 48(4):563–573
Quick JC, Macik-Frey M, Cooper CL (2007) Managerial dimensions of organizational health: the healthy leader at work. J Manage Stud 44(2):189–205
Schmidt KH (1996) Wahrgenommenes Vorgesetztenverhalten, Fehlzeiten und Fluktuation. Zeitschrift für Arbeits- und Organisationspsychologie 40(2):54–62
Skakon J, Nielsen K, Borg V et al (2010) Are leaders' well-being, behaviours and style associated with the affective well-being of their employees? A systematic review of three decades of research. Work Stress 24:107–139
Sosik JJ, Godshalk VM (2000) Leadership styles, mentoring functions received, and job-related stress: a conceptual model and preliminary study. J Organ Behav 21:365–390
Spiess E, Stadler P (2007) Gesundheitsförderliches Führen—Defizite erkennen und Fehlbelastungen der Mitarbeiter reduzieren (Health-promoting management—to recognize deficits and to reduce loads). In: Weber A, Hörmann G (eds) Psychosoziale Gesundheit im Beruf. Genter, Stuttgart, pp 255–274
Stadler P, Spiess E (2003) Psychosoziale Gefährdung am Arbeitsplatz (Psychosocial endangerment in the workplace). Wirtschaftsverlag NW GmbH, Bremerhaven

Health at Work Through Health-Promoting Leadership

Torsten J. Holstad, Astrid Emmerich, Christiane R. Stempel, and Sabine Korek

Abstract

Health-promotion in the work-place is becoming increasingly important in the face of rising absences caused by psychological issues. This chapter proposes health-promoting leadership as a means of improving employees' well-being. Health-promoting leadership is defined as leadership behavior that provides job resources and limits high job demands and thus has the potential to increase followers' well-being and motivation. Moreover, implications for the design of training programs with a focus on health-promoting leadership and guidelines for their successful implementation are presented.

1 Why Do We Need Health-Promoting Leadership?

In Germany, a trend toward increasing absences from work caused by psychological ill health between 1999 and 2010 (Badura et al. 2011) has meant that mental health has been given greater priority on the agendas of companies and organizations. The World Health Organization (WHO 2001) estimates that depression will be the second leading cause of the global disability burden by 2020, implying that this trend will continue. In addition to increasing sickness absence, organizations also foresee difficulties finding qualified personnel because of an expected shortfall of professional experts as a consequence of future demographic changes. This makes maintenance of the present workforce's ability to work even more important. Health-promoting leadership can be a means of furthering

T.J. Holstad (✉) • A. Emmerich • C.R. Stempel • S. Korek
Abteilung für Arbeits- und Organisationspsychologie, Institut für Psychologie, Universität Leipzig, Neumarkt 9-19, 04109 Leipzig, Germany
e-mail: torsten.holstad@uni-leipzig.de; astrid.emmerich@uni-leipzig.de; christiane.stempel@uni-leipzig.de; korek@uni-leipzig.de

M. Wiencke et al. (eds.), *Healthy at Work*, DOI 10.1007/978-3-319-32331-2_9

employees' well-being and thus counteracting a high rate of absences caused by sickness.

2 What Is Health-Promoting Leadership?

Health-promoting leadership, leadership behavior that provides job resources and limits extreme job demands, has been found to significantly impact psychological well-being (Bakker and Demerouti 2007; Mauno et al. 2007). This implies that leaders not only impact their followers' well-being via direct interactions with them but also via the way they arrange task characteristics and resources. According to the Job Demands-Resources Model (Bakker and Demerouti 2007; Demerouti et al. 2001), job demands and resources have a considerable impact on employees' psychological well-being and motivation. Bakker and Demerouti (2007) define job demands as aspects of a job that require increased effort, whereas job resources are defined as job characteristics that facilitate achievement of goals. Examples of job demands are work-load, time pressure, and emotional demands. It is important to note that these do not necessarily have a negative impact on employees' well-being (Bakker and Demerouti 2007). However, excessive job demands can increase strain, especially when few job resources are provided (Bakker et al. 2005, 2007). Examples of job resources are autonomy, social and task-related support, recognition of employees' contributions, meaningful work, a climate of fairness, and appreciation of employees. For example, when employees have sufficient autonomy to accomplish tasks in a way that suits them, they find it easier to cope with high work-loads.

Given that leaders can impact their followers' job demands and job resources, it is logical to postulate that they affect their employees' well-being and motivation. For example, by providing their followers with sufficient autonomy, leaders can help them to master high job demands. Providing resources is thus a means by which leaders may help to improve their employees' well-being. Several studies have demonstrated that leaders affect their followers' motivation by taking care of their job demands and job resources (Holstad et al., unpublished manuscript, "Burned out or engaged? How good and poor leadership impacts follower well-being through job demands and job resources"; Nielsen et al. 2008).

These arguments suggest that leadership has a causal effect on followers' well-being (regular causation). However, because leadership is not a one-way street but rather a bidirectional interaction between a leader and follower(s) (Van Dierendonck et al. 2004), the positive relationship between leadership and well-being may equally well be a result of reverse causation. Hence, followers with poor well-being may tend to evaluate their leaders less favorably than those with good well-being. Longitudinal studies have supported models that imply reciprocal causation; that is, mutual influence processes between leadership and follower well-being (Holstad et al., unpublished manuscript, "Burned out or engaged? How good and poor leadership impacts follower well-being through job demands and job resources"; Nielsen et al. 2008; Van Dierendonck et al. 2004). An experimental study has also confirmed an effect of leadership on followers' well-being (Lyons and Schneider 2009). It can thus be concluded that leaders may promote follower well-being and that, conversely, leadership ratings most likely depend on followers' well-being, too. Moreover, it seems to be especially important that leaders influence followers' well-being via altering job characteristics, such as job resources and demands (Holstad et al., unpublished manuscript, "Burned out or engaged? How good and poor leadership impacts follower well-being through job demands and job resources"; Nielsen et al. 2008). The following sections provide an overview of important job resources and demands through which leaders can help to improve followers' well-being.

3 Important Resources in the Work Environment

This section gives an overview of the most important job resources that can be influenced by leaders and provides some practical guidelines to help leaders to utilize these resources. Please note that this selection of job resources is not exhaustive; for additional factors see Vincent (2012).

Job control refers to the degree of freedom employees experience in their work (Karasek 1979). This concept has also been emphasized by other theorists researching well-being at work. For example, Hackman and Oldham included the concept of autonomy, with virtually the same meaning, in their Job Characteristics Model (Hackman and Oldham 1976). Karasek (1979) suggested that job control is a main driver of well-being in the work-place. The health-promoting effect of job control has been confirmed in many studies (Häusser et al. 2010; Van der Doef and Maes 1999). By assigning their followers a high degree of job control, leaders can improve their followers' motivation (Holstad et al., unpublished manuscript, "Burned out or engaged? How good and poor leadership impacts follower well-being through job demands and job resources"). For example, it can be helpful to give followers information about upcoming tasks in a timely manner to facilitate scheduling and planning, or even to involve followers in the long-term planning of tasks. Moreover, leaders can delegate projects to their followers or give them responsibility for important clients or processes without monitoring them closely. In doing so, leaders demonstrate their trust in followers and their belief in their followers' ability to master these tasks successfully.

"Meaning of work" is defined as employees' belief that their work makes sense and has an important purpose (Arnold et al. 2007). Employees tend to regard their work as meaningful if the values associated with work goals match the individual's goals and ideals (Harris et al. 2007). Research has demonstrated that meaning of work relates positively to job satisfaction and intrinsic motivation (Fried and Ferris 1987). Because they are in a position to communicate the organization's goals to their followers (Rosso et al. 2010), leaders can affect the meaning that followers attribute to their work (Arnold et al. 2007). They can, for example, explain why tasks are necessary and help followers to understand how their colleagues' tasks contribute to achieving higher-level goals. In this way, leaders can help their followers to enlarge their perspectives and appreciate their contributions to business objectives.

Participation means that followers are involved in decisions in the work-place that directly affect them. To avoid followers feeling disrespected or unfairly treated, leaders should inform them of matters that impact on them in a timely manner and provide the opportunity for followers to express their ideas and concerns. Of course, it will not always be possible for leaders and followers to make decisions together. However, if leaders consult with followers prior to making decisions, this will have a positive impact on the perceived fairness and acceptance of the decision (Vroom and Yetton 1973). This may result in the improved well-being of followers. Moreover, in the context of health and well-being, we also consider it important to emphasize that followers have to take responsibility for their own well-being. It is necessary that team members proactively take steps to sustain their well-being, for example by openly addressing problems at work, asking for assistance when they experience work overload, ensuring sufficient time to relax, detaching from work in their free time, taking time to recover from illness, and contributing to a positive team climate. Leaders are encouraged to act as role models in this regard because their behavior has a signaling effect on their team.

The establishment of an appropriate error management culture that regards errors as opportunities for improvement rather than undesirable incidents can help in identifying errors early, thus averting negative consequences for individuals. If leaders increase their followers' job control and participation as suggested in the preceding paragraph, errors may occur more frequently, especially with younger employees who lack relevant job experience. Leaders can help to establish an error-friendly culture by paying attention to the task rather than blaming the person in charge, giving honest and constructive feedback, being loyal to employees, and even taking responsibility for and sharing their own errors. If employees do not have to fear negative consequences when they make errors and emphasis is placed on the learning potential they provide, it is less likely that errors will be covered up. Establishment of an appropriate error management culture may thus help to reduce distress resulting from errors, which can be harmful to employees.

Supervisory social support refers to helpful social interaction between leaders and their followers. Leaders can support their followers both regarding task completion and on a social level, for example by providing affective support and acknowledging followers' decisions (Frese 1999). While task-related support is necessary mainly for inexperienced employees or for handling new complex tasks, social support should regularly be provided to all employees. Strong social support makes individuals feel valued by their supervisors, which has been found to increase both followers' well-being and work engagement (Bakker et al. 2007; Holstad et al. 2014; Maslach and Leiter 2008).

The fair treatment of employees is another important factor for well-being. This concerns various aspects of the work environment, for example fair distribution of tasks, fair decision-making procedures, and a fair process for linking effort with rewards. Fair decision-making processes are especially important for employees (Tyler and Lind 1990), as is reflected in the considerable impact these have on their well-being (Kivimäki et al. 2005). Leaders can improve and sustain followers' well-being by ensuring fair decision-making processes within their teams (Holstad et al. 2013). This can be achieved by letting the team participate in important decisions, making sure that everyone has access to relevant information, and applying high moral standards to their own behavior. However, fair distribution of outcomes, such as pay or recognition, is also important. An imbalance between the efforts employees put into their work and the rewards they receive can have detrimental effects on their well-being (Siegrist 1996).

A leader's appreciation has a powerful impact on followers. Leaders should show their appreciation of employees' contributions to the team by explicitly mentioning these contributions and thanking them for their efforts. Showing interest in followers, for example by asking questions and listening to them, as well as providing social support regarding work-related and private problems, are also ways of expressing appreciation. Of course, friendly non-verbal affirmation like smiling at team members can also make a difference.

It can be concluded that leaders have various opportunities for promoting and sustaining the well-being of their followers by providing job resources. Even if some leaders (especially those at lower organizational levels) lack the influence to redesign work-places and provide high levels of job control, they can still find other opportunities. All leaders are able to provide social support, show their appreciation, explain the importance of tasks, and treat their followers fairly. These kinds of behaviors help to foster their followers' belief in their own occupational efficacy (Nielsen and Munir 2009) and thus can be considered a first step toward a health-promoting work environment.

4 Important Job Demands

In this section an overview of important job demands is provided that, as with job resources, is not intended to be complete (see Vincent 2012). Job demands can be classified as challenge or hindrance demands. The former have the potential to stimulate mastery, personal growth, and future gains, whereas the latter may prevent these (Crawford et al. 2010).

High work-load is considered a challenge demand. Employees facing high work-load may perceive these as opportunities to demonstrate their competence and effectiveness. However, given that high work-load has been found to be associated with high levels of burnout components emotional exhaustion and depersonalization (Lee and Ashforth 1996), the adverse effects of high work-load must also be considered. Leaders are therefore encouraged to keep track of their followers' work-load to avoid negative effects on their well-being.

Time pressure refers to a mismatch between the volume of tasks that have to be accomplished and the time frame in which this must be achieved. Although time pressure can stimulate followers' creativity (Ohly and Fritz 2010), long-term negative effects on health are likely because of continuing overexploitation of physiological and psychological resources. Furthermore, decreases in quality or violations of safety rules may become more likely under conditions of time pressure. Because leaders are also massively affected by time pressure, they may not be able to protect their followers from all forms of it; however, they do have certain opportunities to attenuate its negative effects: they can negotiate realistic work packages and time deadlines with their followers, provide instrumental and social support, arrange tasks that allow for cooperation between team members, or acquire more personnel resources in their departments.

Strong emotional demands are also reportedly associated with increased emotional exhaustion (De Jonge et al. 2008; Van Vegchel et al. 2004; Zapf et al. 2001). Emotional demands are experienced in tense and emotionally charged situations in the work-place, for example when confronted with other people's personal problems or angry customers (Pugliesi 1999). The strong association with emotional exhaustion makes emotional demands a potential hindrance demand and stressor; they should therefore be avoided if possible or kept to a minimum. If emotional demands are unavoidable, employees are encouraged to monitor their degree of strain and actively work on recovering. Leaders of these employees should try to provide additional job resources (see Sect. 3) and discuss emotional demands regularly (Holstad et al., unpublished manuscript, "Burned out or engaged? How good and poor leadership impacts follower well-being through job demands and job resources"). Providing opportunities to discuss emotional demands with colleagues and display connected emotions can reduce adverse effects on followers' well-being. A climate of authenticity in which employees can display their emotions has been found to alleviate the association between emotional work and burnout (Grandey et al. 2012).

Intra-team processes such as conflicts may have considerable impacts on team members' well-being and performance. Task conflicts may even have positive effects on team performance (Alper and Tjosvold 2000); however, relationship conflicts have been found to impair the well-being of team members (De Dreu et al. 2004; De Dreu and Weingart 2003). Leaders can contribute to increased well-being and performance within their teams by establishing a climate of open communication that may facilitate resolution of conflicts. Also, taking a neutral position and playing an active role as a conflict manager when conflicts arise may help to improve followers' well-being. However, when conflicts escalate and more than two team members are involved, it is recommended that leaders seek the help of external specialists like mediators.

Job demands do not necessarily have negative impacts on followers' well-being. On the contrary, provided sufficient resources are in place (Bakker et al. 2007), strong demands can even increase followers' work engagement, which is an indicator of well-being in the work-place (Bakker et al. 2008). However, leaders should monitor job demands and not allow them to become excessive. Leaders should encourage followers who keep thinking about work-related problems in their free time and maintain a high level of arousal even when not working (Mohr et al. 2006) to cultivate greater detachment from their work (Sonnentag and Fritz 2014).

5 How Should Training Programs Regarding Health-Promoting Leadership Be Designed?

So far, we have shown that providing job resources and preventing the negative effects of job demands are valuable means by which leaders can promote their followers' health. It therefore follows that training in these leadership behaviors should be incorporated into leadership development programs.

Programs that aim to develop health-promoting leadership should start with some kind of assessment that aims to identify both beneficial and suboptimal behaviors of leaders (Vincent 2012) and both favorable and unfavorable working conditions. This assessment can include employee surveys, interviews, or structured work-place analysis. We recommend using followers' evaluations of leadership behavior, which should ideally be combined with structured work-place analysis. This approach combines the advantages of drawing on both subjective and objective sources of information. The followers' subjective perspectives are important regarding individual stress reactions (Mohr and Semmer 2002), whereas objective information such as occupational health data (e.g., sickness absence) can be used to validate subjective information. As an option, leaders can be asked to rate their own leadership behavior, thus enabling comparisons between leaders' and followers' perspectives.

As a second step, leaders are encouraged to elaborate action plans with the aim of implementing sustainable changes in leadership behavior and working conditions. Based on their individual leadership profiles and the results of the work-place analysis, leaders select areas in which they want to improve their leadership behavior. Together with a trainer or coach, they can discuss the results of the evaluation by their followers, reflect on possibilities for improvements, and set specific goals. Differences between self-ratings and evaluation by followers can be especially valuable information for leadership development. It is recommended that goals are defined according to the SMART criteria (Locke and Latham 2002). Thus, goals should be specific, measurable, attainable and ambitious, relevant and realistic, and time-framed. All goals should be specified in detail in action plans in which each goal is broken down into sub-goals and the precise steps to be taken to reach these goals are defined. Potential hindrances should also to be taken into account (De Bono 1992).

In parallel with this process, both leaders and followers should receive input about the relationship between work characteristics and well-being. We recommend including both leaders and followers in this step because leadership is a social interaction process (Graen and Uhl-Bien 1995). Providing information about work characteristics and their relationship with well-being has the aim of stimulating leaders and followers to take responsibility for their own well-being. This is necessary because improvements in followers' well-being cannot be achieved without their own contributions (Rigotti et al. 2014). To illustrate how different aspects of the work environment impact the well-being of employees, theoretical models like the Job Demands-Resources Model (Bakker and Demerouti 2007) can be helpful. This model can facilitate understanding the interplay between job resources, job demands, well-being, and motivation. This knowledge can be useful for both leaders and subordinates when developing common goals. Moreover, leaders and subordinates can discuss which job demands and resources are especially important for them and place particular focus on these factors. Such a procedure results in tailor-made interventions that fit the demands and challenges of the organization or team concerned (Rigotti et al. 2014).

We strongly recommend using the powerful tool of role-play in leadership training with the aim of leaders practicing leadership behavior in the company of their peers in practical settings. To ensure realistic and relevant training situations, role-play should be based on critical incidents from the participants' daily work. We recommend that the participating leaders experience both the leaders' and followers' roles for two equally important reasons. First, leaders take an active role in practicing relevant leadership situations, which builds leadership skills and facilitates their transfer to daily work. Second, leaders playing the role of a follower gain an immediate experience of a follower's perspective on and reaction to various leadership behaviors. They may, for example, experience the difference between an appreciative leader and one who is not. To demonstrate the impact of leadership behavior, trainers can instruct the participant playing the role of the leader to adopt dysfunctional leadership styles (for example, not paying attention to followers' issues, making fun of followers or exposing them). Of course, exchanging experience with other leaders is also valuable for leaders, especially if they have no interactions with peers in their daily work.

Moreover, leadership training programs should include coaching for leaders after completion of the training with the aim of facilitating the transfer of newly learned leadership behavior to work settings (Rigotti et al. 2014). Coaching can have an important function in helping participants to transfer such behavior to their daily work. In the coaching sessions, the coach and leader should reflect together on the reasons for transfer problems and explore strategies for overcoming these. Moreover, coaching can help leaders to reflect upon their function as role models for their followers regarding health-conscious behavior in the work-place.

Finally, we strongly encourage thorough evaluations of programs to improve their effectiveness. For a description of an evaluation of a training program that focuses on health-promoting leadership, see Rigotti et al. (2014).

6 Recommendations for the Successful Implementation of Health-Promoting Leadership

The benefits of a healthy and motivated staff, such as reduced costs due to sickness absence, low turnover, and high productivity (Van Dierendonck et al. 2004), should be attractive to most organizations. However, these are long-term rather than short-term benefits. The effects of leadership on follower well-being can take some time to develop (Nielsen et al. 2008; Van Dierendonck et al. 2004). In day-to-day business, short-term goals such as meeting deadlines or short-term economic success are often prioritized over long-term goals. As a consequence, long-term goals may receive less attention or be postponed or downgraded in priority. We therefore recommend that long-term goals like work-place health-promotion be integrated into regular work routines so they will be addressed regularly. This can be achieved, for example, by reserving a fixed slot in weekly team meetings.

Another potential obstacle to the successful implementation of health-promoting leadership is the limited impact and autonomy of first-level leaders. These leaders often have insufficient latitude with decision-making to change working conditions and may be stuck between management's targets and the limitations of organizational practice. For example, such leaders may not be able to reduce the work-load because their teams are understaffed or procedures may be determined by external regulations that are beyond the influence of first-level leaders. In this case, health-promoting leadership needs support from higher organizational levels. We therefore recommend that the organization's management be involved in the implementation of health-promoting leadership.

Finally, a conflict between the objectives of the leader's and followers' well-being may prevent the application of health-promoting leadership. Because health-promoting leadership imposes additional demands on leaders at short notice, this may impair their health and well-being; such unfavorable consequences make frequent displays of health-promoting leadership unlikely. It is therefore important that organizations consider providing leaders with supplementary resources when establishing health-promoting leadership and motivate them to adopt more health-promoting leadership behaviors. For example, additional staff for a limited period and a high degree of autonomy may be helpful. It can also be helpful to make health-promotion a corporate value or to give leaders financial rewards based on leadership evaluations. In the long run, lower levels of staff turnover and sickness absence will reward health-promoting leadership behavior (Rigotti et al. 2014).

7 Conclusions

Leaders can affect the well-being and health of their followers in various ways, for example by providing job resources like autonomy, social support, meaning of work, a climate of fairness, and appreciation. Reduction in extreme job demands may also help to reduce strain and promote well-being. Moreover, empirical evidence shows that health-promoting leadership can be trained and that this may be a way of improving followers' well-being and health in the work-place (Rigotti et al. 2014). To ensure sustainable effects and support their successful implementation, we recommend that such training programs are based on assessment data from which specific goals are derived, that both leaders and followers are addressed, and that programs include behavioral training and coaching for leaders. We regard health-promoting leadership as a promising approach to further health and well-being in the work-place. However, there are some pitfalls that can result in failure of health-promoting leadership interventions. It is therefore important to plan and implement such programs carefully.

References

Alper S, Tjosvold KS (2000) Conflict management, efficacy, and performance in organizational teams. Pers Psychol 53:625–642

Arnold KA, Turner N, Barling J et al (2007) Transformational leadership and psychological well-being: the mediating role of meaningful work. J Occup Health Psychol 12:193–203

Badura B, Ducki A, Schröder H, Klose J, Macco K (eds) (2011) Fehlzeitenreport 2011: Führung und Gesundheit. Springer, Berlin (in German)

Bakker AB, Demerouti E (2007) The job demands-resources model: state of the art. J Manage Psychol 22:309–328

Bakker AB, Demerouti E, Euwema MC (2005) Job resources buffer the impact of job demands on burnout. J Occup Health Psychol 10:170–180

Bakker AB, Hakanen JJ, Demerouti E et al (2007) Job resources boost work engagement, particularly when job demands are high. J Educ Psychol 99:274–284

Bakker AB, Schaufeli WB, Leiter MP et al (2008) Work engagement: an emerging concept in occupational health psychology. Work Stress 22:187–200

Crawford ER, Lepine JA, Rich BL (2010) Linking job demands and resources to employee engagement and burnout: a theoretical extension and meta-analytic test. J Appl Psychol 95:834–848

De Bono E (1992) Serious creativity. Using the power of lateral thinking to create new ideas: a systematic approach to take you beyond the power of lateral thinking. Harper Business, New York

De Dreu CKW, Weingart LR (2003) Task versus relationship conflict, team performance, and team member satisfaction: a meta-analysis. J Appl Psychol 88:741–749

De Dreu CKW, Van Dierendonck D, Dijkstra MT (2004) Conflicts at work and individual well-being. Int J Confl Manage 15:6–26

De Jonge J, Le Blanc PM, Peeters MC et al (2008) Emotional job demands and the role of matching job resources: a cross-sectional survey study among health care workers. Int J Nurs Stud 45:1460–1469

Demerouti E, Bakker AB, Nachreiner F et al (2001) The job demands-resources model of burnout. J Appl Psychol 86:499–512

Frese M (1999) Social support as a moderator of the relationship between work stressors and psychological dysfunctioning: a longitudinal study with objective measures. J Occup Health Psychol 4:179–192

Fried Y, Ferris GR (1987) The validity of the job characteristics model: a review and meta-analysis. Pers Psychol 40:287–322

Graen GB, Uhl-Bien M (1995) Relationship-based approach to leadership: development of leader-member exchange (LMX) theory of leadership over 25 years: applying a multi-level multi-domain perspective. Leader Q 6:219–247

Grandey A, Foo SC, Groth M et al (2012) Free to be you and me: a climate of authenticity alleviates burnout from emotional labor. J Occup Health Psychol 17:1–14

Hackman JR, Oldham GR (1976) Motivation through the design of work: test of a theory. Organ Behav Hum Perform 16:250–279

Harris KJ, Kacmar KM, Zivnuska S (2007) An investigation of abusive supervision as a predictor of performance and the meaning of work as a moderator of the relationship. Leader Q 18:252–263

Häusser JA, Mojzisch A, Niesel M et al (2010) Ten years on: a review of recent research on the job demand-control (-support) model and psychological 'well-being'. Work Stress 24:1–35

Holstad TJ, Rigotti T, Otto K (2013) Prozedurale Fairness als Mediator zwischen transformationaler Führung und psychischer Beanspruchung am Arbeitsplatz. Z Arb Organ 57:163–176 (in German)

Holstad TJ, Korek S, Rigotti T et al (2014) The relation between transformational leadership and follower emotional strain: the moderating role of professional ambition. Leadership 10:269–288

Karasek RA (1979) Job demands, job decision latitude, and mental strain: implications for job redesign. Admin Sci Q 24:285–308
Kivimäki M, Ferrie JE, Brunner E et al (2005) Justice at work and reduced risk of coronary heart disease among employees: the Whitehall II study. Leader Q 165:2245–2251
Lee RT, Ashforth BE (1996) A meta-analytic examination of the correlates of the three dimensions of job burnout. J Appl Psychol 81:123–133
Locke EA, Latham GP (2002) Building a practically useful theory of goal setting and task motivation: a 35-year odyssey. Am Psychol 57:705–717
Lyons JB, Schneider TR (2009) The effects of leadership style on stress outcomes. Leader Q 20:737–748
Maslach C, Leiter MP (2008) Early predictors of job burnout and engagement. J Appl Psychol 93:498–512
Mauno S, Kinnunen U, Ruokolainen M (2007) Job demands and resources as antecedents of work engagement: a longitudinal study. J Vocat Behav 70:149–171
Mohr G, Semmer NK (2002) Arbeit und Gesundheit. Psychol Rundsch 53:77–84 (in German)
Mohr G, Müller A, Rigotti T et al (2006) The assessment of psychological strain in work contexts. Eur J Psychol Assess 22:198–206
Nielsen K, Munir F (2009) How do transformational leaders influence followers' affective well-being? Exploring the mediating role of self-efficacy. Work Stress 23:313–329
Nielsen K, Randall R, Yarker J et al (2008) The effects of transformational leadership on followers' perceived work characteristics and psychological well-being: a longitudinal study. Work Stress 22:16–32. doi:10.1080/02678370801979430
Ohly S, Fritz C (2010) Work characteristics, challenge appraisal, creativity, and proactive behavior: a multi-level study. J Organ Behav 31:543–565
Pugliesi K (1999) The consequences of emotional labor: effects on work stress, job satisfaction, and well-being. Motiv Emot 23:125–154
Rigotti T, Holstad TJ, Mohr G et al (2014) Rewarding and sustainable health-promoting leadership. Bundesanstalt für Arbeitsschutz und Arbeitmedizin (BAuA), Dortmund
Rosso BD, Dekas KH, Wrzesniewski A (2010) On the meaning of work: a theoretical integration and review. Res Organ Behav 30:91–127
Siegrist J (1996) Adverse health effects of high-effort/low-reward conditions. J Occup Health Psychol 1:27–41
Sonnentag S, Fritz C (2014) Recovery from job stress: the stressor-detachment model as an integrative framework. J Organ Behav 36:72–103
Tyler T, Lind E (1990) Intrinsic versus community-based justice models: when does group membership matter? J Soc Issues 46:83–94
Van der Doef M, Maes S (1999) The job demand-control (-support) model and psychological well-being: a review of 20 years of empirical research. Work Stress 13:87–114
Van Dierendonck D, Haynes C, Borrill C et al (2004) Leadership behavior and subordinate well-being. J Occup Health Psychol 9(2):165–175
Van Vegchel N, De Jonge J, Söderfeldt M et al (2004) Quantitative versus emotional demands among Swedish human service employees: moderating effects of job control and social support. Int J Stress Manage 11:21–40
Vincent S (2012) Gesundheits- und entwicklungsförderliches Führungsverhalten: Gendertypische Differenzen. Gruppendynamik Organ 43:61–89 (in German)
Vroom VH, Yetton PW (1973) Leadership and decision-making. University of Pittsburgh Press, Pittsburgh
World Health Organization (2001) The world health report 2001: mental health: new understanding, new hope. WHO, Geneva
Zapf D, Seifert C, Schmutte B et al (2001) Emotion work and job stressors and their effects on burnout. Psychol Health 16:527–545

Managing (in) Times of Uncertainty: The Effects of Leadership Sensemaking on Employee Well-Being in Dynamic Business Environments

Fokko J. Eller, Sebastian Fischer, and Laura Diedrich

Abstract

Leadership behavior is an important factor for employee well-being. Research about occupational health promotion has shown that the support of managers is crucial to the success of measures for improving employee health and well-being. However, little interest has been paid to the psychological effects of dynamic environments on employee and management health. To address this gap, we will review publications concerning the interplay of environmental dynamism, employee uncertainty, leadership sensemaking, and employee well-being. We argue that change(s) in economic systems promotes personal uncertainty, which affects employee well-being. More specifically, we focus on leadership behaviors that diminish uncertainty in times of dynamic environments. In this chapter, we will describe two alternative routes for leadership influence on employee well-being and argue that environmental change has to be actively managed. Possible implications that can be derived from our concepts are: (1) that preserving employee well-being through leadership sensemaking in turbulent economic times could contribute to maximizing employee performance; and (2) that especially designed leadership trainings could provide organizations with means of dealing with employee uncertainty stemming from rapidly changing economic environments.

1 Introduction

According to Gates and Hemingway (1999) business is going to change more in the next decade than it has in the last 50. In these times of globalization and technologization, changes in the economic environment can be rapid and hard to

F.J. Eller (✉) • S. Fischer • L. Diedrich
Leuphana University of Luneburg, Luneburg, Germany
e-mail: Eller@leuphana.de; sebastian.fsc@gmail.com

M. Wiencke et al. (eds.), *Healthy at Work*, DOI 10.1007/978-3-319-32331-2_10

predict. Branches that rely on macro-economic, social, and political developments, in which rapid changes may occur more frequently, may be more severely affected. Recent examples of rapidly changing economic environments include the turnaround in some countries' energy policies from nuclear power to renewable energy following the Fukushima catastrophe of 2011, and the economic downturn following the global financial crisis in 2009, which strongly affected some countries' economic systems while leaving others relatively unaffected. Even with less intense changes, increasing interdependence and internationalization accompanied by constant changes will progressively play a more important role in almost all economic sectors. Some might perceive these changes as opportunities and take initiatives, whereas others might be paralyzed by them. In the latter case, change could constrain a person from taking action and adapting to new situations (Barr and Huff 1997; Driskell and Salas 2013). Until now, little interest has been paid to the psychological effects of such turbulent economic environments on employees and management. Conventional wisdom suggests that rapid change creates uncertainty among employees, possibly resulting in psychological strain (Bordia et al. 2004). However, this process is not universal. In fact, it has been shown that too little uncertainty and too much control can decrease well-being, possibly pointing to a curvilinear rather than linear relationship between the exertion of control over the working environment and employee well-being (Kubicek et al. 2014).

Furthermore, we may only speculate exactly how turbulent economic environments influence employees. Change in economic systems may directly influence employee uncertainty, for example via public debate. However, this effect may be limited because such change is so removed from the situation of employees. Alternatively, change in economic systems may first affect businesses and their leaders, who subsequently transfer their personal uncertainty to their employees: This route may be more promising. Therefore, we suggest that the underlying processes that connect uncertainty and well-being within employees may be understood from a multilevel perspective (Johns 2006), our specific focus being on leadership behaviors that diminish uncertainty in times of dynamic environments.

In this chapter, we aim to build theory by proposing a model of the relationships between environmental dynamism (Dess and Beard 1984; Ensley et al. 2006), employee uncertainty (Bordia et al. 2004), leadership sensemaking (Ancona 2012; Weick 1995) and employee well-being (Diener 2005, 2009). After defining the relevant constructs, we will review publications concerning the interplay of environmental dynamism, employee uncertainty, leadership sensemaking, and employee well-being, and outline a conceptual model of the relationships between environmental dynamism, employee uncertainty, leadership sensemaking, and employee well-being. We propose two routes for leadership influence on employee well-being in dynamic environments. Third, we will discuss implications for corporate practice and leadership research.

2 Environmental Dynamism, Uncertainty, Sensemaking, and Well-Being

Organizations' environments differ in their degree of dynamism. Environmental dynamism represents the degree of unpredictable change in an environment to which companies and individuals have to adapt (Dess and Beard 1984). Highly dynamic environments are characterized by unpredictable and rapid change, which increases uncertainty about the future for companies operating in those environments (Dess and Beard 1984; Duncan 1972; Ensley et al. 2006).

Uncertainty can be conceptualized on two levels: The business level (including the management level) and the employee level.

For businesses, uncertainty about the future is defined as (1) the absence of information concerning environmental factors associated with a given decision making situation, (2) the lack of knowledge with regard to the outcome of a specific decision in terms of how much the business would lose if the decision were incorrect, and (3) the inability to assign probabilities with any degree of confidence with regard to how environmental factors are going to affect the success or failure of the decision (Duncan 1972). Because the future is hard to predict, uncertainty is normal in business life; however, specifications of products or services need to be defined with the future customer in mind. Businesses employ numerous means to reduce uncertainty about the future. For example, they do market research on their products, take out insurance, analyze business data, and employ consultants. Although these measures do not eliminate business uncertainty, they provide a basis for managers' decisions.

For employees, as for businesses, a certain degree of uncertainty about the future is a normal part of life. However, uncertainty about the future may transform personal uncertainty into a negative psychological state characterized by a sense of doubt or instability in self- or world-views, associated with feelings of decreased manageability and control over one's own life (van den Bos 2009). Personal uncertainty is more disturbing than uncertainty about the future because it reduces belief in a good future and therefore provides a basis for reduced well-being or even depression.

So, what is the relationship between personal uncertainty and uncertainty about the future? Uncertainty about the future does not per se increase personal uncertainty. Everybody experiences some uncertainty about the future, usually without suffering from it. Depending on personality and cultural orientation (Boelen and Carleton 2012; Boelen et al. 2010; Hofstede and McCrae 2004; Zuckerman 1994), little to moderate uncertainty is generally tolerated well, or even experienced as stimulating, according to the dictum "no risk, no fun". However, extreme uncertainty can reduce well-being, which can be defined as how people evaluate their lives when they ask themselves whether their lives are going well according to their own standards (Diener 2005, 2009; Dugas et al. 2004).

When people experience considerable uncertainty about the future in a central domain of their life, personal uncertainty occurs and this is highly relevant to well-being. Reduced well-being because of extreme uncertainty is a well-known phenomenon from research on job insecurity (Nelson et al. 1995; Sverke and Hellgren 2002). Butterworth et al. (2011) have found that people who work in uncertain jobs have worse mental health than those who are unemployed. Uncertainty is regularly connected with feelings of discomfort, fear, and alarm and with mental health problems (Elovainio et al. 2005).

This negative relationship between uncertainty and well-being is not limited to employees at low levels of an organizational hierarchy, but occurs on all levels. For example, when the future is very uncertain, decision-makers in companies have to process complex information, which can lead to cognitive overload (Tushman 1979). As a result, these individuals are at risk of experiencing greater stress and anxiety (Waldman et al. 2001), because environments perceived as highly uncertain are frequently seen as very risky, whereby a few flawed decisions could result in severe trouble, and possibly risk the survival of the organization (Waldman et al. 2011).

The main psychological effects of uncertainty and reduced well-being are to produce feelings of stress and anxiety which can lead to rigidity, denial, and avoidance (Staw et al. 1981). Therefore, the work performance of employees with diminished well-being suffers (Lim et al. 2000; Wright and Cropanzano 2000; Wright and Staw 1999), especially in the case of those whose job is to devise innovations to equip the company for the future. Therefore, employee well-being is particularly relevant for organizations. Based on our line of reasoning we postulate the following:

Proposition 1: High environmental dynamism in an organization's environment is positively related to employee uncertainty.

Proposition 2: High employee uncertainty caused by environmental dynamism is negatively related to employee well-being.

These issues require counteraction by firms. So, what can be done to overcome the negative effects of extreme uncertainty about the future? Well-being can be influenced by leadership behavior (Ilies et al. 2005); we will therefore explore leadership influences in more detail in the next sections.

Power centralization is commonly seen as one outcome of dynamic environments (Eisenhardt 1989; Staw et al. 1981). Power centralization means that in the course of mechanistic shifts (Burns and Stalker 1961) in highly dynamic environments, decision-making processes tend to be transferred from team members to the team leader. Leadership is described as a process of social influence in which one person can request assistance and support of others in order to accomplish a common goal (Chemers 1997). Thus, leadership can be understood as a behavioral characteristic rather than a certain hierarchical status within a company (even though for many leaders status and behavior do go hand in hand). Irrespective of their seniority, leadership influences can come from leaders in middle to top management when they actively influence the behaviors, cognitions, and emotions of their subordinates.

The phenomenon of transferred decision-making competencies can be understood from the perspective of evolutionary psychology. As mentioned above, many individuals perceive extreme uncertainty as a threat (Bredemeier and Berenbaum 2008; Izard 1991; Mandler 1981). In threatening situations, dominant leaders (e.g. those who are seen as confident decision-makers) are preferred by their subordinates because they seem to know what they are doing (Little et al. 2007; McCann 2001). Accordingly, in times of high uncertainty, employees tend to rely more strongly on their leaders' guidance (Schein 2004; Tushman 1979). Thus, leaders who are confident about the future seem to make good decisions and seem strong in the face of adversity, which may buffer some of the negative effects of dynamic environments. This leads to the following proposition:

Proposition 3: High environmental dynamism leads to power centralization, which puts decision-making in the hands of leaders and may buffer some of the negative effects of dynamic environments.

However, transferring decision-making responsibility to a leader who is strong and confident may be insufficient to buffer the detrimental effects of high uncertainty on employee well-being. For example, turbulence may be too strong, or confidence in the leader may be eroded by incorrect interpretations of external events, incorrect decisions, or other problems.

This is where the process of sensemaking provides an alternative, second route to leader influence (Ancona 2012; Gioia and Chittipeddi 1991; Pye 2005; Weick 1995). Weick (1995) defines sensemaking as making something sensible; that is, understandable and meaningful. The construct of sensemaking thus resembles the cognitive and motivational elements of Antonovsky's sense of coherence (e.g. Antonovsky 1996); however, rather than representing an individual and global orientation, it describes a social and active process of structuring the unknown (Waterman 1990). Thus, leaders help others to make sense out of an ambiguous situation in order to attain a shared understanding (Plowman et al. 2007). For example, if turbulence is getting worse, leader will have to define actions for handling the new situation. This process can be understood as sensemaking by leaders. They then engage in sensegiving by communicating their new strategies to their teams, who in turn start their own processes of sensemaking by interpreting the information and drawing their own conclusions. The teams can then engage in the process of sensegiving by expressing their opinions about the new strategies. Such interactions lead to a shared understanding of the situation and thus to collective action.

Sensemaking is mainly about modeling team members' perceptions of organizational identity and the desired future image (Gioia and Thomas 1996). According to Weick (1995), effective sensemaking has the following characteristics (adapted from Lipshitz et al. 2004, p. 100): It is grounded in identity construction, retrospective, enactive of sensible environments (i.e. it happens in interaction with the surroundings), social, ongoing, and driven by environmental cues and plausibility rather than accuracy.

For successful sensemaking, leaders' interpretations do not have to be correct, but they do have to provide purpose and an image of where the team stands and where it is going (Ancona 2012). Sensemaking allows people to craft, understand, and accept new conceptualizations of their organization (Smircich 1983, cited after Kezar and Eckel 2002) and consequently to behave in ways consistent with those new interpretations and perceptions (Gioia et al. 1996; Weick 1979). Sensemaking has also been shown to be an important factor in transformational change (Kezar and Eckel 2002). Thus, if the leader develops an inner map of the events in his environment (Ring and Rands 1989) and then attempts to make sense of it for his followers, he may, if not resolve it, at least keep fear at a distance (Ancona 2012). Therefore, we suggest that sensemaking may be superior to power centralization as a route of influence because it does not require the leader to be correct in all interpretations and decisions, but rather influences the organization to come up with its own, new interpretations of the situations and develop its own solutions from there. We thus postulate the following proposition:

Proposition 4: Strong sensemaking by leadership weakens the negative relationship between employee uncertainty and employee well-being and should thus buffer against the potentially detrimental effects of high employee uncertainty on employee well-being.

We employed a three stage search strategy in order to identify pertinent articles for our review. First, we searched the web of science based on the variation and combination of the following key words: Uncertainty, leadership sensemaking, employee well-being and environmental dynamism. In the second step, we reviewed the articles' citations and abstracts. In the final step, we hand searched the remaining articles for thematically applicable ones. Table 1 summarizes the main findings of relevant published reports.

3 Discussion

In summary, we suggest that there are two routes via which leaders influence employee well-being in dynamic environments. First, mechanistic shifts in highly dynamic environments may enhance leaders' influence on employees' uncertainty. When these occur, they enable leaders to influence employee behaviors by providing strong and powerful evidence against uncertainty. Second, leaders may help employees to make sense of their surroundings, thereby potentially reducing employee uncertainty, and thence increasing their well-being and performance.

In practice, organizations with leaders who are either both strong and powerful or who excel in making sense of the environment for their employees perform the best under highly dynamic circumstances. They are successful because their employees are enabled to react to changes. This indicates that businesses that succeed are not those with the best knowledge or those that best predict future trends. Rather, good sensemaking enables employees to work with little doubt or

Table 1 Review of published reports

Author(s), year	Relevant predictor(s)	Relevant outcome(s)	Main finding(s)
Environmental dynamism/employee uncertainty			
Duncan (1972; empirical)	Dynamic/complex environments	Team uncertainty in decision-making	Individuals in decision units with dynamic/complex environments experience the greatest uncertainty in decision-making
Dess and Beard (1984; theoretical)	Environmental dynamism	Uncertainty	Environmental dynamism heightens uncertainty for key organizational members
Tushman (1979; empirical)	Environmental dynamism/stability and technical support/ research teams	Team behavior	Strong environmental dynamism leads to greater reliance on supervisory direction in technical support teams, whereas it leads to better intra-team communication in research teams
Employee uncertainty/employee well-being			
Staw et al. (1981; empirical)	Perceived threat	Restriction in information processing and constriction of control	Evidence from multiple levels of analysis is summarized, showing a restriction of information processing and constriction of control under threat conditions
Nelson et al. (1995; empirical)	Company privatization	Employee well-being	Decline in job satisfaction in the period during which privatization took place
Leadership sensemaking/relationship between employee uncertainty and employee well-being			
Gioia and Thomas (1996; empirical)	Strategic (internal) change in colleges and universities in the USA	Employee interpretation of processes	Top management members' perceptions of identity and desired (future) image serve as important links between change and team members' interpretations thereof (=sensemaking)

instability in their self- or world-views, to experience life as manageable, and to feel in control of their lives. Accordingly, they perform well.

We sought to integrate diverse published reports to inform construction of a framework for how leaders may buffer the possibly detrimental effects of strong environmental dynamism and resulting employee uncertainty on employee well-being. We argue that although leaders cannot control what happens in the environment of their organization ('what'), they can actively structure and make sense of these events for their subordinates ('how'), in concordance with the proverb "we cannot direct the wind, but we can adjust the sails".

So, what implications for research and practice does our conceptual model suggest? First, our research emphasizes the importance of leadership behavior for employee well-being. Research about occupational health promotion has documented that the support of managers is crucial for the success of measures to improve employee health and well-being (Dollard and Bakker 2010; Dollard and McTernan 2011; Hall et al. 2010; Law et al. 2011). Our suggestions illustrate that even if employee well-being is not an explicit goal of leadership behavior (which would, in our example, be sensemaking), it may still preserve or enhance employee well-being. Indeed, our model calls for an active leadership approach (rather than laissez-faire behavior) to handling environmental dynamism: Environmental change thus has to be actively *managed.*

To enable implementation of this notion, businesses must prioritize leadership well-being. As outlined in Sect. 2, leaders as organizational decision-makers may also be affected by environmental dynamism, resulting in cognitive overload and stress. Therefore, a focus on promoting leadership well-being would support managers to be effective sensemakers. Additionally, in organizational practice, organizations might opt for special leadership trainings on how to manage environmental change through sensemaking. To enhance leaders' abilities to make sense of the environment, it might pay for organizations to equip leaders with solid economic, geopolitical, and business knowledge, for instance through regular in-house training or cooperation with academia. Furthermore, organizations could guide leaders in how to package this knowledge and link it to the organization's past and current situation through organizational storytelling (Abma 2003; Boyce 1996; Taylor et al. 2002). Such comprehensive content-wise and methodological knowledge would support leaders to excel at sensemaking. From clinical settings we know that for the mind to know, it must actively give meaning and order to the reality to which it is responding (Raskin 2002).

Second, our conceptions might mean that preserving employee well-being through leadership sensemaking in turbulent economic times could also contribute to maximizing employee performance (Wright and Cropanzano 2000; Wright and Staw 1999)—an asset dearly needed by businesses in times of economic turmoil.

To conclude, we suggest that in practice our model would provide organizations with means of dealing with employee uncertainty stemming from rapidly changing economic environments, especially designed leadership trainings. Our model aims to build theory by linking constructs that have not yet been comprehensively studied. A possible risk of our model could arise from the incorrect interpretation that, when facing uncertainty, employees can passively rely on their leaders to structure the environmental dynamics for them. In the era of the entrepreneurial employee (Cardon 2008; Stam 2013), this is how employees perceive themselves and are perceived by their leaders. Rather, effective sensemaking by leaders in times of high economic dynamism should prevent the petrifying of employees that otherwise often results from massive uncertainty, leading to sustained action and initiative in the face of the unknown.

References

Abma TA (2003) Learning by telling—storytelling workshops as an organizational learning intervention. Manage Learn 34(2):221–240

Ancona D (2012) Sensemaking: framing and acting in the unknown. In: Snook SA, Nohria N, Khurana R (eds) The handbook for teaching leadership: knowing, doing, and being. Sage, Thousand Oaks, CA

Antonovsky A (1996) The salutogenic model as a theory to guide health promotion. Health Promot Int 11(1):11–18

Barr PS, Huff AS (1997) Seeing isn't believing: understanding diversity in the timing of strategic response. J Manage Stud 34(3):337–370

Boelen PA, Carleton RN (2012) Intolerance of uncertainty, hypochondriacal concerns, obsessive-compulsive symptoms, and worry. J Nerv Ment Dis 200(3):208–213

Boelen PA, Vrinssen I, van Tulder F (2010) Intolerance of uncertainty in adolescents correlations with worry, social anxiety, and depression. J Nerv Ment Dis 198(3):194–200

Bordia P, Hobman E, Jones E et al (2004) Uncertainty during organizational change: types, consequences, and management strategies. J Bus Psychol 18(4):507–532

Boyce ME (1996) Organizational story and storytelling: a critical review. J Organ Change Manage 9(5):5–26

Bredemeier K, Berenbaum H (2008) Intolerance of uncertainty and perceived threat. Behav Res Ther 46(1):28–38

Burns T, Stalker GM (1961) The management of innovation. Tavistock, London

Butterworth P, Leach LS, Strazdins L et al (2011) The psychosocial quality of work determines whether employment has benefits for mental health: results from a longitudinal national household panel survey. Occup Environ Med 68(11):806–812

Cardon MS (2008) Is passion contagious? The transference of entrepreneurial passion to employees. Hum Resour Manage Rev 18(2):77–86

Chemers MM (1997) An integrative theory of leadership. Lawrence Erlbaum, Mahwah, NJ

Dess GG, Beard DW (1984) Dimensions of organizational task environments. Admin Sci Q 29 (1):52–73

Diener E (2005) Guidelines for national indicators of subjective well-being and ill-being. University of Illinois, Urbana-Champaign

Diener E (2009) Frequently answered questions. Available from http://internal.psychology.illinois.edu/~ediener/faq.html#happiest. Accessed 1 Sept 2014

Dollard MF, Bakker AB (2010) Psychosocial safety climate as a precursor to conducive work environments, psychological health problems, and employee engagement. J Occup Organ Psychol 83(3):579–599

Dollard MF, McTernan W (2011) Psychosocial safety climate: a multilevel theory of work stress in the health and community service sector. Epidemiol Psychiatr Sci 20(4):287–293

Driskell JE, Salas E (eds) (2013) Stress and human performance. Psychology, Mahwah, NJ

Dugas MJ, Schwartz A, Francis K (2004) Brief report: intolerance of uncertainty, worry, and depression. Cogn Ther Res 6:835–842

Duncan BW (1972) Characteristics of organizational environments and perceived environmental uncertainty. Admin Sci Q 17(3):313–327

Eisenhardt KM (1989) Making fast strategic decisions in high-velocity environments. Acad Manage J 32(3):543–576

Elovainio M, van den Bos K, Linna A et al (2005) Combined effects of uncertainty and organizational justice on employee health: testing the uncertainty management model of fairness judgments among Finnish public sector employees. Soc Sci Med 61(12):2501–2512

Ensley MD, Pearce CL, Hmieleski KM (2006) The moderating effect of environmental dynamism on the relationship between entrepreneur leadership behavior and new venture performance. J Bus Venturing 21(2):243–263

Gates B, Hemingway C (1999) Business @ the speed of thought: using a digital nervous system. Warner, New York

Gioia DA, Chittipeddi K (1991) Sensemaking and sensegiving in strategic change initiation. Strateg Manage J 12(6):433–448

Gioia DA, Thomas JB (1996) Identity, image, and issue interpretation: sensemaking during strategic change in academia. Adm Sci Q 41(3):370–403

Hall GB, Dollard MF, Coward J (2010) Psychosocial safety climate: development of the PSC-12. Int J Stress Manage 17(4):353–383

Hofstede G, McCrae RR (2004) Personality and culture revisited: linking traits and dimensions of culture. Cross Cult Res 38(1):52–88

Ilies R, Morgeson FP, Nahrgang JD (2005) Authentic leadership and eudaemonic well-being: understanding leader-follower outcomes. Leadersh Q 16(3):373–394

Izard CE (1991) The psychology of emotions. Plenum, New York

Johns G (2006) The essential impact of context on organizational behavior. Acad Manage Rev 31 (2):386–408

Kezar A, Eckel P (2002) Examining the institutional transformation process: the importance of sensemaking, interrelated strategies, and balance. Res High Educ 43(3):295–328

Kubicek B, Korunka C, Tement S (2014) Too much job control? Two studies on curvilinear relations between job control and eldercare workers' well-being. Int J Nurs Stud 51 (12):1644–1653

Law R, Dollard MF, Tuckey MR et al (2011) Psychosocial safety climate as a lead indicator of workplace bullying and harassment, job resources, psychological health and employee engagement. Accid Anal Prev 43(5):1782–1793

Lim D, Sanderson K, Andrews G (2000) Lost productivity among full-time workers with mental disorders. J Ment Health Policy Econ 3:139–146

Lipshitz R, Ron N, Popper M (2004) Retrospective sensemaking and foresight: studying the past to prepare for the future. In: Tsoukas H, Shepherd J (eds) Managing the future: foresight in the knowledge economy. Wiley, Hoboken, NJ

Little AC, Burriss RP, Jones BC et al (2007) Facial appearance affects voting decisions. Evol Hum Behav 28(1):18–27

Mandler G (1981) Mind and body. W.W. Norton, New York

McCann SJH (2001) Height, societal threat, and the victory margin in presidential elections (1824–1992). Psychol Rep 88(3):741–742

Nelson A, Cooper CL, Jackson PR (1995) Uncertainty amidst change: the impact of privatization on employee job satisfaction and well-being. J Occup Organ Psychol 68(1):57–71

Plowman DA, Solansky S, Beck TE et al (2007) The role of leadership in emergent, self-organization. Leadersh Q 18(4):341–356

Pye A (2005) Leadership and organizing: sensemaking in action. Leadership 1(1):31–49

Raskin JD (2002) Constructivism in psychology: personal construct psychology, radical constructivism, and social constructionism. Am Commun J 5(3):1–26

Ring PS, Rands G (1989) Sensemaking, understanding and committing: emergent transaction processes in the evolution of 3M's microgravity research program. In: Van de Ven AH, Angle H, Poole MS (eds) Research on the management of innovation: the Minnesota studies. Ballinger/Harper & Row, New York, NY, pp 337–366

Schein EH (2004) Organizational culture and leadership, 2nd edn. Jossey-Bass, San Francisco, CA

Stam E (2013) Knowledge and entrepreneurial employees: a country-level analysis. Small Bus Econ 41(4):887–898

Staw BM, Sandelands LE, Dutton JE (1981) Threat rigidity effects in organizational behavior: a multilevel analysis. Admin Sci Q 26(4):501–524

Smircich L (1983) Organizations as shared meanings. In: Pondy LR, Frost PJ, Morgan G, Dandridge TC (eds) Organizational symbolism. JAI, Greenwich, CT, pp 55–65

Sverke M, Hellgren J (2002) The nature of job insecurity: understanding employment uncertainty on the brink of a new millennium. Appl Psychol 51(1):23–42

Taylor SS, Fisher D, Dufresne RL (2002) The aesthetics of management storytelling—a key to organizational learning. Manage Learn 33(3):313–330
Tushman ML (1979) Work characteristics and subunit communication structure: a contingency analysis. Admin Sci Q 24(1):82–98
Van den Bos K (2009) Making sense of life: the existential self trying to deal with personal uncertainty. Psychol Inq 20(4):197–217
Waldman DA, Ramirez GG, House RJ et al (2001) Does leadership matter? CEO leadership attributes and profitability under conditions of perceived environmental uncertainty. Acad Manage J 44(1):134–143
Waldman DA, Carmeli A, Halevi MY (2011) Beyond the red tape: how victims of terrorism perceive and react to organizational responses to their suffering. J Organ Behav 32(7):938–954
Waterman RH Jr (1990) Adhocracy: the power to change. Whittle Direct, Memphis, TN
Weick KE (1979) The social psychology of organizing. Random House, New York, NY
Weick KE (1995) Sensemaking in organizations. Sage, Thousand Oaks, CA
Wright TA, Cropanzano R (2000) Psychological well-being and job satisfaction as predictors of job performance. J Occup Health Psychol 5(1):84–94
Wright TA, Staw BM (1999) Affect and favorable work outcomes: two longitudinal tests of the happy-productive worker thesis. J Organ Behav 20(1):1–23
Zuckerman M (1994) Behavioral expressions and biosocial bases of sensation seeking. Cambridge University Press, Cambridge

The Need for Healthy Leadership in the Health Care Sector: Consideration of Specific Conditions for Implementation

David Horstmann and Hanna Lisa Eckerth

Abstract

Because of various demographic changes, the health care sector is currently facing increasing demands and decreasing resources, which is leading to increasingly greater demands on nurses and compromises in patients' quality of care. These upcoming changes as well as sector-specific frameworks are affecting leadership in the health care sector and resulting in certain requirements for leaders. As there is considerable evidence that leadership has a positive influence on employees' health, it seems reasonable to take leaders into account when considering the increasing demands on nurses. Healthy leadership aims to support employees' health and be an adequate instrument for health promotion in the work-place. In addition to a healthy leadership style, leaders have to consider the specific circumstances of the sector, mainly legal regulations and limited resources. New techniques for leadership development and implementation of health promotion in the work-place may be necessary to enable health care companies to combat these restrictions. This chapter gives an overview of current and future challenges in the health care sector and describes how leaders can address these challenges by having healthy leadership styles and taking sector-specific conditions into account.

1 Challenges in the Health Care Sector

The following overview promotes understanding of the current situation in the health care sector, which is characterized by a high physical and mental demands and impairment of the health status of nurses. These circumstances point to the needs that should be addressed by leadership in the health care sector. Additionally,

D. Horstmann (✉) • H.L. Eckerth
Leuphana University Luneburg, Luneburg, Germany
e-mail: david.horstmann@leuphana.de; hanna_lisa.eckerth@inkubator.leuphana.de

M. Wiencke et al. (eds.), *Healthy at Work*, DOI 10.1007/978-3-319-32331-2_11

specific aspects of the legal framework and limited time and financial resources also require consideration for successful implementation of leadership.

1.1 Demands on Nurses and Their Health Status

Interactions with patients are both the main and the most demanding field of action for nurses. Nurses frequently have to deal with suffering or aggressive patients and are confronted with death and sickness every day. These intense pressures are emotionally demanding for nurses, especially for those in the geriatric care sector (Simon et al. 2005). In addition to this mental stress, geriatric care nurses must also perform physically demanding work such as lifting and moving patients, especially if no work aids are available (Estryn-Behar et al. 2005). Furthermore, nurses' overall working conditions are demanding: they have to document every work step, which leads to a lot of paperwork and numerous interruptions to their workflow (Metz et al. 2009). Providing continuous care for patients makes shift work necessary; it is widely known that working shifts is associated with poorer health status (e.g., Von Treuer et al. 2014).

These demands have consequences for geriatric care nurses. The nursing profession is characterized by a poorer health status than other professions, which leads to a higher rate of absence and inactive periods due to sick leave (Kordt 2014). In a recent study, the authors examined the organizational health in 28 nursing homes in Lower Saxony.[1] This questionnaire-based study surveyed 874 employees in the geriatric care sector to identify the most common strains. Half of the employees (49.6 %) experienced musculoskeletal symptoms at least every few days, 48.4 % fatigue at least every few days, 42.7 % insomnia, and 18.7 % frequent feelings of depression. The high rate of presenteeism is also noteworthy because it results in sick nurses exposing their patients and colleagues to the risk of infection. About one third of these employees go to work even when they feel sick and against to their doctors' advice (Remdisch et al. 2013). This pattern of symptoms reflects the specific demands on geriatric care nurses: heavy physical work and overall workload on the one hand and emotional demands and mental stress on the other. The most common pressures reported by the nurses are a low patient–staff ratio, a great deal of paperwork, quality inspections by the medical service, and the overall workload. These results are in accordance with the findings of previous studies (e.g. Metz et al. 2009; Simon et al. 2005).

However, these employees also reported resources in their work-places: they derived benefit from their responsibilities, the meaningfulness of their work, social support from colleagues, and taking care of their patients. Similar resources have been reported by previous researchers (Simon et al. 2005). Furthermore, we found that these employees have varying perceptions of their supervisors. Whereas they inform their supervisors about common work-related demands and stressors, they

[1] The study was part of the European Union-funded project "Innovation-Incubator Luneburg".

do not perceive them as appropriate or motivating role models concerning health. On an individual level, regular feedback rarely occurs. The acceptance of health-oriented actions is often short term, because they are not integrated into their daily routines (Remdisch et al. 2013).

Because of the upcoming shortage of skilled workers, nurses' workloads and therefore their work-related health issues will increase further. Nurses' working conditions will be exacerbated by future societal developments. Demographic changes will lead to changes in Western society's age structure in the coming decades (European Commission 2011). Declining birth-rates and improvements in life expectancy will lead to an increasing proportion of older persons in society. Therefore, the number of people requiring professional care will continue to increase. In Germany, 2.5 million people currently need care; this number is expected to increase by 30 % to 3.4 million by the year 2030 (Statistisches Bundesamt 2010). Similar trends have been identified in other European countries (European Commission 2011). The more older adults who need care, the more nursing staff will be required. Additionally, increasingly fewer people will be available for the labor market (Nowossadeck 2013). Researchers predict that the requirement for a minimum of 250,000 full-time workers that will be needed in the geriatric care sector in Germany by the year 2030 may not be met (Rothgang et al. 2012).

As a result, geriatric care nurses will face an overall increase in workload, which will lead to even greater demands on them (Metz et al. 2009). The health status of nurses must be addressed because high work-related demands can lead to decreased quality of health care (Von dem Knesebeck et al. 2009). Leadership will play a central role in counteracting this development and maintain high-quality nursing care (Papathanasiou et al. 2014). Further, reviews of published reports have shown that leadership is an important factor in reducing employees' demands and thus supporting their health and well-being (Skakon et al. 2010; Gregersen et al. 2011).

1.2 Legal Framework and Limited Resources

In addition to societal changes and health-related issues, the geriatric care sector is constrained by a specific legal framework and limited resources.

First, the health care sector has a strict legal framework. To guarantee a high quality of caregiving, care facilities must follow a strict set of rules and laws. Working processes have to be documented, independent quality checks are mandatory, and the patient–staff ratio and proportion of qualified staff are statutory (Köhler and Goldmann 2010). These regulations lead to additional demands on nurses. As mentioned in Sect. 1.1, the high volume of paperwork and quality checks by medical service are among the most frequently cited demands (Remdisch et al. 2013). Although these requirements aim to ensure quality care, they also restrict each company's freedom and possible courses of action. In addition, the political framework hinders initiation of change by companies. The numerous relevant political and social factors, such as health and nursing care insurance, and authorities for social assistance, rehabilitation, and medical services, which each have their own interests and stipulations, make it necessary to accommodate a variety of requirements and reduce the flexibility with which health care facilities may act (Köhler and Goldmann 2010). In general, politically driven decisions lead to uncertainty in the health care market, because every new government may make new decisions. Therefore, reliable planning is quite difficult: environmental stability is necessary to initiate processes for change and innovation (Greenhalgh et al. 2004).

Second, there are often few resources for preventing development of health issues in the health care sector. For example, many nursing homes are rather small facilities with few human or financial resources for health promotion and personnel development. In Germany, the majority of the 12,000 geriatric nursing homes have 60 patients or fewer (Statistisches Bundesamt 2013). The smaller the company, the smaller the investment in qualifications and further education (Vollmar 2013). However, the pressure of competition is also constantly increasing in hospitals and the majority of the hospitals are in debt. Consequently, resources are limited, making further education in the health care sector a low priority (Blum et al. 2013). Although promotion of employees' health is needed in health care companies, the companies' foci remain on keeping everyday business up and running.

Because leaders play an important role in creating healthy work environments in the health care sector (Shirey 2006), the limitations mentioned in this section also affect leadership. Leaders have to work within this operational framework and find feasible ways of achieving their goals.

2 Leadership in the Health Care Sector

The current situation in the health care sector requires leadership to address certain challenges and circumstances. Because employees in the health care sector face high demands and experience specific strains, leaders should focus on having healthy leadership styles for addressing these demands. Additionally, consideration of specific conditions, such as the legal framework and limited resources, is helpful when attempting to address the needs of the health care sector. In the following paragraph we present findings on the role of leadership in employees' health and suggest several ideas for successfully implementing leadership in the health care sector.

2.1 Focus on Healthy Leadership

Supporting employees' health and well-being is a major challenge in the health care sector (Hasselhorn et al. 2005; Remdisch et al. 2013). In this context, leadership plays an important role in health promotion in the work-place. Leadership can be understood as a process of influencing group members to contribute to the achievement of group or organizational goals (Rost 2008). There is considerable evidence that leadership can positively influence employees' health, well-being, and performance. Reviews have identified that positive leadership styles have significant effects on several health-related outcomes, including employees' job satisfaction, work-related strain and burnout, and absentee rates (Skakon et al. 2010; Gregersen et al. 2011). To understand these effects, it is helpful to introduce the concept of healthy leadership. Healthy leadership can be defined as any leadership technique, style, or behavior that focuses on employees' well-being and health and thereby fosters efficiency and motivation in employees (Spiess and Stadler 2007, p. 258). According to this definition, the goal of healthy leadership is to foster employees' well-being in the first place rather than focusing on employees' performance, customer service or other matters. Therefore healthy leaders focus on health issues, inform employees about health protection, implement health promotion in the work-place, and take care of their own health. By doing so, leaders can create an organizational climate of health (Gurt et al. 2011).

Franke et al. (2014) define three central aspects of healthy leadership, thus highlighting that health-orientation occurs not only on a behavioral level but also on a cognitive level. Leaders should therefore not only act in health-supporting ways but also place high value on health and be aware of health-related issues and interventions. Valuing health means that leaders are interested in, and feel responsible for, their own health and the health of their employees. This sets a motivational basis for leaders to engage in health issues and is necessary for implementing health-oriented actions. Wilde et al. (2009) found that supervisors' personal attitude toward healthy leadership is the most relevant factor in their motivation to lead in a healthy way. Additionally, leaders need to show health awareness. They should be aware of job-related demands and resources and potential interventions and need to notice changes in followers' health status (Franke et al. 2014). Supervisors' perceived influence on employees' health and their competencies fosters their motivation to practice healthy leadership (Wilde et al. 2009). Therefore, increasing leaders' awareness of health-related issues and their possible course of action could facilitate them to positively influence their employees' health. In this context the specific pattern of physically demanding work combined with mental stress (see Sect. 1.1) in the health care sector must be mentioned. Leaders should be aware of these demands and possible interventions aimed at minimizing musculoskeletal complaints, psychological stress, burnout, and work–life conflicts.

Being motivated and knowing about employees' health statuses and issues is not enough in itself. Leaders also have to perform health-related actions, such as providing information on health promotion, giving feedback, or changing work conditions (Franke and Felfe 2011). Concerning such health-oriented behavior, supervisors can influence employees' health via several pathways (Wegge et al. 2014).

On the individual level, a dyadic relationship between supervisor and subordinate is particularly relevant to employees' health and well-being (Gregersen et al. 2014). Direct contact with their subordinates enables leaders to support employees' heath, for instance by motivation and communication. Employees who have high-quality relationships with their leaders tend to be more satisfied and have a greater sense of well-being (Hooper and Martin 2008). However, leaders can also harm their subordinates' health. Destructive leadership behavior tends to impact employees' health negatively. Tepper (2007) reported that abusive leadership behavior leads to psychological distress and job strain for subordinates.

On an environmental level, leaders can implement actions and policies that support employees' health and well-being. The relevance of work-place design, such as ergonomics and working tools for work-related health, is widely known (e.g. Aarås 1994; Lee et al. 2014). Leaders can also influence working conditions (e.g. autonomy, task variety, job complexity, or team constellations) to foster employees' health (Humphrey et al. 2007). By changing such environmental factors, leaders indirectly influence their employees' health. For example, Nielsen et al. (2008) showed that work characteristics mediate the effects of leadership on employees' job satisfaction and well-being.

Furthermore, leaders act as role models for their subordinates by taking care of their own health. Leaders' impact on employees' behavior and satisfaction depends on whether the subordinates perceive their supervisors as role models (Ogunfowora 2014). If supervisors are not aware of their own resources and needs and do not take care of their own health, they are unlikely to do so for their subordinates. A leader's lack of self-care leads to their employees perceiving them to be inauthentic and reluctant to accept their leader's health-oriented behaviors (Franke et al. 2011). Reflecting on their own health behavior, resources, and stressors provides supervisors with an opportunity for self-guidance and can establish a basis for health-related role modeling. Employees who perceive that their leaders have a high level of health awareness and behavior take more care of their own health and have better health statuses (Franke et al. 2014). The authors' own study—presented in Sect. 1.1—revealed a gap between leaders' self-awareness and their employees' perceptions concerning health-related role modeling. When leaders over-rate their own health behavior and their influence as role models, employees do not necessarily see them as positive examples concerning health-related issues (Remdisch et al. 2013).

However, any of the actions, leaders intend to support their employees' health must be matched to their specific needs. There is no silver bullet for health promotion in this sector: the individual circumstances of each company and their employees must be considered. An individualized assessment of health complaints and work-related demands and resources can help identify where action needs to be taken, after which customized concepts and measures can be developed to address the needs of each individual employee and company.

2.2 Considering Specific Conditions

Besides focusing on employees' health and healthy leadership, leadership in the health care sector has to consider the specific framework of the sector in general and the individual circumstances of each company. As shown above, healthy leadership can help to support employees' health and well-being. However, limited resources and restrictive legal regulations not only affect employees' working condition, but also limit the possibilities for leaders to take action and implement healthy leadership. We will now present specific conditions that have to be considered as well as central suggestions for dealing with these limitations.

First, legal regulations and requirements limit health care companies' possible courses of action. A strict set of rules and laws aimed at guaranteeing a high quality of work regulate working processes and patient–staff ratios (Köhler and Goldmann 2010). This framework may prevent leaders from taking certain actions aimed at optimizing working conditions. For instance, reducing the requirements for paperwork would be a rational way of reducing one of the most frequent demands on nurses. However, for legal reasons this requirement cannot be ignored; thus, alternative solutions have to be found. Strengthening individual resources is one possible solution. Building up individual coping strategies for dealing with work-related stressors that cannot be reduced on a situational level can support employees to cope with such stressors and foster work engagement (Xanthopoulou et al. 2007). Interventions on an individual level are economic and effective in reducing stress-related mental strains, particularly with high-risk employees (Sockoll et al. 2008). Leaders therefore need to identify individual needs and support the development of individual resources to cope with situational stress. These individual resources can also affect matters on the organizational level. For example, leaders could assign tasks on the basis of each individual's strengths and personal interests.

Second, limits in resources of health care facilities need to be considered. Because many nursing homes in Germany are small companies (Statistisches Bundesamt 2013), they do not have access to adequate time and financial resources. These circumstances both emphasize the importance of leaders and affect leadership in several ways (Ulich and Wülser 2010). There is little or no capacity for systematic leadership development and everyday business is prioritized over implementing health-oriented actions. To address these issues, it seems reasonable to consider new personnel development methods, especially for smaller companies. Corporate networks could potentially play a central role in the implementation of health promotion by small and medium-sized enterprises in the health care sector. By these means, small companies can profit from the knowledge and resources of other companies (Dehning and Dostalek 2012). For instance, because group-based training is more cost-effective than individual training, combining with other small companies can decrease the costs of leadership development programs or training sessions. In addition to corporate networks, universities and other actors in the labor market, such as chambers and the Health Insurance or Employer's Liability Insurance Association, are potential points of access to health promotion resources and personal development. Furthermore such programs should be characterized by strong practical relevance and aim to prioritize health-related topics. Therefore programs should not only transfer knowledge about possible interventions for supporting employees' health, but also increase leaders' valuing and awareness of health-related issues.

3 Conclusions

The aim of this chapter was to present the current situation in the health care sector and outline the implications for leadership in this sector. Demographic changes and upcoming societal challenges in health care mandate increasing workload and job strains for nurses. These developments and other specific characteristic of the health care sector lead to specific requirements for leadership. Leadership in the health care sector should focus on healthy leadership to counteract impairment of the health status of nurses: research has shown that certain aspects of leadership are helpful for employees' health.

Health care companies do not have the same resources and freedom of action as other companies to address the major challenge of impairment of the health status of nurses. Thus, leaders in this field may not develop their full potential for supporting employees' health and well-being. However, there are some guidelines that compensate for the identified hindering factors. Findings of previous research suggest some central ideas. Focusing on innovative learning concepts, supporting individual resources, increasing leaders valuing and awareness of health, and engaging in network cooperation would likely facilitate more effective implementation of healthy leadership in the health care sector. There is no silver bullet for the successful implementation of healthy leadership in such companies. To create customized interventions, the purposes and individual needs of each company must be considered. Future research should provide empirical evidence for the stated suggestions and thereby set a basis for successful implementation of healthy leadership in the health care sector.

References

Aarås A (1994) The impact of ergonomic intervention on individual health and corporate prosperity in a telecommunications environment. Ergonomics 37:1679–1696

Blum K, Löffert S, Offermanns M et al (2013) Krankenhausbarometer 2013 [Hospital barometer 2013]. Deutsches Krankenhaus Institut, Düsseldorf (in German)

Dehning W, Dostalek K (2012) Regioanle Kooperationen—ein Netzwerk von Netzwerken: Das Beispiel "Gesund Richtung Zukunft" [Regional cooperations—a network of networks: the example "healthy towards the future"]. In: Glückler J, Dehning W, Janneck M, Armbrüster T (eds) Unternehmensnetzwerke. Springer, Berlin, pp 297–312 (in German)

Estryn-Behar M, Le Nézet O, Laine A et al (2005) Körperliche Belastungen bei Pflegepersonal [Physical demands on nursing staff]. In: Hasselhorn H-M, Müller BH, Tackenberg P et al (eds) Berufsausstieg bei Pflegepersonal. Arbeitsbedingungen und beabsichtigter Berufsausstieg bei Pflegepersonal in Deutschland und Europa. Bundesanstalt für Arbeitsschutz und Arbeitsmedizin (BAuA), Dortmund, pp 101–108 (in German)

European Commission (2011) Demography report 2010. Publication Office of the European Union, Luxembourg

Franke F, Felfe J (2011) Diagnose gesundheitsförderlicher Führung—Das Instrument "health-oriented leadership" [Diagnosing health-supporting leadership—the instrument "health-oriented leadership"]. In: Badura B, Ducki A, Schröder H, Macco K (eds) Fehlzeiten-Report 2011. Springer, Berlin, pp 3–13 (in German)

Franke F, Vincent S, Felfe J (2011) Gesundheitsbezogene Führung [Health-related leadership]. In: Bamberg E, Ducki A, Metz A-M (eds) Gesundheitsförderung und Gesundheitsmanagement in der Arbeitswelt. Hogrefe, Göttingen, pp 371–391 (in German)

Franke F, Felfe J, Pundt A (2014) The impact of health-oriented leadership on follower health: development and test of a new instrument measuring health-promoting leadership. Z Pers 28:139–161

Greenhalgh T, Robert G, Macfarlane F et al (2004) Comment on "Diffusion of innovations in service organizations: systematic review and recommendations". Milbank Q 82:581–629

Gregersen S, Kuhnert S, Zimber A, Niehaus A (2011) Führungsverhalten und Gesundheit—Zum Stand der Forschung [Leadership behavior and health—state of research]. Gesundheitswesen 73(1):3–12 (in German)

Gregersen S, Vincent-Höper S, Nienhaus A (2014) Health–relevant leadership behaviour: a comparison of leadership constructs. Z Pers 28:117–138

Gurt J, Schwennen C, Elke G (2011) Health-specific leadership: is there an association between leader consideration for the health of employees and their strain and well-being? Work Stress 25:108–127

Hasselhorn H-M, Müller BH, Tackenberg P et al (2005) Berufsausstieg bei Pflegepersonal. Arbeitsbedingungen und beabsichtigter Berufsausstieg bei Pflegepersonal in Deutschland und Europa [Career endings for nurses. Working conditions and intended career ending for nurses in Germany and Europe]. BAuA, Dortmund (in German)

Hooper DT, Martin R (2008) Beyond personal leader–member exchange (LMX) quality: the effects of perceived LMX variability on employee reactions. Leader Q 19:20–30

Humphrey SE, Nahrgang JD, Morgeson FP (2007) Integrating motivational, social, and contextual work design features: a meta-analytic summary and theoretical extension of the work design literature. J Appl Psychol 92:1332–1356

Köhler K, Goldmann M (2010) Soziale Innovation in der Pflege—Vernetzung und Transfer im Fokus einer Zukunftsbranche [Social innovation in health care—networking and transfer in the focus of a future-sector]. In: Howaldt J, Jacobsen H (eds) Soziale Innovation. Auf dem Weg zu einem postindustriellen Innovationsparadigma. VS Verlag für Sozialwissenschaften, Wiesbaden, pp 253–270 (in German)

Kordt M (2014) Gesundheitsreport 2014 [Health report 2014]. DAK Gesundheit, Hamburg (in German)

Lee EWC, Fok JPC, Lam AT et al (2014) The application of participatory ergonomics in a healthcare setting in Hong Kong. Work 48:511–519

Metz A-M, Kunze D, Hamann L et al (2009) Demographischer Wandel in der Pflege. Konzepte und Modelle für den Erhalt und die die Förderung der Arbeits- und Beschäftigungsfähigkeit von Pflegekräften [Demographic change in the health care sector. Concepts and models for preventing and supporting workability for nurses]. BAuA, Berlin (in German)

Nielsen K, Yarker J, Brenner S-O et al (2008) The importance of transformational leadership style for the well-being of employees working with older people. J Adv Nurs 63:465–475

Nowossadeck S (2013) Demografischer Wandel, Pflegebedürftige und der künftige Bedarf an Pflegekräften. Eine Übersicht [Demographic change, people needing long-term care, and the future need for carers. An overview]. Bundesgesundheitsbla 56:1040–1047 (in German)

Ogunfowora B (2014) It's all a matter of consensus: leader role modeling strength as a moderator of the links between ethical leadership and employee outcomes. Hum Relat 67:1467–1490

Papathanasiou IV, Fradelos EC, Kleisiaris CF et al (2014) Motivation, leadership, empowerment and confidence: their relation with nurses' burnout. Mater Sociomed 26:405–410

Remdisch S, Gerstenberg R, Storm V (2013) Ergebnisbericht zur Studie "Organisationale Gesundheit in der Pflegebranche" [Result report on the study "Organizational health in the health care sector"]. Leuphana University Luneburg, Luneburg (in German)

Rost JC (2008) Leadership definition. In: Marturano A, Gosling J (eds) Leadership: the key concepts. Routledge, New York, pp 94–99

Rothgang H, Müller R, Unger R (2012) Themenreport "Pflege 2030" [Thematic report "Health care 2030"]. https://www.bertelsmann-stiftung.de/de/unsere-projekte/pflege-vor-ort/projektthemen/pflegereport-2030. Accessed 20 Mar 2015 (in German)

Shirey MR (2006) Authentic leaders creating healthy work environments for nursing practice. Am J Crit Care 15:256–268

Simon M, Tackenberg P, Hasselhorn H-M et al (2005) Auswertung der ersten Befragung der NEXT-Studie in Deutschland [Results from the first survey of the NEXT-study in Germany]. Universität Wuppertal, Wuppertal (in German)

Skakon J, Nielsen K, Borg V, Guzman J (2010) Are leaders' well-being, behaviours and style associated with the affective well-being of their employees? A systematic review of three decades of research. Work Stress 24:107–139

Sockoll I, Kramer I, Bödeker W (2008) iga-Report 13: Wirksamkeit und Nutzen betrieblicher Gesundheitsförderung und Prävention. Zusammenstellung der wissenschaftlichen Evidenz 2000 bis 2006 [iga-Report 13: effectiveness and profit of work-place health promotion and prevention. Summary of scientific evidence 2000 to 2006]. BKK Bundesverband, Essen (in German)

Spiess E, Stadler P (2007) Gesundheitsförderliches Führen—Defizite erkennen und Fehlbelastungen der Mitarbeiter reduzieren [Health-supportive leading—identifying deficits and reducing employees' strains]. In: Weber A, Hörmann G (eds) Psychosoziale Gesundheit im Beruf: Mensch, Arbeitswelt. Gesellschaft. Genter-Verlag, Stuttgart, pp 255–274 (in German)

Statistisches Bundesamt (2010) Demografischer Wandel in Deutschland. Auswirkungen auf Krankenhausbehandlungen und Pflegebedürftige im Bund und in den Ländern [Demographic change in Germany. Effects on hospital treatment and people in need of care at federal and state level]. Statistisches Bundesamt, Wiesbaden (in German)

Statistisches Bundesamt (2013) Pflegestatistik 2011—Pflege im Rahmen der Pflegeversicherung—Deutschlandergebnisse [Health care statistics 2011—health care in the scope of health care insurance—results for Germany]. Statistisches Bundesamt, Wiesbaden (in German)

Tepper BJ (2007) Abusive supervision in work organizations: review, synthesis, and research agenda. J Manage 33:261–289

Ulich E, Wülser M (2010) Gesundheitsmanagement in Unternehmen. Arbeitspsychologische Perspektiven [Health management in companies. Work-psychological perspectives], 4th edn. Gabler, Wiesbaden (in German)

Vollmar M (2013) Berufliche Weiterbildung in Unternehmen 2010—Methodik und erste Ergebnisse [Professional development in companies 2010—method and initial results]. In: Statistisches Bundesamt (ed) Wirtschaft und Statistik. Statistisches Bundesamt, Wiesbaden, pp 276–287 (in German)

Von dem Knesebeck O, Blum K, Grosse K et al (2009) Arbeitsbedingungen und Patientenversorgung [Working conditions and patient care]. Arzt Krankenhaus 7:210–213 (in German)

Von Treuer K, Fuller-Tyszkiewicz M, Little G (2014) The impact of shift work and organizational work climate on health outcomes in nurses. J Occup Health Psychol 19(4):453–461

Wegge J, Shemla M, Haslam SA (2014) Leader behavior as a determinant of health at work: specification and evidence of five key pathways. Z Pers 28:6–23

Wilde B, Hinrichs S, Pavez CB et al (2009) Führungskräfte und ihre Verantwortung für die Gesundheit ihrer Mitarbeiter—Eine empirische Untersuschung zu den Bedingungsfaktoren gesundheitsförderlichen Führens [Leaders and their responsibility for employees' health—an empirical study on factors influencing health-supporting leadership]. Wirtschaftspsychologie 2:74–89 (in German)

Xanthopoulou D, Bakker AB, Demerouti E et al (2007) The role of personal resources in the job demands-resources model. Int J Stress Manage 14:121–141

Part III

Public and Private Policies Promoting Occupational Health

Improved Health and Safety at Work by Cooperation with (Social) Partners: A European and Global Perspective

Swen Malte John and Patricia Weinert

Abstract

The present article aims at showcasing how cooperation with important stakeholders and building linkages with key policy actors can increasingly bring safety and health to the forefront of the OSH agenda. Despite of the disease burden, it is unfortunately not yet well known amongst policy-makers that occupational skin diseases (OSD) are the most frequently recognized occupational diseases in Europe with a share of 35 % of all work-related illnesses. While recent studies have shown that preventive interdisciplinary intervention can significantly reduce the burden, there is still a lack of targeted preventive measures and coordinated efforts to improve options for risk assessment and specific dermatological interventions. Through awareness raising and building partnerships improvements can however be reached and lead to a win-win situation for all actors involved.

1 Introduction

It is undisputed that the promotion of social dialogue and the role of social partners are important factors in achieving healthy workplaces. While not always binding, social dialogue provides an important orientation for government policy. Indeed, as emphasized by the Director of Eurofound, "Research shows that the role of social

S.M. John
Department of Dermatology, Environmental Medicine, Health Theory, Faculty of Human Sciences, University of Osnabrück, Osnabrück, Germany
e-mail: sjohn@uos.de

P. Weinert (✉)
Institute for Interdisciplinary Dermatological Prevention and Rehabilitation (iDerm), University of Osnabrück, Osnabrück, Germany
e-mail: patricia.weinert@uni-osnabrueck.de

M. Wiencke et al. (eds.), *Healthy at Work*, DOI 10.1007/978-3-319-32331-2_12

dialogue and social partners is relevant to raise awareness and implement interventions" (Menéndez-Valdés 2014).

Hence, since its launch in 1985, which had the aim of ensuring a social dimension in the development of the European single market, social dialogue has increasingly been a driving force at both national and European levels for promoting better working conditions and social well-being. Thus, social dialogue has become an essential element of the European social model (European Commission 2015). Obviously, the organization of social partner involvement at national level differs because of varying socioeconomic and political contexts in which national schemes have developed. However, at EU level social dialogue has received strong institutional recognition and developed well-established structures.

Many European social dialogue committees are working to improve occupational safety and health (OSH) in their own sectors. They have influenced European OSH regulations and have been able to set some minimum standards through agreements implemented by EU legislation. Moreover, EU cross-industry social partners have signed autonomous agreements that are implemented by national social partner organizations in accordance with national procedures and practices (European Agency for Safety and Health at Work 2014).

In addition to the above, advocacy and raising awareness are also necessary elements for prompting action at policy level because they increase the scope of influences on improving policies. As our own experiences have shown, a key to advocacy is building linkages among stakeholders and gaining support, including by means of awareness-raising activities. Hence, the approaches of finding appropriate allies, building on existing institutions, lobbying, using examples, linking to key partners, and building strategic partnerships can all facilitate achieving the objective of healthier workplaces.

This chapter thus aims to showcase how the interplay between these approaches and associated actions can help to move OHS matters higher on the policy agenda. As an example, it will focus on specific initiatives taken by European social partners with the aim of improving health and safety conditions for high-risk professional groups that are particularly prone to occupational skin diseases (OSDs). It will furthermore demonstrate how building partnerships with common goals, identifying allies, and linking to key institutions can be of extreme benefit and lead to win-win situations for all actors involved.

2 Improved Safety and Health by Cooperation with (Social) Partners

We have chosen to focus on OSDs because it is still not sufficiently well known that OSDs are the most frequently recognized occupational diseases in Europe, constituting up to 35 % of all work-related illnesses and causing considerable morbidity in affected workers. OSDs are often underreported because of failure to recognize their association with the work environment. Individuals affected by OSDs are of all ages and both sexes and have a large variety of occupations. High-risk industries for OSDs include health-care, hairdressing/beauty salons, metal work, construction, manufacturing, food production, agriculture, printing, and janitorial services.

The burden of OSDs is quite significant. Annual costs for medical treatment, sick leave, and loss of productivity incurred by OSDs in the EU are estimated to vastly exceed 5 billion €. Most of these costs fall on small and medium-sized enterprises, which employ two thirds of EU workers (Sartorelli et al. 2011). The implications of demographic ageing and increasing challenges of finding skilled labour has resulted in such businesses attempting to improve their competitiveness by reducing costs, which often negatively impacts OSH. For affected individuals, the chronic course of OSDs can have detrimental socioeconomic and psychological consequences, including absenteeism, need for retraining, job losses and long-term unemployment.

One sector that has received considerable attention over the past decade for its work on OSH is hairdressing. This industry is of great importance in Europe because it comprises some 400,000 salons and employs over 1.5 million workers, which is approximately 10 % of the total service sector in Europe (Observatoire Social Européen 2010). Accordingly, the European social dialogue committee in this sector has worked for well over 15 years to reduce work-related risks to safety and health and thus benefit hairdressers and hairdressing businesses. This work has resulted in the production and adoption of numerous joint texts ranging from common opinions and codes of conduct to tool kits, frameworks for action, and autonomous agreements (Andor 2013).

With the Declaration of Dresden, signed in September 2010, an important breakthrough was achieved by the social partners of this sector. This declaration was the main outcome of the EU-funded "SafeHair 1.0" project, in which employers' and workers' representatives, suppliers, and safety engineers reached an autonomous agreement on the importance of implementing state-of-the-art measures to prevent OSDs in the hairdressing trade. The agreement contains key recommendations on how to impart available scientifically acquired health and safety knowledge in the hairdressing sector to self-employed hairdressers, employees and apprentices through appropriate educational and teaching measures and thus facilitate implementation of such knowledge (Safehair 2010a). The SafeHair project was continued in 2012 by the social partners to put theory into practice by implementing the Dresden recommendations. As a result, a comprehensive virtual modular toolbox that is considered a very valuable tool for educating workers and employers about OSH has been made available in over ten European countries (Safehair 2010b). This virtual toolbox is now used in a number of European countries, for instance in Austria in vocational schools, with specific training for multipliers (e.g. vocational school teachers) and hairdressers' apprentices.

An additional step was initiated by the social partners in this sector; namely, the first autonomous European framework agreement on the protection of OSH in the hairdressing sector was reached and signed in April 2012 in Brussels in the presence of Laszlo Andor, the then EU Commissioner in charge of Employment, Social Affairs and Inclusion (Coiffure EU and UNI Europa Hair & Beauty 2012).

The agreement is an example of a successfully integrated policy achieved through social dialogue. Indeed, the European Agency for Safety and Health at Work (EU-OSHA) states that "Experiences of countries that have implemented measures similar to those contained in the agreement point to clear benefits for employers and workers alike in lowering sick leave and absence, reducing staff turnover and reducing treatment and follow-up costs for health systems in treating occupational disease. Evidence suggests that the costs of implementing the agreed prevention measures are low—less than EUR 0.5 per customer, or just over 1 % of the annual turnover of an average salon" (Eeckelaert et al. 2014).

3 Advocacy and Raising Awareness Is Never Enough

With this in mind, in 2009, the European Association for Dermatology and Venereology (EADV) launched the pan-European "Healthy skin@work" campaign. Following a "One-size-fits-all" concept, the campaign has been seeking to draw the attention of key stakeholders (EU Parliament, international organizations e.g. the World Health Organization [WHO], and social partners) to the economic burden of OSDs and the need to implement improved measures to prevent them. The campaign has subsequently been addressed at a number of European events over past years, including the Healthy Workplaces Summit of the EU-OSHA in 2013 and various workshops organized by the EU-Directorate General Employment, Social Affairs and Inclusion on future health and safety concepts within the EU social model. Safety and health issues pertaining to skin diseases were also discussed at meetings between European experts and representatives from the European Centre for Vocational Training (CEDEFOP), EU-OSHA, European Foundation for the Improvement of Working and Living Conditions (Eurofound), and European Training Foundation (ETF). At these meetings, the latest trends in social policy and how the EU social model contributes to improvements in the current social and economic landscape by ensuring a sustainable and competitive future were debated. Additionally, "Healthy skin@work" has become an official partner of the EU-OSHA for campaigning activities.

In parallel with efforts to gain the commitment of key stakeholders at the European level, national awareness-raising activities under the umbrella of the "Healthy skin@work" campaign have been developed since the inception of this campaign (European Initiative for the Prevention of Occupational Skin Diseases 2010) under the slogan "Your skin: the most important 2 m^2 in your life". This slogan has been translated into eight languages and a range of activities are regularly organized in Germany, Bulgaria, the Czech Republic, Croatia, Denmark, Romania, Slovakia, and Serbia. More countries continue to join the campaign. The manifold activities carried out at national, regional and local level include workshops for high-risk professions, yearly weeks of occupational dermatitis (WOOD), implementation of a self-diagnosing app for determining risk exposure and skin damage, provision of free dermatological consultations and patch-testing, and holding press conferences with key stakeholders, including social partners and policy makers. The impact has been notable: for example, in Germany WOOD has led to a 30 % increase in OSD notifications and thus decreased the number of underreported cases. Additionally, costs of vocational rehabilitation have substantially decreased as a result of earlier dermatological interventions, resulting in fewer resignations. In Romania, thanks to the national campaign activities, additional financial support for acquiring further technical equipment for diagnosing OSDs has been awarded.

Although OSDs are currently receiving increasing attention, there is still one neglected skin disease; namely, non-melanoma skin cancers caused by long-term exposure to solar ultraviolet (UV) light. Indeed, recent research has demonstrated that sun exposure in workplaces of outdoor workers (e.g. construction workers, farmers, fishers) is a highly relevant occupational hazard in Europe (Bauer et al. 2015). Thus, from a socioeconomic point of view, prevention must play a major role in minimizing occupational risks and reducing both personal and financial costs. This has been recognized by some EU member states, including Austria, Croatia, Denmark, Portugal and, most recently, Germany, which have implemented appropriate legislation. However, because there is still a significant gap between national regulations and their implementation, more needs to be done to raise awareness at EU and national levels and consequently improve prevention measures for this particular group of workers. In many instances, this discrepancy is attributable to a lack of awareness and evidence-based intervention concepts.

Accordingly, for the past 4 years the EADV, Association of European Cancer Leagues, and Euromelanoma (the official skin cancer prevention campaign initiated by the European Association of Dermato-Oncology) have been organizing an annual Skin Cancer Awareness Day (EADV: European Academy of Dermatology and Venereology 2015). Representatives of the EU Parliament, EU-OSHA, WHO, International Labour Organization (ILO), and social partners have been invited to discuss how outdoor workers can be better protected from exposure and how the European Commission can contribute to addressing this problem. In 2014, a Call to Action on the protection of outdoor workers from occupational skin cancer in the form of an open letter was addressed to the European Commission. This letter was signed by speakers at the 2014 Skin Cancer Awareness Day from within and outside the European Parliament and social partners (John et al. 2014). Remarkably, this call received sustained attention and was officially answered by the President of the European Commission and by the Commissioner of Employment, Social Affairs and Inclusion. It was made clear that OSDs are a high priority for the European Commission. This subsequently fueled a range of activities that aimed to put OSDs, including skin cancer, on the map of health and safety. One most recent important milestone has been reached with the draft of a scientific interpretative document on non-melanoma skin cancer to the EU Occupational Safety and Health Strategic Framework 2014–2020. This document will be published on the homepage of the EU-OSHA with the support of that organization.

Additional meetings have occurred between Members of the European Parliament and the EU Commission with the aim of attempting to improve surveillance and prevention of OSDs. Actions taken have included improving Directives 2001/

83 EC and 2001/20 EC, which relate to the implementation of good clinical practice in the conduct of clinical trials on medicinal products for human use, the aim of these improvements being to allow undertaking of patch-testing trials to identify new occupational type 4 allergens.

At international level, strategic partnerships have been established with the WHO and ILO. A workshop involving renowned experts worldwide was organized by the EADV and WHO and resulted in defining strategic directions for prevention and early detection of OSDs. Dermatological expert support has also been requested for revision of the WHO International Classification of Diseases list 11. Widely accepted clinical definitions on a range of skin diseases, including occupational allergic contact dermatitis and contact urticaria, have been integrated into this list. This is a major achievement because it will help capture more comprehensive epidemiological data on OSDs at both national and international levels, which is greatly needed. It will thus contribute substantially to identifying the true dimensions of the OSD problem. Since 2013, the WHO further seeks expert dermatological advice at its yearly INTERSUN meetings. This WHO program promotes and evaluates research on the effects on health of UV radiation with the aim of developing appropriate responses through guidelines, recommendations, and dissemination of information. The chair of EADV's "Healthy skin@work" meanwhile serves as a regular scientific adviser to both the WHO and the ILO and advises on aspects of occupational skin hazards.

These illustrative examples have together created win-win situations for all actors involved, as have a growing number of joint research projects. One such example is the EU-funded COST Action "StanDerm" (Development and Implementation of European Standards on Prevention of Occupational Skin Diseases) under the EU-HORIZON 2020 Agenda (COST: European Cooperation in Science and Technology 2012). With a network of almost 140 experts from 28 EU countries, StanDerm strives for increased transnational intervention studies and the development of evidence-based common European standards on OSD prevention and patient management. The project has linked up with the EADV "Healthy skin@work" campaign and the WHO and provides expertise to different forums when requested.

Another example pertains to the GENeration and Extraction System for Individual expoSure (GENESIS-UV) project, which was first initiated in 2014 in Germany and involved measuring daily UV exposure of 300 volunteers from all relevant outdoor professions throughout the country with a device named GENESIS-UV. This project has been recognized to be very relevant not only at national but also at EU and global levels. Indeed, gathering scientific data on a global scale on the degree and quality of exposure to non-ionizing radiation and thence creating a job exposure matrix is a unique undertaking. The acquired knowledge about risk profiles will form an essential basis for primary and secondary prevention of skin cancer and sound education of outdoor workers on a large scale.

4 Summary

Awareness by various stakeholders of the challenges associated with OSDs, including skin cancer, has been growing over recent years. Persevering efforts to network, raise awareness, build strategic partnerships, and gain the commitment of social partners have yielded positive results at many levels, from changes in individual behaviors up to adoption of agreements at EU level. While evidence-based data is an important element, practical achievements in safety and health at work very much depend on the cooperation of participating (social) partners. However, this process is time-consuming and requires additional capacities that need to be taken into consideration; nevertheless, in retrospect the outcomes are worthwhile.

References

Andor L (2013) European strategies for health and safety at work: the role of social partners. Paper presented at the European Trade Union Institute conference on "Trade unions and civil society for a strong and ambitious EU strategy for health and safety at work 2013–2020", Brussels, Belgium, 27 Mar 2013

Bauer A, Beissert S, Knuschke P (2015) [Prevention of occupational solar UV radiation-induced epithelial skin cancer]. Hautarzt 66(3):173–178 (in German)

Coiffure EU, UNI Europa Hair & Beauty (2012) Common statement on "European framework agreement on the protection of occupational health and safety in the hairdressing sector—26 April 2012". Available from http://www.dfkf.dk/xdoc/153/1_Common_statement.pdf. Accessed 22 Apr 2015

COST: European Cooperation in Science and Technology (2012) Development and implementation of European Standards on Prevention of Occupational Skin Diseases (StanDerm). Available from http://www.cost.eu/COST_Actions/isch/Actions/TD1206. Accessed 22 Apr 2015

EADV: European Academy of Dermatology and Venereology (2015) Skin cancer awareness day. Available from http://www.eadv.org/press-corner/skin-awareness-day/. Accessed 22 Apr 2015

Eeckelaert L, Dontas S, Georgiadou E et al (2014) Occupational health and safety in the hairdressing sector. Publications Office of the European Union, Luxembourg

European Agency for Safety and Health at Work (2014) Psychosocial risks in Europe: prevalence and strategies for prevention. Publications Office of the European Union, Luxembourg

European Commission (2015) A new start for social dialogue. European Union, Belgium

European Initiative for the Prevention of Occupational Skin Diseases (2010) European Initiative for the Prevention of Occupational Skin Diseases—"healthyskin@work". Available from http://www.epos2010.eu/. Accessed 23 Apr 2015

John SM, Augustin M, Hercegova J et al (2014) Call to action to the European Commission on the protection of outdoor workers from occupational skin cancer. Open letter. http://www.standerm.eu/fileadmin/user_upload/documents/StanDerm/Publications/2014/John_Augustin_CallToActionSkinCancer_2014.pdf. Accessed 22 Apr 2015

Menéndez-Valdés J (2014) Event brochure for "Occupational Health and Safety in Europe: Fostering Healthy Workplaces for Sustainable and Inclusive Growth". Available from http://www.publicpolicyexchange.co.uk/events/FB24-PPE2. Accessed 22 Apr 2015

Observatoire Social Européen (2010) European sectoral social dialogue. Factsheets. Project coordinated by Christophe Degryse, www.worker-participation.eu/EU-Social-Dialogue/Sectoral-ESD. Accessed 22 Apr 2015

Safehair (2010a) Common skin protection recommendations for the hairdressing sector in Europe: declaration of Dresden. Available from http://safehair.loungemedia.de/safehair/safehair-10/declaration-of-dresden.html. Accessed 29 Apr 2015

SafeHair (2010b) Systematic development and validation of a modular target specific instrument for the design of teaching-learning arrangement for the prevention of occupational skin diseases in the hairdressing industry (2.0). Available from http://safehair.loungemedia.de/safehair/safehair-20.html. Accessed 23 Apr 2015

Sartorelli P, Kezic S, Larese Filon F et al (2011) Prevention of occupational dermatitis. Int J Immunopathol Pharmacol 24(1 Suppl):89S–93S

Occupational Health Services in Scotland

Markus F.C. Steiner and Katherine Targett

Abstract

There is now considerable evidence that work is good for you. Positive effects include improved financial prosperity, social participation, self-esteem and improved physical and mental well-being. Overall, the benefits of work appear to outweigh the risks and to be much greater than the detrimental effects of prolonged sickness absence or long-term unemployment. Given the cost of sickness absence to business and that of worklessness to the tax-payer, there has been a drive in the UK to reform provision of welfare and occupational health (OH). Most health care within the UK is provided by the tax-funded National Health Service (NHS), free at the point of delivery. Historically, however, the NHS did not provide OH; rather, it was provided either as in-house OH services by larger organizations (public and private sector) or externally by specialist OH companies. Consequently, many small and medium-sized enterprises received no such services. The result has been patchy OH provision across the UK and a lack of a "joined-up approach". This chapter describes the changing face of OH provision in the UK over recent years, using the OH service provided by GO Health Services, Grampian, as an exemplar model.

1 UK Welfare Reforms since 2007

The UK has relatively large numbers of people out of work and claiming incapacity benefits. There were 2.5 million incapacity claimants in 2012; a million more than those out of work and receiving unemployment benefits even in the wake of a

M.F.C. Steiner (✉) • K. Targett
GO Health Services, NHS Grampian, Foresterhill Lea, Foresterhill Site, Aberdeen, UK
e-mail: m.steiner@nhs.net; k.targett@nhs.net

M. Wiencke et al. (eds.), *Healthy at Work*, DOI 10.1007/978-3-319-32331-2_13

recession (Beatty et al. 2013). This compares to 1.72 million in Germany, where the population is almost 25 % larger than in the UK and the numbers receiving unemployment benefits have decreased by 5 % over the past 20 years (Deutsche Rentenversicherung Bund 2014).

In the UK, over the last 30 years the number of incapacity claimants has increased dramatically, despite a gradual improvement in the health of the working-age population. There appears to be a significant number of people claiming a sick role that is legitimized by sick certification, often in the absence of significant disease or impairment. Evidence suggests that the large numbers claiming incapacity benefits reflect both job shortages, particularly in older industrial areas, and a tendency for those claimants to have poor skills and qualifications and therefore to not be chosen by employers. Nonetheless, health problems shape the way that incapacity claimants see their future employment prospects: survey data show that only around a third of these claimants would like a job now or in the future and over 90 % state that their reason for not wanting a job is that their health is not good enough. Similarly, even among those who would like to work, 90 % cite ill health, injury or disability as a barrier to finding work and 75 % believe that employers would regard them as too ill or disabled or too big a risk (Beatty et al. 2013). There is however considerable evidence that work is good for you. Not only include positive effects improved financial prosperity, social participation, and self-esteem but also improved physical and mental well-being (Waddel and Burton 2006). Overall, the benefits of work appear to outweigh the risks and to be much greater than the detrimental effects of prolonged sickness absence or long-term unemployment.

Not surprisingly, given the substantial cost to the government, reducing incapacity claimant numbers has become an important goal and has driven a series of changes in welfare provision.

The Welfare Reform Act 2007 led to the replacement of previous incapacity benefits by the Employment Support Allowance and a new medical test, the Work Capability Assessment, in 2008 (Parliament of the United Kingdom 2007). Following this assessment claimants are allocated to one of two groups: a "Work-Related Activity Group", comprising those who are considered potentially able to work and a "Support Group" consisting of individuals whose health condition or disability is so severe that they are considered unable to work. Those in the work-related activity group attend work-focused interviews with a personal adviser to ensure they receive appropriate training and condition management support.

In "*Working for a healthier tomorrow*", Black reviewed the health of the UK's working-age population and estimated that the costs to the country's economy of working days lost and worklessness caused by ill health of working-age individuals was over £100 billion annually (Black 2008). She emphasized the importance of access to expert advice to enable individuals to gain and maintain employment with an expanded role for occupational health (OH) to make it "*available to all, whether they are entering work, seeking to stay in work, or trying to return to work without delay in the wake of illness or injury*".

In 2008 the UK government responded with "*Improving health and work: changing lives*" in which it committed to the development of a fit-for-work service (Great Britain: Department for Work and Pensions and Department of Health 2008). In Scotland this service is named *Working Health Services* and uses a case-managed, multidisciplinary approach to provide treatment, advice and guidance for people in the early stages of sickness absence. It is aimed at employees of small and medium-sized enterprises (SMEs) who would not otherwise have access to specialized Occupational Health Services (OHS).

Complementary to this, in 2010 the *Fit Note* (statement of fitness for work) replaced the old "sick note", which stated either "you should refrain from work" or "you need not refrain from work". In contrast, the Fit Note focuses on what an employee may be able to do at work rather than what they cannot do. One of the Fit Note's options is "may be fit for work taking account of the following advice" and the doctor is encouraged to suggest work-place adaptations.

More recently, in 2011 "*Health at Work—an independent review of sickness absence*" was published, prompting the development of *Fit for Work*, which is known in Scotland as "*Fit for Work Scotland*" (Black and Frost 2011) and was launched in December 2014. This program provides supportive work-focused assessment for patients who could be absent from work for 4 or more weeks. Early intervention is a key feature because it is widely acknowledged that the longer an employee is off work, the smaller their chances of ever returning to work. The program is free, may be accessed by employers, general practitioners (GPs), and employees and offers access to quality, independent advice and a specialist work-focused assessment; recommendations on work-place adjustments; and support for absence prevention. Additionally, employers may claim a tax exemption on costs of up to £500 (680 € a year per employee) to fund any recommended adjustments. The service is intended to complement rather than replace existing OHS provided by employers and is likely to be of particular benefit to smaller employers who do not have access to in-house OHS. Although there is clearly some overlap between *Fit for Work Scotland* and *Working Health Services*, there is currently no plan to merge them.

2 The Organization of OHS in Scotland

Most public sector organizations and many large private firms provide an in-house or external work-place OHS. However, fewer than one in seven SMEs provide OHS for their employees. OHS are predominantly offered by private providers, as is the case in other European countries such as Germany.

In Scotland the National Health Service (NHS) has an in-house OHS for its own employees. In addition, the NHS OHS offers services to external companies and delivers *Fit for Work Scotland* and *Working Health Services*. The following section describes these services for NHS Grampian, a rural region in northeastern Scotland that includes the Orkney and Shetland Islands.

3 Example of Provision of OHS: Grampian

Grampian is on the northeast coast of Scotland and has an average population of 530,000, approximately 10 % of the population of Scotland; half of this population (280,000) is of working age. It covers an area of about 110 × 150 km on the mainland plus the Shetland and Orkney Islands. In its capital, Aberdeen, which is regarded as the center of the European oil and gas industry, the unemployment rate is lower than the national average (National Statistics (NOMIS) 2015). Grampian is home to around 30,000 enterprises, 95 % of which are SMEs; the other 5 % comprises the NHS, higher education (two universities), councils, and employers in the oil and gas sector. About 151,000 (54 %) of those of working age are working in SMEs.

Table 1 gives an overview of the services provided by GO Health Services, the OHS provider of NHS Grampian, which serves a total workforce of 28,000 employees in the Grampian area, including 16,000 staff from the NHS.

It offers an exemplar model for OH provision in that it provides a comprehensive range of complementary services. We will describe the points named in bold font in more detail in Sects. 3.2 through 3.6.

3.1 The Multidisciplinary Team

GO Health Services is delivered by four whole-time equivalent (WTE) doctors, 16 WTE OH nurses, two WTE counsellors and 11 WTE administration/support staff. It is managed by a head of service who is supported by a clinical leader, nurse manager, business manager, and improvement and development manager. Members of this multidisciplinary team work closely and share learning to promote an optimal approach to case management. Continuous professional development is encouraged and staff members are funded to undertake specific training in OH. Members of the OH team work closely with employers, health and safety specialists, and partnership organizations to promote employee well-being.

3.2 Management of Referrals

Employees may be referred to GO Health Services by their managers or may self-refer. Requests for management referrals are assessed and prioritized daily to facilitate provision of the appropriate time, date and level of assessment.

In the financial year 2013/2014, GO Health Services received 3866 management referrals, predominantly from NHS Grampian. All referrals (both from management and self-referrals) are triaged by an OH nurse adviser who makes initial telephone contact and agrees on an action plan with the employee. The referral may be managed entirely by telephone; however, if appropriate, an appointment is made for a face-to-face nurse assessment. More complex cases may be forwarded for medical triage by an OH physician. Medical input in terms of triage and/or

Table 1 Overview of services provided by GO Health Services, the OHS provider of NHS Grampian

• Pre-employment health screening
• **Management referrals and self-referrals**
• Case management
• Advice on work-place adjustment programs
• Immunization and screening
• Risk assessments
• Health surveillance
• Counselling
• Pensions applications
• **Sickness absence management (iAM)**
• Travel clinic
• External contract work
• **Healthy Working Lives**
• **Working Health Services**
• **Fit for work Scotland**

assessment was required for 626 of all cases (17 %); 495 of these cases (13 %) subsequently require medical telephone or face-to-face assessments and possibly follow-up review appointments. Case conferences between OH, the employee, their manager, and Human Resources may be arranged to discuss the case, share understanding of work-place issues, and facilitate the development of an appropriate action plan.

GO Health Services provides a counselling service for employees with mental health conditions, mainly anxiety and depression. During 2013/2014, the internal counselling service received referrals for 220 such clients and provided them with an average of six counselling sessions each.

3.3 Attendance Management

The iAM (intelligent Attendance Management) service was developed by GO Health Services and provides employees with OH contact via phone on the first day of absence from work to offer advice and support as appropriate. A pilot study was undertaken to evaluate the impact of proactive OHS on sickness absence rates and clients' and managers' satisfaction with the service. This pilot study was well received and iAM has subsequently been rolled out to a number of different sectors within NHS Grampian with a focus on those with a higher than average short-term sickness absence rate.

In the financial year 2013/2014, GO Health Services received 1305 referrals from the participating sectors for iAM. The number of absences in the project group decreased by 19 % compared with a decrease of 3.7 % in the number of absences in the remainder of the organization. Similarly, hours lost decreased in this group by 18 % compared with only a 0.2 % reduction in the remainder of the organization over the same time period. A subgroup analysis comparing the project group with a similar control group that did not take part in iAM revealed that 8.7 % of individuals had absences in the project group compared with 24.6 % in the control group; 13.9 % of hours were lost in the project group and 26.5 % in the control group.

During the pilot phase, participants completed a satisfaction questionnaire and a EuroQol-5D (quality of life) questionnaire (Brooks et al. 2003) at 1 week, 3 months and 12 months to assess the acceptability of the program and collect feedback. An improvement in quality of life according to this measure was reported by 34 % of the clients and most viewed the intervention positively.

3.4 Healthy Working Lives

Healthy Working Lives, an initiative led by the Scottish Centre for Healthy Working Lives, aims to work with employers to enable them to understand, protect, and improve the health of their employees, thus promoting a healthier and more motivated workforce. Free services are provided through a network of advisers, a website (NHS/Health Scotland 2011) and a telephone advice line.

A work-place team in GO Health provides the Health Promotions component of Healthy Working Lives and is the main point of contact and advisory center for the Healthy Working Lives award scheme. The Healthy Working Lives award scheme includes a requirement for OH and Safety, rehabilitation, and environmental criteria to be met by clients. Support is provided by advisers with specialist expertise about OH and safety. Award support includes providing clients with training on criteria at kick-start events and assessing the OHS criteria element of client submissions for awards.

This service provides OH and safety advice to SMEs and assists client companies to implement and develop safety management systems. Following their initial visit, clients receive a bespoke report with recommendations and supporting documentation. Further support is provided via follow-up visits, telephone and email.

Advisers contribute to the development of Healthy Working Lives services by participating in national groups.

3.5 Working Health Services

Working Health Services uses a case-managed, multidisciplinary approach to provide treatment, advice and guidance for people whose health is affecting their ability to work. It is aimed at employees of SMEs who would not otherwise have access to specialized OHS. The service is free, having been funded by the Scottish Government since 2010. This service was initially aimed at SMEs with 250 or fewer employees and was for both those on sickness absence from work and those at work. However, since April 2015 funding has been reduced, prompting a restriction of eligibility to SMEs with 50 or fewer employees and only employees

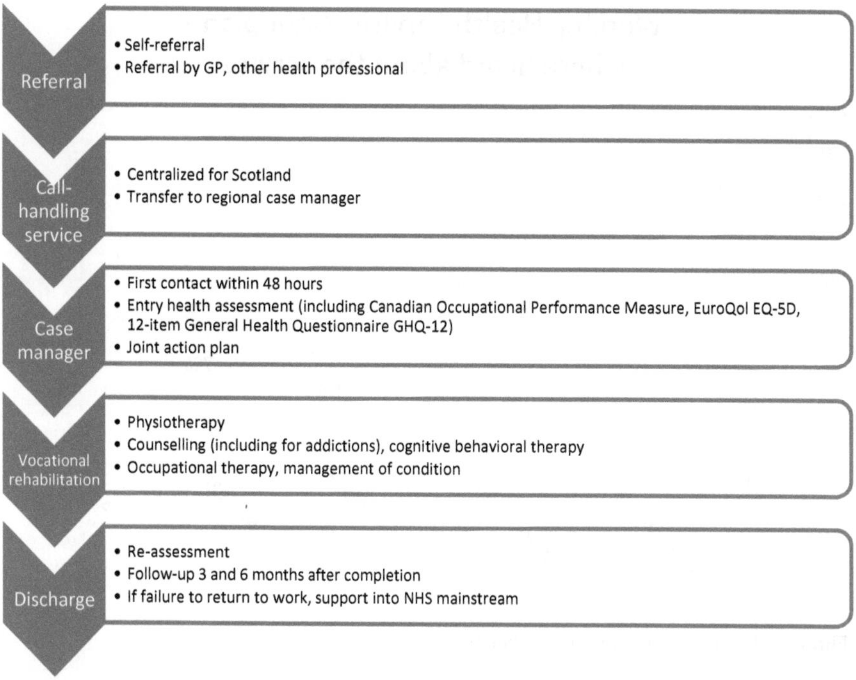

Fig. 1 Flow chart showing how the Working Health Services provide free OH support to SMEs

who are at work. The service may be accessed by employer referral, self-referral, or referral by the family doctor or another health professional. It offers a dedicated case manager, access to OH nurses and advisers, OH physicians and consultants, counsellors and physiotherapists, a wide variety of support groups, advice lines, community projects, social work services, and local charity groups. Because of the rural geography, physiotherapy and counselling are often contracted out to local providers rather than being provided in-house.

Figure 1 illustrates the Working Health Services referral process.

A central call-handling service refers clients to the relevant regional case manager (e.g., the case manager at GO Health Services). The regional case manager contacts the client within 48 h; then an initial assessment takes place via telephone and an action plan, which might involve referral to an OH physician or referral to other health services such as physiotherapy or counselling, is agreed upon. Follow-ups occur 3 and 6 months after review and discharge of the client from the service; if the health condition has persisted and prevented a return to the work-place, the client is supported into an appropriate mainstream treatment route for further treatment.

Over the financial year 2013/14, 140 clients contacted GO Health Services via the Working Health Services scheme, the main proportion referred by their GP or other health professionals (Fig. 2).

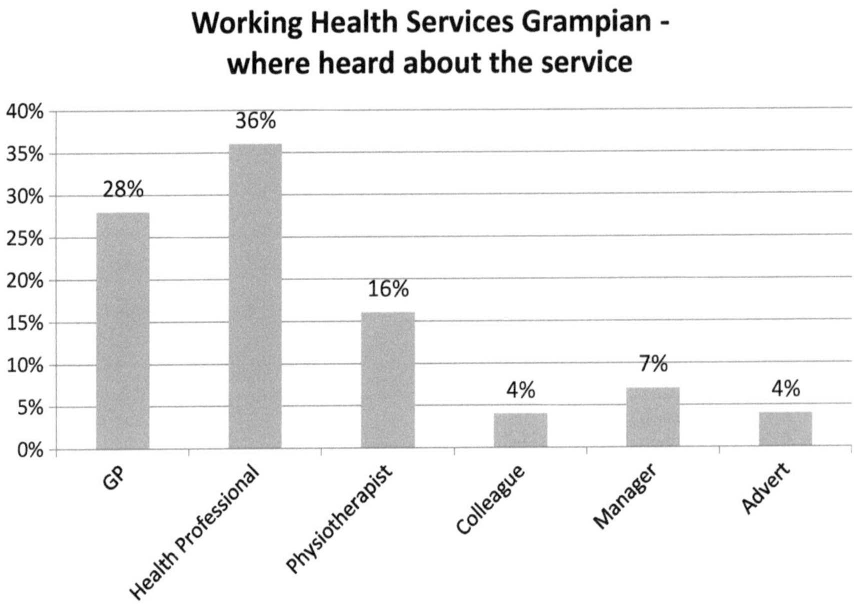

Fig. 2 Advertising and referral distribution

Of the clients who were referred to this service, 81 % reported musculoskeletal problems, 16 % had mental health problems, mainly anxiety and depression, and only 5 % had other health conditions (Fig. 3).

Forty-three percent of the clients were aged over 50 years; the remaining 57 % were distributed more or less evenly over the younger age groups (Fig. 4).

Because a significant proportion of them were directly from physiotherapy waiting lists, referrals may have been skewed towards musculoskeletal conditions.

All participants in the scheme are asked to complete an initial questionnaire that includes the EuroQol-5D quality of life questions, follow-up questionnaires being completed at discharge and again after 3 and 6 months. Scores for well-being increased from an average of 62 on a visual analog scale at the initial assessment to 77 at discharge with an overall very positive response to the service.

Despite advertising of the service by several means, only 0.01 % of those working in SMEs in Grampian have used or been referred to the Working Health Services annually, which is only about a quarter of the numbers expected. Is the explanation for this discrepancy that employees of SMEs are much healthier than those of large companies in the public and private sector, in which absence rates of 4 % and more are not uncommon, or is it attributable to a reluctance to use these services? Further health economic analysis into the costs of £420 per client for such OHS should give more insight into the efficacy and potential savings for an employer when sickness absence can be avoided and productivity increased.

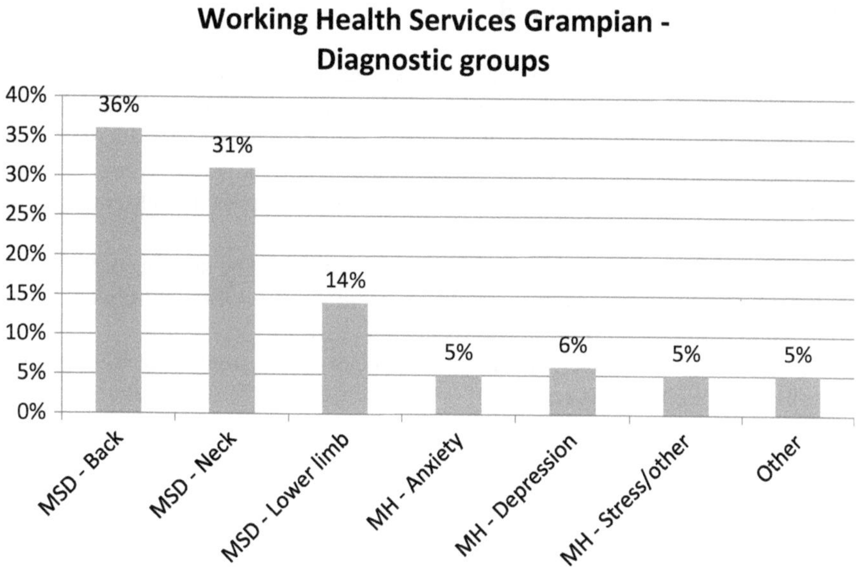

Fig. 3 Diagnoses of individuals referred to Working Health Services. *MH* mental health, *MSD* musculoskeletal disorder

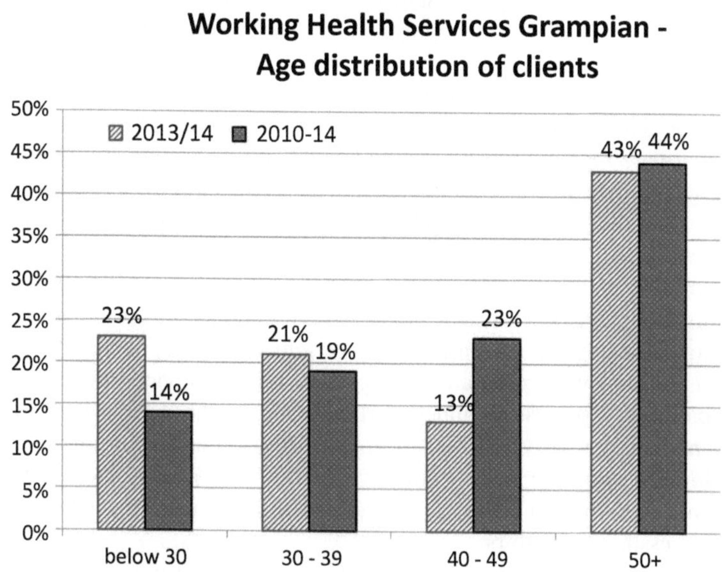

Fig. 4 Age (years) distribution of clients referred to Working Health Services from SMEs

3.6 Fit for Work Scotland

Fit for work Scotland (Department for Work and Pensions and Scottish Government 2015), a free advice and assessment service supporting employees, employers, and GPs to manage return to work, is still in its infancy; GO Health started to undertake this work in April 2015. Its stated aim is to "support a reduction in the length of sickness absence from work and reduce the impact that absence has on individuals, employers and the State".

To be eligible for an assessment, an individual must: (1) be in paid employment; (2) have been absent, or likely to be absent, from work for 4 or more weeks; and (3) agree to being referred for an assessment.

Individuals are ineligible if they: (1) have been absent from work for less than 4 weeks; (2) are either self-employed or unemployed; or (3) are re-referred within 12 months of a previous referral.

4 Comparison of OHS in Scotland and Germany

Here follows a comparison of provision of OHS in Scotland and Germany, where OHS is split into different elements; accordingly, in the latter country, employers need to contact different organizations depending on their needs.

In Germany, if an employee is signed off work (a medical sick note is required after the third day off work), an employer can request information about their fitness to work via the employee's health insurance company, which then arranges a medical assessment by (one department of) the medical advisory board of the statutory (or private) health insurance (Medizinische Dienst der Krankenkassen—MDK), these are regionally organized in every state. The MDK only provides the requesting employer with a recommendation about the employee's fitness to work, and if or when that employee might return to work. They do not offer advice on return-to-work programs.

OHS or OH medical departments in Germany offer the following services: pre-employment assessments, advising and setting up health surveillance, dealing with the identification, management, and prevention of occupational diseases, and advice on return-to-work programs. Some of the services concerning identification of occupational diseases go much further than in the UK, where specialist assessments are normally referred to appropriate specialist consultants (e.g., to dermatologists, respiratory physicians, orthopedists, etc.) and some OH medicine departments do their own allergy testing (patch and skin prick testing), pulmonary function assessment and so on.

According to statute §84 Abs. 2 SBG IX, in Germany employers are legally obliged to offer "occupational integration management" ("betriebliches Eingliederungsmanagement") if an employee has been off work continuously or intermittently for more than 6 week in any 1 year. In larger organizations this service may be provided at least partially in-house; however, numerous charitable and commercial organizations offer such services together with individual coaching of the employee. These services are implemented specifically when dealing with employees with chronic diseases, cancer, mental health problems, or depression.

The fourth type of institution that employers may want to contact are those for statutory accident insurance for the industrial and public sectors (the 'Berufsgenossenschaften' (BGs) and the public-sector accident insurers respectively)They provide resources for the prevention of work accidents, occupational diseases, and work-related health risks, and to ensure effective first aid. They are also responsible for rehabilitation through comprehensive medical care of an employee after a work-related injury, and social integration into society which might include re-training of an employee when redeployment is necessitated by work-related ill health and disability (vocational rehabilitation), or compensation including injury benefits and pension in case of ill-health retirement due to work-related health effects. In some ways the BGen also has a similar function to the Health and Safety Executive in the UK, inspectors from which visit employers to check health and safety provisions for prevention of accidents in the work-place.

The main distinction between these two systems is that employers in the UK, and specifically in Scotland, can deal with a single provider for all work-related health issues and do not have to navigate through different providers with overlapping duties. The second distinction is that in Scotland this service is predominantly run by OH nurses rather than occupational physicians (in a ratio of 80/20). This allows for more cost-efficient provision of OHS, making it more affordable, specifically for SMEs.

Nurses in Scotland who wish to specialize in OH take a postgraduate degree in OH Practice. This Bachelor of Science course is studied over 2 years part-time and is accredited by the Nursing and Midwifery Council, the nursing regulatory body in the UK. Students who take this route have access to a practice teacher and complete a range of modules, including managing risks in the work-place, attendance management, leadership in public health, wellness and work, and key concepts in decision making in public health; they also complete a research-based project and undertake a practice placement (Royal College of Nursing 2011).

References

Beatty C, Duncan K, Fothergill S et al (2013) The role of health interventions in reducing incapacity claimant numbers. Sheffield Hallam University, Sheffield

Black C (2008) Working for a healthier tomorrow. The Stationery Office, London

Black C, Frost D (2011) Health at work—an independent review of sickness absence. The Stationery Office, London

Brooks R, Rabin R, de Charro F (eds) (2003) The measurement and valuation of health status EQ-5D: a European perspective. Springer, Dordrecht

Department for Work and Pensions and Scottish Government (2015) Fit for work Scotland. Available from http://www.fitforworkscotland.scot/. Accessed 28 Apr 2015

Deutsche Rentenversicherung Bund (DRV) (2014) Rentenversicherung in Zeitreihen. DRV-Schriften Band 22. Deutsche Rentenversicherung Bund, Berlin. Available from http://www.deutsche-rentenversicherung.de/Allgemein/de/Inhalt/6_Wir_ueber_uns/03_fakten_und_ zahlen/03_statistiken/02_statistikpublikationen/03_rv_in_zeitreihen.pdf?__blob=publicationFile&v=15. Accessed 29 Apr 2015 (in German)

Great Britain: Department for Work and Pensions, Department of Health (2008) Improving health and work: changing lives. The Stationery Office, London

National Statistics (NOMIS) (2015) Aberdeen City and Shire key statistics—unemployment rates ranked by local authority area June 2015, updated July 2015. Available from http://www.acsef.co.uk/uploads/fileUploads/ACSEF%20Unemployment%20-%20Rates%20Ranked%20by%20Local%20Authority%20Area%202015%20%28Jul%2015%29.pdf. Accessed 15 Sept 2015

NHS/Health Scotland (2011) Healthy working lives. Available from http://www.healthyworkinglives.com/. Accessed 18 May 2015

Parliament of the United Kingdom (2007) Welfare Reform Act 2007. The Stationery Office, London

Royal College of Nursing (2011) Occupational health nursing: career and competence development—RCN guideline. Available from https://www.rcn.org.uk/__data/assets/pdf_file/0010/409438/004123.pdf. Accessed 5 Apr 2015

Waddel G, Burton AK (2006) Is work good for your health and well-being? The Stationery Office, London

New Ways of Providing Occupational Health Management via a Network for Small- and Medium-Sized Enterprises

Sabine Nitsche

Abstract

This chapter provides an insight into the research project InnoGema that has been undertaken at the University of Applied Science Berlin (Hochschule für Technik und Wirtschaft Berlin) since 2007. The overall goal of InnoGema is to develop a specific support structure for small and medium-sized enterprises (SME) with the aim of facilitating the implementation of sustainable occupational health management. The main concept of InnoGema and its network approach are outlined, as is the offline and online support being provided to SME by InnoGema's supraregional service center. In addition, information about the expansion of the InnoGema network from Berlin to Brandenburg is presented.

1 New Ways of Providing Occupational Health Management via a Network for Small- and Medium-Sized Enterprises

Because of demographic changes in Germany and the resulting shortage of skilled workers, the need for occupational health management (OHM) has become increasingly important in recent years. Companies need to use their human resources more efficiently (see Bundesagentur für Arbeit 2013; ISO 2009); in particular, strategic human resource management needs to place more focus on the ability of existing workers, especially older ones, to work (Ilmarinen 2011). In addition, companies can attract talented workers by offering a good OHM system to their employees, thus presenting themselves as good employers (Nitsche 2012). Numerous measures for preventing employees' ill-health and maintenance or restoration of their health

S. Nitsche (✉)
University of Applied Science Berlin (Hochschule für Technik und Wirtschaft Berlin), Ostendstr. 1, 12459 Berlin, Germany
e-mail: nitsche@htw-berlin.de

M. Wiencke et al. (eds.), *Healthy at Work*, DOI 10.1007/978-3-319-32331-2_14

(Badura et al. 2011) have been developed in recent decades and there are numerous books and articles (Bamberg et al. 2011; Uhle and Treier 2013; Ulrich and Wülser 2012 etc.) that provide a huge amount of information about processes, tools, and techniques in OHM. According to an online survey of the Association of German Chambers of Commerce and Industry (Deutscher Industrie- und Handelskammertag, DIHK) in 2013, almost two-thirds (65 %) of 1500 companies in Germany have an increased commitment to promoting occupational health. One-third of the surveyed companies planned to increase their OHM-related activities in terms of providing healthy work environments (70 %), vaccination and preventive medical check-ups (45 %), physical activities (43 %), and healthy canteen food or nutritional advice (24 %) over the coming 5 years (DIHK 2014). Overall, the need for OHM has increased and companies who want to be competitive will have to be active in this field.

As research shows, small and medium-sized enterprises (SME) experience significant constraints in implementing OHM systems and there are few structures for supporting them. Because classical OHM instruments cannot be transferred without major adjustments from large- to small-scale enterprises, there is a particular need for concepts that are specifically tailor-made for implementing OHM procedures by SME. Compared with large-scale enterprises, SME mainly do not have the structural, financial, or personal resources to implement complex OHM systems. In particular, their limited resources must be taken into account and reflected in OHM systems for SME.

Against this background, an OHM network for SME, InnoGema, was developed at the University of Applied Science in Berlin (Hochschule für Technik und Wirtschaft Berlin (HTW Berlin)) in association with the health insurance company Kaufmännische Krankenkasse (KKH) and the statutory accident insurance company Gesetzliche Unfallversicherung (VBG). The network has been jointly developed with SME (Simon und Heger 2009) in Berlin[1] since 2007 and in Brandenburg[2] since 2011 with the aim of addressing their particular needs. In this context OHM activities were identified as more attractive amongst employees when (1) they are close to the work-place (nearby and easy to reach); (2) they can be used flexibly (at various times of the day, flexible booking, user card); (3) colleagues can join (motivation); (4) there is choice between various courses and providers; (5) information about providers and the quality of their work is transparent (including evaluation); and (6) when the company and the health insurance are responsible for some of the costs.

[1] This project was funded by the Bundesministerium für Bildung und Forschung (Federal Ministry of Education and Research) from 2007 to 2010 (FKZ: 01FM0709).

[2] This project was funded by the Ministerium für Arbeit, Soziales, Gesundheit, Frauen und Familie (Ministry of Labor, Social Affairs, Health, Women and Family) with grants from the Land Brandenburg and European Social Fund from 2011 to 2014.

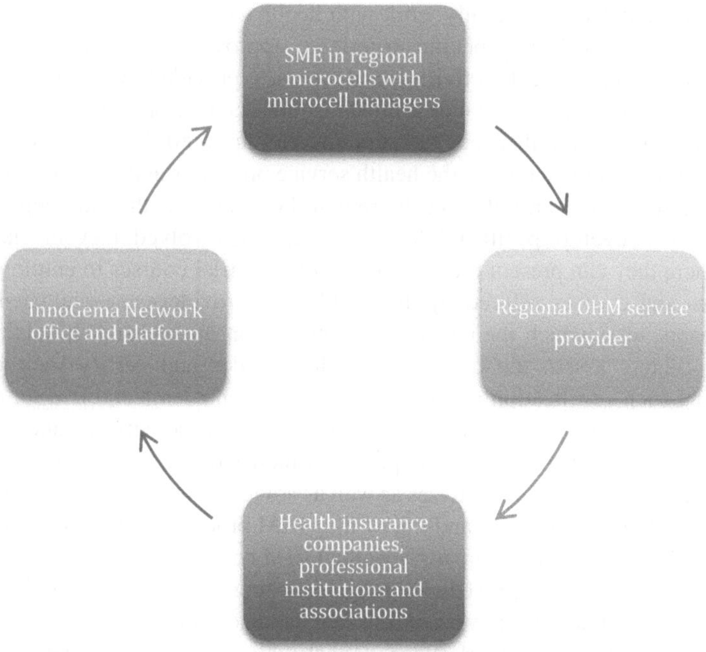

Fig. 1 The regional network approach

These findings led to the development of a network structure with regional microcells that is supported by both online and offline activities; these will now be explained in more detail.

InnoGema stands for "Innovatives Gesundheitsmanagement", meaning innovative health management. InnoGema is a network of SME, health service providers, health insurance companies, and other professional associations and institutions in the field of occupational health as well as a supraregional service center at HTW Berlin with the portal www.innogema.de. InnoGema brings SME together in regional "microcells" that connect them and are supported by local service units, so-called microcell managers (see Fig. 1).

The main idea of the regional network approach is to bring SME in one region together to enable exchange between them on all topics related to OHM. Various activities such as health days, business breakfasts/lunches, and after-work discussions around the theme of OHM are offered within these microcells. The network structure enables the establishment of groups of economic size and provision of courses close to work-places. Rooms and equipment can be shared amongst the SME in a microcell, allowing them to offer a wider selection of activities and OHM providers to their employees as well as tailor-made activities as needed. The cooperation with other companies also facilitates better cooperation with health insurance companies and other professional associations in the field of occupational health.

Because InnoGema will only work with health service providers of guaranteed quality, it has developed specific criteria to approve health service providers' qualifications and accept activities and courses provided by them. Activities approved by the health insurance companies are denoted on the website with a § 20 emblem, indicating that they are supported by up to 80 % by those companies. Furthermore, InnoGema trains the health service providers and encourages them to cooperate with other providers in the region. For example, they are supported by InnoGema to develop specific OHM activities and be involved in events like health days where they can present their services and offer trial courses to employees. All activities and courses offered by the health care providers are evaluated by the participants, after which their feedback is placed on the website. In addition, the service centers assess the quality of each provider and certify their courses, indicating such certification on the portal with the InnoGema emblem for quality.

InnoGema offers professional advice to all SME in the network and steers the process by which each company implements their specific OHM. In this context, companies receive detailed analyses as well as advice on issues like communication within the organization and handling of the OHM process by the organization. To facilitate development of company-specific OHM concepts, InnoGema offers various analyses such as "Arbeitsbewältigungs-coaching" (AB-Coaching), Work Ability Index, work situation analysis, and strategy and goal-finding workshops. Further analyses such as work-place analysis, ergonomic advice, and "Gesundheitsbarometer" (a health barometer) are offered by health insurance companies and the professional bodies via the network. SME are supported by InnoGema in the implementation of their OHM by the process described below, which is summarized in a specific guideline for SME (Simon et al. 2011). This process comprises (1) interviews with key persons in the organization like the CEO, human resource manager and heads of units; (2) an employee survey and AB-Coaching or work situation analysis depending on the size of the organization; (3) workshops aimed at identifying the specific needs of the organization and the individual employees; (4) organization of specific health days and tailor-made OHM activities; and (5) quality control and evaluation of all OHM activities and adaptation if needed.

Individual service packages are being developed on this basis. The service center provides a wide range of information about OHM and supports the network by providing services from a single source. It offers a pool of regional health service providers and a wide range of courses on topics like physical activities, nutrition, prevention of stress, drug dependence, leadership, and organization. InnoGema provides contacts with health insurance companies and health service providers. The advantage for the SME is that they save time and resources by having a single central point of contact that takes care of organizational handling and finding information.

The InnoGema service is supported online by the portal www.innogema.de (Fig. 2), which offers companies access to a wide range of information about OHM.

Through it, they can find regional health service providers as well as other companies in their region with whom they can build partnerships to accomplish

Fig. 2 www.innogema.de home page

certain OHM activities. They can also find courses and dates of health management activities, directly book courses, and generate vouchers for their employees that they can redeem for OHM activities. The aim of the network providing the service center and the various functions of the website is to minimize the effort required by the involved companies.

2 Network Expansion

From 2012 to 2014, InnoGema was further developed and transferred to Brandenburg, where three microcells were established. In addition, the portal www.innogema.de was renewed and enriched with web2.0 elements.

Three specific regions for developing new microcells within Brandenburg were identified in cooperation with the partners KKH and VBG. Ninety-seven potential companies in central Potsdam were contacted via letter or email about expanding the network into their region and an associated offer of support for implementing OHM systems in their companies; 98 companies in the "Potsdamer Centrum für Technologie" (PCT [Potsdam Center for Technology]) were similarly visited and informed about the project, as were 91 companies in Teltow. Eventually, three microcells comprising nine companies and 150 employees overall were developed in Brandenburg: two in Potsdam and one in Teltow. A manager for each microcell was found and trained to serve as a local point of contact for providing regional support to the companies.

In addition, cooperation contracts with 30 regional health service providers offering OHM activities in the fields of physical activities, nutrition, stress and drug dependence prevention, leadership, and organization were signed and added to the network.

All three microcells were represented on the website, thus linking the companies, regional health care providers, local network manager, health insurance companies, other professional associations and institutions, and the supraregional service center InnoGema. All have individual profiles on the website (Fig. 3).

The process of developing the new microcells was diverse. It was very difficult to determine who was in charge of OHM in most of the companies; however, in most companies the CEOs were found to be responsible for this issue because SME are small and have flat hierarchies and limited personnel resources. The initial

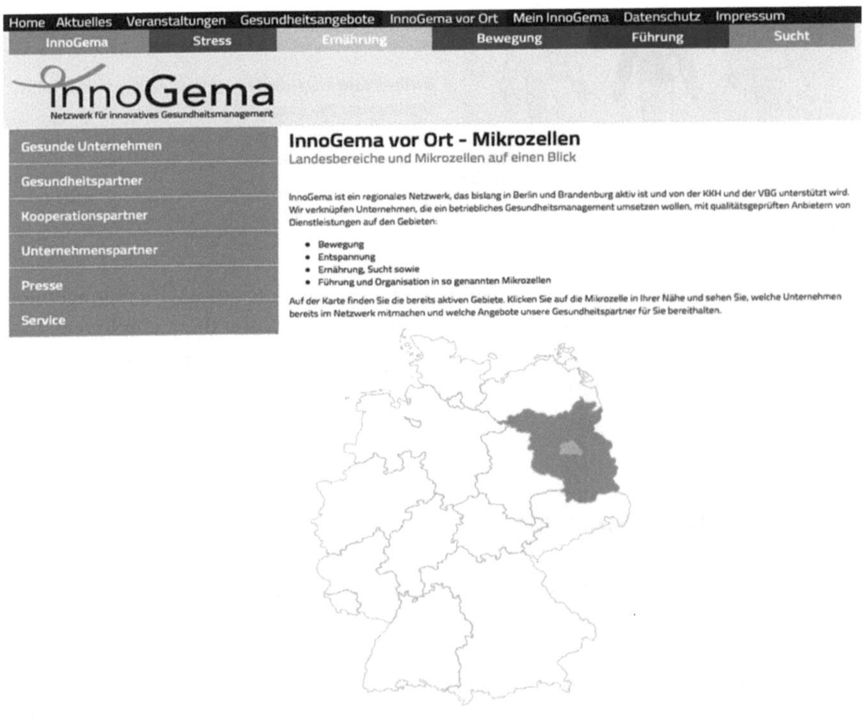

Fig. 3 www.innogema.de regional microcells

letters and emails were followed up by telephone conversations and ideally by direct contacts on site. The large workload and limited time resources within the SME made it even more difficult to make appointments for personal visits. It is particularly important for the person in charge to see the need for OHM and understand the benefits for the company. Although this topic was already on the agendas of some SME, they nevertheless expressed concerns; namely, that the process might move beyond their control and create expectations amongst the employees they could not fulfill. Furthermore, they anticipated that the process and organizational handling of OHM would require additional resources that were not available within the companies. In this context, the flexibility and holistic approach of InnoGema were seen as advantages by the companies who finally joined the network.

In addition, the parallel acquisition of the companies and regional health services slowed the development of the new microcells. To acquire the companies, it was important to have already established a good network of regional health service providers and ideally to have their profiles, courses, and activities on the website. Conversely, to acquire the regional health service providers, it was important to have already established a good network of companies and hence potential clients to whom they could offer their activities and courses. In the future, one solution to this problem may be to open the network to large-scale companies and thus more quickly reach a critical mass of employees and activities. Another solution may develop when InnoGema eventually becomes popular.

In 2013 and 2014, the specific circumstances and needs of nine SME in Brandenburg were identified and individual action plans concerning their OHM developed. To this end, AB-Coachings, including the Work Ability Index (BAuA 2011; Gruber and Frevel 2009; Tempel and Ilmarinen 2012), were conducted with a total of 82 employees in four companies, the "Gesundheitsbarometer", a specific online-analysis tool offered by KKH, was used with 76 employees of three companies, the work situation was analyzed in four companies, and several workshops were conducted with management and employees with the aim of identifying and developing specific OHM activities for the involved companies. The purpose of the various analyses was to identify factors influencing the employees' health and ability to work such as coping strategies, social support from colleagues, leadership, work organization, working environment and conditions, compatibility of work and family life, information and organizational support, and the development of work-related competencies.

By the end of the analysis phase, more than 150 employees had participated in the analysis process and nine company-specific guidelines had been developed and transferred into practice with the support of the InnoGema network and platform. The employees then began to use the platform: in the first quarter of 2014, 53 employees (29 male, 24 female) booked various activities via the platform (Fig. 4).

After the analysis phase, the microcells became livelier. Interestingly, employees from five SME that were not part of the network but were located close to the microcells started to book courses via the platform, demonstrating

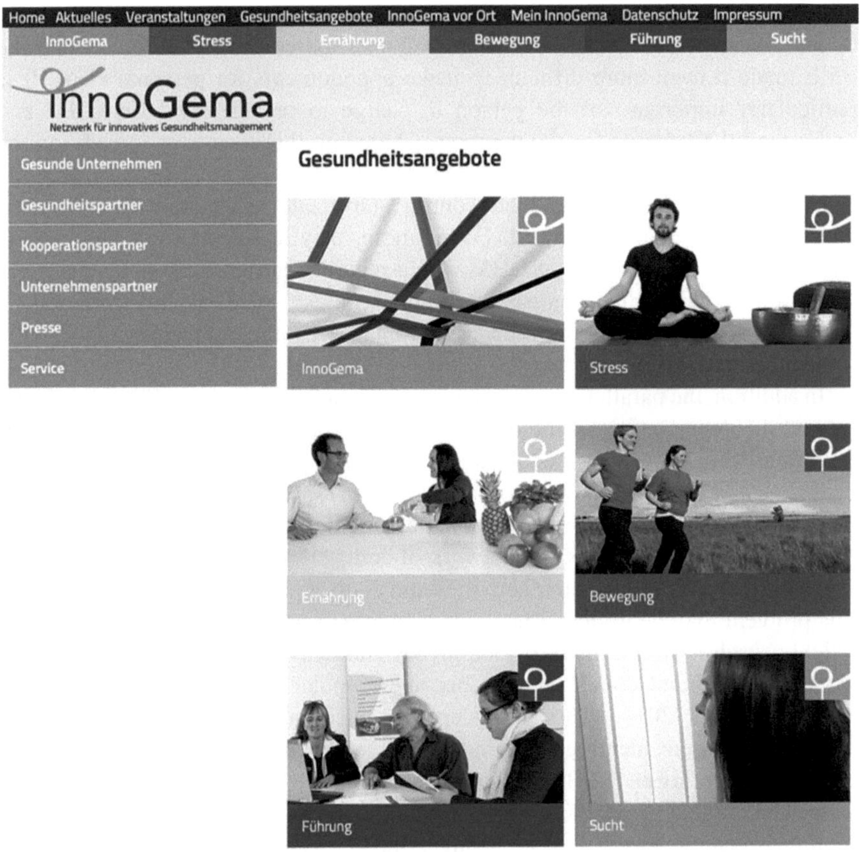

Fig. 4 www.innogema.de courses and activities

that the regional network approach works for SME and that microcells can start to grow on their own. This indicates that companies whose management has thus far not joined the network may do so eventually via participation of their employees in InnoGema activities.

Overall, InnoGema offered a variety of activities to the companies and their employees during the course of the project. The microcell managers worked regular office hours; business breakfast, lunch, or after-work meetings on selected topics like motivation of employees for OHM, addiction, healthy leadership and men's health were regularly offered; and health days and in-house workshops were conducted on demand by InnoGema.

In addition to a wide range of offline activities, online activities were offered by InnoGema via the portal www.innogema.de, which was continuously optimized and enriched with short articles, information, short film clips, and expert chats during the course of the project. Live chats with experts on various topics regarding OHM were offered and employees and health service providers could use the

InnoGema forum and contribute to discussions. A web2.0 concept was written and specific social media like Twitter and Facebook were selected and put into practice. Regular newsletters were sent to over 200 subscribers and could also be downloaded via the platform. By May 2014, InnoGema had 56 followers on Twitter and 17 Facebook Likes. In general, a continuous increase in the use of these applications was evident; they were mainly used to share information. A proposal to tape functions such as business lunches and show the resultant videos via the portal or similarly display material and films (e.g. provided by our cooperation partners) was not achieved because of technical and legal issues. Furthermore, employees had limited access to the internet, Facebook and so on at their workplaces; in some companies employees did not even have computers or were not allowed to use the internet at work. Nevertheless, the portal had a progressively increasing number of users; from 746 per month in January 2014 to 975 in April 2014.

The transfer of InnoGema to Brandenburg and the microcell approach can be evaluated as successful overall because both the organizations involved in the project and employees from other organizations located close to the microcells are increasingly taking up InnoGema's offers. Feedback from the SME who have participated in the project has been positive; in particular, the intensive and holistic offline and online support provided by InnoGema is valued. It was found that the analysis and implementation process take time. The limited time available to SME must be taken into account and this factor is largely responsible for the high valuing of the flexible online and offline support offered by InnoGema.

The overall goal of InnoGema is to develop a specific support structure for SME that facilitates implementation of their OHM and is sustainable. The network structure facilitates cooperation between SME in a region, which can substitute, for example, for structural disadvantages caused by a lack of the resources needed for implementing OHM. The connection of SME with local service providers and other organizations offering support in OHM allows companies to contact a pool of various qualified service providers and directly book required courses via the portal, thus reducing the resources each enterprise has to provide. Detailed information about each service provider can be found on the system, together with the results of evaluation and quality control conducted by the InnoGema team. Experience with the network has shown that it compensates for the limited resources of SME and can support the organizational handling and implementation of OHM systems within SME.

References

Badura B, Ducki A, Schröder H, Klose J, Macco K (eds) (2011) Fehlzeitenreport. Führung und Gesundheit. Springer, Berlin (in German)

Bamberg E, Ducki A, Metz A-M (2011) Gesundheitsförderung und Gesundheitsmanagement in der Arbeitswelt—ein Handbuch. Hogrefe, Göttingen (in German)

Bundesagentur für Arbeit (2013) Der Arbeitsmarkt in Deutschland. Fachkräfteengpassanalyse, Nürnberg (in German)

Bundesanstalt für Arbeitsschutz und Arbeitsmedizin (BAuA) (2011) Why WAI—Der Work Ability Index im Einsatz für Arbeitsfähigkeit und Prävention. Erfahrungsberichte aus der Praxis. Available from: http://www.baua.de/de/Publikationen/Broschueren/A51.pdf?__blob=publicationFile. Accessed 5 Feb 2014 (in German)

DIHK (2014) An Apple a Day... Gesundheitsförderung im Betrieb kommt an, DIHK-Unternehmensbarometer zur Gesundheitsvorsorge. Available from: http://www.dihk.de/presse/meldungen/2014-01-08-unternehmensbarometergesundheitsfoerderung. Accessed 28 Jan 2014 (in German)

Gruber B, Frevel A (2009) Arbeitsbewältigungscoaching. Neue Herausforderungen erfordern neue Beratungswerkzeuge. In: INQA-Bericht, vol 38 Available from: http://www.inqa.de/SharedDocs/PDFs/DE/Publikationen/inqa-38-arbeitsbewaeltigungscoaching.pdf?__blob=publicationFile. Accessed 5 Feb 2014 (in German)

Ilmarinen J (2011) Das Haus der Arbeitsfähigkeit—der finnische Blick auf den demografischen Wandel. Available from: http://www.inqa.de/SharedDocs/PDFs/DE/Gute-Praxis/Vortrag-von-Juhani-Ilmarinen-zum-Haus-der-Arbeitsfaehigkeit.pdf?__blob=publicationFile. Accessed 5 Feb 2014 (in German)

InnoGema. Netzwerk für innovatives Gesundheitsmanagement. Available at: http://www.innogema.de/

ISO (2009) Pflege 2030: Chancen und Herausforderungen! Dokumentation der gemeinsamen Fachtagung am 1. Juli 2009 in Berlin (Documentation of GE joint conference on 1 July 2009 in Berlin). Institut für Sozialforschung und Sozialwirtschaft, Saarbrücken (in German)

Nitsche S (2012) Gesundheitsmanagement als kompetitiver Arbeitgebervorteil beim Employer Branding. In: Praeview—Zeitschrift für innovative Arbeitsgestaltung und Prävention, pp 22–24 (in German)

Simon D, Heger G (2009) Kundenintegration bei der Entwicklung innovativer Dienstleistungen zur betrieblichen Gesundheitsprävention. In: Henning K, Leisten I, Hees F (eds) Innovationsfähigkeit stärken—Wettbewerbsfähigkeit erhalten. Präventiver Arbeits- und Gesundheitsschutz als Treiber. Wissenschaftsverlag Mainz, Aachen, p 362 f. (in German)

Simon D, Heger G, Reszies S (eds) (2011) Praxishandbuch Betriebliche Gesundheitsförderung. Ein Leitfaden für kleine und mittlere Unternehmen. Kohlhammer, Stuttgart (in German)

Tempel J, Ilmarinen J (eds) (2012) Arbeitsleben 2025. Das Haus der Arbeitsfähigkeit im Unternehmen bauen. VSA, Hamburg (in German)

Uhle T, Treier M (2013) Betriebliches Gesundheitsmanagement—Gesundheitsförderung in der Arbeitswelt—Mitarbeiter einbinden, Prozesse gestalten, Erfolge messen. Springer, Berlin (in German)

Ulrich E, Wülser M (2012) Gesundheitsmanagement in Unternehmen—Arbeitspsychologische Perspektiven. Springer, Gabler (in German)

"More than the Sum of Its Parts": Demographic Change and the Future of Outpatient Nursing and Long-Term Care: *Zukunft:Pflege* (Future:Care) as an Example of Innovation Through Organizational Learning and Regional Network Development

Wolfgang Ritter, Isabella Schimitzek, and Sina Lürssen

Abstract

The collaborative project *Zukunft:Pflege* seeks to promote long-term employability through the development of skills and competences, enhancing motivation and preserving the health of outpatient care workers. To this end, work process-related and organizational learning strategies for small- and medium-sized enterprises were drawn up and systematically linked together and enhanced by developing and mobilizing resources at a network level. This chapter deals with the coordination aspect of the *Zukunft:Pflege* network. This network comprises outpatient nursing, long-term care services, and so-called value partners such as continuing education providers and outpatient care associations in the Bremen region. In general, inter-organizational networks tend to waver between cooperation and competition. The primary objective of cooperation can be attained provided the coordinators succeed in promoting the development of human and relational potentials and ensuring relationships of trust. To this end, the network agency places emphasis on interactive learning processes among the network partners. Building on experiences from such cooperation, knowledge transfer via the network can be encouraged, potentials and expectations promoted, and the basis for cooperation further strengthened. The main functional focus of the network agency lies in coordination of the various

W. Ritter (✉) • S. Lürssen
Zentrum für Sozialpolitik, Abteilung Gesundheitsökonomie, Gesundheitspolitik und Versorgungsforschung, Mary-Somerville-Straße 5, 28359 Bremen, Germany
e-mail: writter@uni-bremen.de; s.luerssen@uni-bremen.de

I. Schimitzek
APOLLON Hochschule der Gesundheitswirtschaft, Universitätsallee 18, 28359 Bremen, Germany
e-mail: isabella.schimitzek@apollon-hochschule.de

M. Wiencke et al. (eds.), *Healthy at Work*, DOI 10.1007/978-3-319-32331-2_15

interests of the network members, identification of synergy potentials and thence development of possible courses of action for promoting long-term employability in outpatient nursing and long-term care work.

1 Background Information

In times of tumultuous markets and socio-political change, particularly when demographic changes make issues concerning ambulant care more visible, many organizations seek collective strategies to acquire and develop new knowledge with the aim of improving their ability to react adequately to requirements through comprehensive decision-making. Demographic changes are especially visible in ambulant care because of the ageing population (= more persons in need of care) and low birth rates (= less young and specialized health professionals). In general, including for organizations that provide ambulant care, it can be assumed that more members of an organization contribute their individual knowledge to a knowledge pool, the larger is the knowledge output of the organization. This helps to ensure the economic survival of the organization. It is postulated that group strategies lead to a broader dissemination of knowledge and control within an organization, thus better ensuring its survival. "Learning organization" is considered an efficient strategy in knowledge-based economic environments because it can adapt more quickly to changing circumstances (Argyris and Schön 1999; Willke 1998).

Against this background, interactive cooperation and exchange of knowledge between businesses is of increasing importance, particularly in long-term care. Within the next decade, more and more people will potentially need care because of the increasing age of the population. This increase in demand for care will have to be met by a skilled workforce that is itself ageing. As a consequence, small and medium-sized companies active in nursing and long-term care are facing pressing challenges. These challenges include enhancing the attractiveness of the nursing occupation, reducing high physical and mental work pressures, securing a highly qualified workforce, and reducing the high early exit rates. In particular, in long-term care there are still few satisfactory strategies for overcoming the growing problem of staff shortages.

This is where the strategy of the network *Zukunft:Pflege* comes into play. A regional network has been set up by the Centre for Social Policy Research (Zentrum für Sozialpolitik[ZeS]) at the University of Bremen to support small- and medium-sized long-term and nursing care companies. To this end, organizational learning developed by ambulatory care companies is systematically linked and enhanced by the development and mobilization of resources at a network level (Becke and Bleses 2013). The aim of the *Zukunft:Pflege* project is to promote long-term employability in outpatient nursing and long-term care work based on competence and skill development, enhancement of motivation and preservation of health. At the same time, the network is intended to serve as a learning and communication platform as well as a "market" for the exchange of knowledge.

Regional networks can be understood as a specific type of purposeful cooperation or exchange between companies, expert organizations and trade associations or lobby groups who pursue common aims and objectives (Weyer 2000, p. 16). In these networks, organizations join forces and, for example, generate innovative knowledge or common strategies they would not have managed on their own. The effectiveness of regional networks results specifically from the fact that they provide especially favorable opportunities for mutual learning and innovative processes (Krugman 1991).

Studies (e.g., INQA 2011) on network research have shown that regional networks seem suitable for dealing with issues such as staff orientation that have hitherto received only marginal attention through conventional channels of communication. However, these studies have also reported some negative aspects of regional network cooperation, such as competition, conflicts of interest and distrust between network participants, and insufficient financial and personnel resources, which is particularly problematic in terms of sustaining such networks (e.g., for regular network coordination). Why and to what extent small and medium-sized companies take part in such regional networks remains largely unexplored. The published studies also lack concrete evidence concerning the creation and consolidation of such networks.

Because of their more or less loose form of organization, networks provide a flexible basis for describing hybrid forms of business cooperation as learning, changing and adapting organizations in which the joint activities of the cooperating businesses give them a competitive advantage over individual businesses. Cooperation within such networks should be coordinated to facilitate generation of synergy potentials.

Section 2 of this chapter deals with the theoretical foundations of these networks in terms of the essential elements required to foster innovation. These elements comprise the development of trust, human and relational resources, and the development of learning from the individual to network level. In Sect. 3, the coordination strategy of the *Zukunft:Pflege* network is described, with a focus on certain characteristics of company cooperation and the development of a learning strategy. Section 4 of this chapter summarizes the most important results in terms of cooperation within the *Zukunft:Pflege* network and indicates the extent to which the exchange of knowledge can be regulated in terms of a coordination strategy. Thus, our main questions are as follows: (1) What influences cooperation in the *Zukunft:Pflege* network? (2) On what does the exchange of knowledge and information between the project partners in the network depend? (3) How can the exchange be coordinated?

2 Theoretical Foundations

2.1 The Role of Trust and as well as Human and Relational Potential Within a Cooperation Network

The actions and activities of two or more stakeholders or organizations reflect their behavior patterns and hence their manner of cooperation. Good cooperation, compared with the solo activities of an individual business or organization, is characterized by the following: (1) objectives are more readily achieved; (2) experienced partners support the innovative process; (3) competences and resources that are needed but not available in every organization are easily accessible to all participating organizations, thus saving time and money; (4) ideas and suggestions are shared between participating organizations; and (5) outcomes that would be impossible for a solo organization are achieved (Becker et al. 2007, p. 5; Schubert 2008, p. 14).

What makes rival companies cooperate? As early as the late 1970s, economists like Kenneth Arrow pointed out that trust facilitates cooperation (Arrow 1974), in particular when it comes to sharing and revealing sensitive internal knowledge and information. Trust is based on common experiences and actions as well as mutual understanding. The more the network participants disclose and share knowledge and information, the stronger mutual trust becomes. Trust arises through close collaboration, openness and communication on a regular basis and changes as a relationship develops. In the course of cooperating, the partners come to identify with each other's values, goals and needs. Development of trust-based cooperation seems to be a fundamental requirement for successful cooperation at a network level.

Stirzel and Armbrüster (2012, p. 61) described human and relational potential as the basis for successful innovation in a network. Human potential comprises individual resources such as skills, know-how, experience, and innovative ability. Relational potential is characterized by the ability and motivation to exchange knowledge and information with other network partners, for instance by openly disclosing one's company's internal knowledge about certain processes or structures. It includes disclosing untapped tacit knowledge that has been gained through practice and experience(Nonaka 1994) and can be used and combined with other information to form a basis for generating new concepts. This implied knowledge is both a, not easy to pass on and not easy to take in (Simonin 1999). These transactions involve social and psychological processes such as intuition, interpretation, integration, and institutionalization (Crossan et al. 1999). Such processes should "flow" through the organization (Bapuji and Crossan 2007). The attributes of knowledge transfer that are necessary for cooperation should be identified and efforts should be made to strengthen the transfer of knowledge and draw up a coordination strategy.

The relational potential of a network is thus characterized by the ability of individual network participants to exchange ideas and describes indirectly the use of human capital, such as knowledge and information about clients, suppliers, competitors, and other companies. For example, regarding procedures for specific staff-related measures, it is important to know the employees' preferences, desires and requirements. Regarding with regard to improve the quality of certain care procedures, it is important to know the patients' requirements and preferences. With the placement of patients, contact with clinics, hospitals and medical practitioners play a critical role. Exchange of knowledge requires a foundation of trust, particularly if, as in the case of a network, horizontal cooperation partners and therefore direct competitors are involved. However, trust is a resource that is not available in a network from the outset, but rather develops in the course of cooperation.

The human and relational potential of network participants thus form the basis for cooperation and hence for network cooperation. The cooperative process is often equated with the term "synergy", synergetic effects being the effects of a process and potential synergies of the as yet unrealized synergetic effects (Ike 2008, p. 184). Potential synergies distinguish successful cooperation from the solo activities of an individual company. If a coordination strategy succeeds in securing a foundation for cooperation and hence potential synergies, then the cooperative partners can use their human and relational potential in such a way that each individual company can achieve their objectives more successfully than if they worked on their own. Efficiency and effectiveness criteria can be used to ascertain the degree of success. Cooperative partners who recognize the benefits of greater efficiency and effectiveness will apply their human and relational potential in such a way that, for instance, the promotion of long-term employability can be developed via network cooperation.

Social mechanisms like trust and economic considerations like effectiveness and efficiency considerations have effects on the cooperative activities of network actors. A coordination strategy within a network should be geared to promote interactions between network partners, create confidence-building cooperation, and thus facilitate maximal use of human and relational potential. Organizational learning is required to facilitate the use of human and relational potential needed within participating companies and even at network level. The project *Zukunft:Pflege* aims to strengthen and develop organizational learning in participating companies and promote organizational learning at a network level. The following aspects of organizational learning will now be considered more closely.

2.2 Network Cooperation Through the Joint Development and Coordination of Learning Strategies: From Organizational Learning to Network Learning

Outpatient nursing and long-term care are a service industry that is particularly dependent on social interactions and conformity with external regulations and conditions. The effective care of patients already requires that information be are widely shared, for instance among physicians, hospitals, day care facilities, and patients. Additionally, new legal and medical regulations constantly need to be considered. Outpatient nursing and long-term care agencies and their employees have to assimilate continually accumulating and changing knowledge. Relevant information is gathered and passed on within an agency or company to the staff concerned, who are expected to be prepared for changes and modifications. This is what is known as a "learning organization". However, what happens if, as already described, small- and medium-sized companies are unable to cope with the sheer volume and complexity of new and relevant information? Can networks support "learning organizations" involved in outpatient nursing and long-term care by more effectively structuring the exchange of information and knowledge?

We believe this is possible, provided such a network also initiates structures and processes that facilitate their members' learning. Moreover, the learning process of the individual members of such a network is an important prerequisite for learning at a network level. We will now outline the various levels of learning, from the individual to group and company level, as aspects of organizational learning, and describe the transition to network learning.

A "learning organization" begins with individuals. The knowledge of each individual, acquired through the mutual exchange and coordination of knowledge and through experience with other individuals, is the starting point for establishing an organizational knowledge base that can also be used to refresh the participants' knowledge (Probst and Büchel 1994). As already indicated with respect to human and relational potential, improving the efficiency or learning ability of an organization requires that tacit knowledge be transformed into explicit knowledge (Geissler 1998, p. 196). Group or team discussions or rather interchange during network meetings play an important role in terms of the interdependence between tacit and explicit learning because they are the fora in which important decisions are made by learning units in organizations and in which individual decisions are transformed (Willke 1998, p. 50). Only a systemic perspective that provides a conceptual framework with common patterns and structures for communication between individuals, yet is detached from individuals, is able to integrate individual learning processes into organizational learning processes. Changes in the organizational knowledge base can contribute to the adaptation of a system or to an increased capacity to solve problems (Argyris and Schön 1999, p. 36).

To implement such a learning strategy, a knowledge infrastructure that allows new learning experiences to be shared easily is required (Grossmann and Scala 1994). In Sect. 3, some features and attributes of business cooperation are considered, after which we describe a coordination strategy in more detail.

3 Coordinating Network Cooperation: The Example of *Zukunft:Pflege*

In general, coordination is understood to be the agreement over or alignment of individual activities within a system of labor division or an organization with a view to achieving a higher objective. The coordination of individual activities is considered to be a central management function (Freese 2000, p. 69; Staehle 1999, p. 555).

The coordination of a system in general and of economic units in particular is goal-oriented: in this sense coordination can be described as a regulatory instrument aimed at achieving a common objective. In a network, congruence between the individual participating businesses' objectives and the objectives of the network is essential for the continued existence of the network. If the objectives do not match, the network is in danger of collapsing (Siebert 1991, p. 307).

The common aim of the *Zukunft:Pflege* network is the promotion of long-term employability in terms of competence and skills development, motivation and health preservation in nursing staff in ambulatory long-term care.

3.1 Characteristics of the *Zukunft:Pflege* Cooperation Network

The coordination of business cooperation is dependent on specific characteristics and features. Killich (2005, p. 18) has compiled a list of the key features of cooperation (see Table 1). According to these features, cooperation differs in terms of coordination.

As well as other participants, the *Zukunft:Pflege* network includes ambulatory care services and hence businesses that operate in the same sector and on the same value creation level. In contrast with vertical cooperation (between partners that operate in the same sector but work on different value creation levels), this is known as horizontal cooperation and entails a relatively competitive situation (Killich 2005, p. 18). This competition between participating ambulatory care services can have different effects on cooperation within the network.

In surveys on the *Zukunft:Pflege* network, members of the ambulatory care services commented on the competition in their sector more or less as follows: competition and rivalry occur on a regional level, but there are enough patients to prevent competition over the acquisition of patients becoming overly aggressive. However, competitive pressure is perceived to be strong because of the current job market structures. The network partners are worried that their respective staff might be recruited away, which has consequences for the exchange of information on the network. According to Killich (2005, p. 18), network cooperation should at least

Table 1 Key features of cooperation according to Killich (2005, p. 18)

Feature	Parameter					
Dimension	Horizontal		Vertical		Diagonal	
Outreach	Local		Regional		National	Global
Linkage	Low		Moderate		High	
Commitment	Agreement		Contract		Contribution of capital	
Time frame	Temporary				Indefinite	
Aim identity	Redistributive				Reciprocal	
Department of cooperation	R&D	Distribution	Purchasing	Marketing	Production	Other

R&D: research and development

include “the elimination of any existing competition between the cooperating businesses”.

Furthermore, the project is supported by so-called value partners, who actively participate in shaping the network and network cooperation. These value partners include continuing education as well as associations of private social services providers, clinic associations, and the trade association. In relation to the care services, the value partners can be understood as cooperation partners from different sectors and with different value creation levels, which results in diagonal cooperation (Killich 2005, p. 18). Up to now, cooperation with these partners has been limited to recommendations for lecturers to provide in-house training seminars in the ambulatory care companies. In addition to diagonal cooperation between the care companies and training institutions, there is horizontal cooperation between the different training institutions that manifest as follows. The ambulatory care companies are not reimbursed through their care contracts for the cost of training their staff; health insurance only covers expenses arising directly from the care of patients. The companies must cover the (usually additional) expenses for further training, and money for training and education is not readily available. The training institutions are in a competitive situation in as much as they have the same target group and make very similar offers. According to the training institutions, it would make sense for them to come to an agreement about continuing and further education curricula, specialization subjects, and the quality and scope of the courses offered.

The linkage between the cooperation partners of the *Zukunft:Pflege* network can be described as moderate to high. If a network is only used to exchange information and opinions, the linkage of the partners can be said to be low. However, the commitment and respective tasks and activities of the participants of the *Zukunft: Pflege* network indicate that they are very willing to cooperate and that all network partners benefit to an equal degree (win-win situation) (Killich 2005, p. 19).

The following factors provide evidence for the ambulatory care companies' commitment to cooperation on the network: (1) participation in work science analyses and development of design concepts; (2) operational testing and participation in the evaluation of a design concept for “sustainable employability”; (3) participation in network meetings; (4) participation in mutual learning in the network; and (5) participation in transfer events organized by the network.

Cooperation with so-called value partners is summarized as follows. The value partners facilitate a broad exchange with the companies and representatives of interest groups within the care sector through the network. They bring their experiences into the research and development process and actively support the transfer of project results and their practical application in the sector.

As to aim identity, Killich (2005, p. 20) describes redistributive and reciprocal forms of cooperation. Redistributive cooperation is when the cooperation partners try to compensate for their common weaknesses by pooling their resources. Reciprocal cooperation is when the companies acknowledge that their cooperation partners are best suited to solving each other's problems.

In the *Zukunft:Pflege* network, each business is represented by a member of the managerial staff who is the business owner or their deputy and responsible for corporate development, sales or marketing; these members therefore have a broad perspective on the purposes of cooperation. Their interest in cooperation is to orient their business more to the demands of the market and to make the whole sector more competitive.

3.2 From an Individual Problem to Network Learning: Coordination of the Learning Process by the Network Agency

What does the coordination of cooperation look like in a network for sharing and generating knowledge with the aim of ensuring sustained employability in ambulatory long-term care? Based on what has been described in Sect. 2.2, the *Zukunft: Pflege* network can be described as a platform for joint, coordinated learning processes.

Possible starting points are diverse and include, for example, a business wishing to create a systematic collection of procedures for dealing with ongoing issues, managing the qualifications of personnel and handling problems in the daily routine of care work (e.g., dealing with difficult patients, ergonomic issues). Carers encountering such problems can discuss difficult cases with colleagues or managers and ask them about their experiences. If this informal approach does not result in a solution, the problems can be discussed in institutionalized team discussions (or a group learning context) in which moderators help team members contribute towards finding a solution by describing their individual experiences. Alternatively, experts from outside the team can be invited to join such a case discussion; this process is described as learning at the enterprise level, so-called organizational learning. It is helpful to keep a record of the problem and the solution, for example by using a protocol. Of note, this form of group and organizational learning is not systematic, but subject to chance. However, it makes sense to systematically accumulate specific guidelines for knowledge transfer, particularly in the case of recurrent or permanent problems (e.g., new nursing regulations or patients with certain chronic diseases) that have to be overcome daily. Such a systematic set of guidelines can help with more reliably identifying qualification requirements, adding them to training programs and measuring the resultant outcomes.

Systematic accumulation of guidelines can occur while network partners follow the cycles of a learning spiral supported by appropriate operational structures and leadership qualities (Fig. 1).

Every learning spiral begins with a diagnosis or analysis of the status quo from which priorities, aims, and parameters can be established. After implementation of the measures, their effectiveness is evaluated, marking the beginning of a new cycle. The systematic knowledge management have then been developed in cooperation with the network partners; that is, with the potential future users.

Assistance from the network for systematically establishing such an education/knowledge management process may be needed if, even after holding individual sessions with employees to consolidate the knowledge acquired, evaluation indicates that practical implementation of course content is not reasonably assured, in

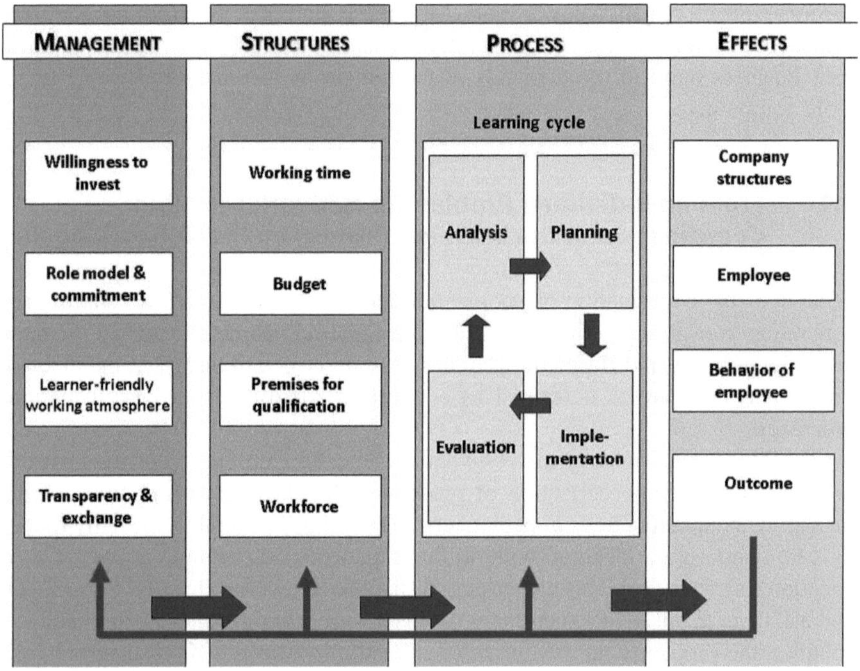

Fig. 1 Diagram depicting the development of knowledge management by network partners in Zukunft:Pflege

which case the newly acquired know-how would likely evaporate. The different levels of transferring the knowledge management (e.g., the individual, patients' needs and route planning levels) should be discussed by a team from the institution concerned and solutions developed and documented. If a solution is not found within that company, it is necessary to go one level further.

The *Zukunft:Pflege* network offers partners the opportunity to seek and discuss solutions at network level with other companies facing similar challenges. Relevant value partners (e.g., teachers of geriatric care, professional associations for careers, the Federal Association of Private Social Service Providers) can attend network meetings, share experiences, organize, or attend professional events to tackle individual problems and discuss possible solutions (second exchange on the network in Fig. 2).

The network is coordinated by a network agency. This agency is located at the University of Bremen and employs three research assistants who, because of their science-based knowledge, are responsible for listing challenges and problems facing the outpatient care sector and facilitating solutions. The network agency should not be understood as an "expert" or "problem solver"; rather, its purpose is to help organizations to help themselves, for instance by forming small groups to work on specific themes.

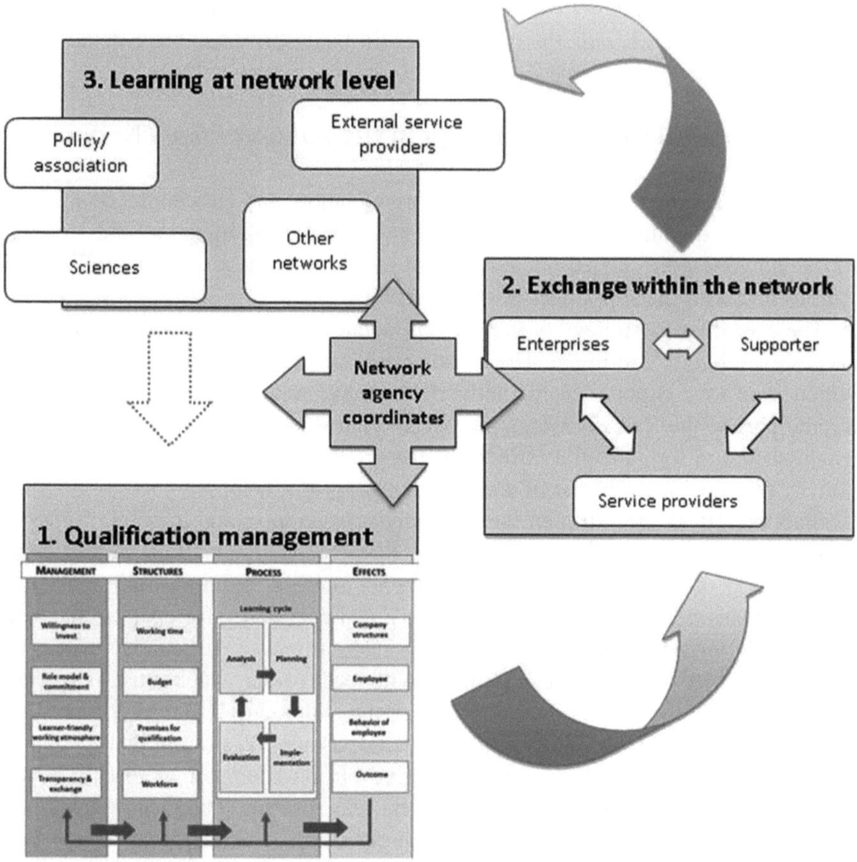

Fig. 2 Transfer of knowledge in *Zukunft:Pflege*

The network agency coordinates and brings partners together to find the right solutions through interactive processes. If no appropriate solution to the problem of an individual company is reached within the network, the network agency can look beyond the network for other solution strategies (e.g., in science, politics, or other external systems). Possible solutions can then be tailored to suit the specific needs of the company in question (third exchange in the network in Fig. 2). The coordination function of the network agency involves arranging interactions on the network aimed at adapting possibly abstract strategies to meet the demand of the company in question (Fig. 2).

The above illustrates coordination by a network agency to develop individual learning processes at the network learning level: this is the hypothetical network strategy for a consensual exchange. The focus of the network is coordination of the different learning processes and learning levels.

In Sect. 3.3, the theoretical significance and practical implementation of this coordination in the *Zukunft:Pflege* network will be described in more detail.

3.3 Practical Network Strategy in the *Zukunft:Pflege* Network

To achieve a broad consensus on the organization and procedure of network coordination, qualitative analyses and surveys on the organization and structure of the network were carried out at different levels of the network partners' organizations and with the supporters of the project.

The coordination strategy of the network agency refers primarily to the network level shown in box 3 of Fig. 2. The learning process in the participating companies, which is of key importance for network learning, was coordinated by the operational part of the *Zukunft:Pflege* project. Here, for example, control circles were established and formalized as structures for group learning. The next step was to convey the specific problems of a company to network level.

Stepping out of the safety of the company surroundings and internal discussion of problems requires a relatively binding form of documentation. In particular, coordination of the various network partners' interests requires trust and confidence; it takes time to develop sufficient trust and confidence for open exchange to occur with unknown partners concerning new projects, particularly when the companies involved are in competition with each other. Such trust is a basic requirement for openness and a balanced "giving and taking" (sharing information, commitment, etc.) by all participating network partners. We found that having all partners in the *Zukunft:Pflege* network agree on a written set of directives and guidelines for reaching agreed objectives was useful. These directives and guidelines also included agreements on participation, confidentiality and openly addressing conflicts and are documented in a paper entitled "Principles of Good Cooperation" that was compiled and agreed to by all network participants.

Overall, a network should be open to new partners and issues; that is, any new network member should feel able to openly introduce certain issues. We therefore preferred a flat hierarchy in which topics are moderated only by the coordination agency.

Because their resources are tight in terms of time and personnel, the network partners prefer to organize quarterly network meetings and monthly working groups. The meetings are moderated and the minutes recorded by the network agency to facilitate effective joint discussions and coordinated exploration of common solution strategies at network level. The network members believe that clearly laying out the objectives of the network and reporting its long-term benefits in the minutes, newsletters and reports are crucial to keeping all participants motivated.

To facilitate informed and purposeful discussions, a creative team has been set up to propose and prepare topics for the network meetings. Priority is given to following up issues that have already been addressed and identifying future areas of activity and prospects in ambulatory long-term care. The search for common solutions is not confined to the network. In addition to including new partners in the region, contact is also made with subject-related networks (care, health promotion, and welfare associations). The network partners appreciate this strategy because it helps generate new ideas on long-term care. To enhance the visibility, appeal and momentum of the network both internally and for the public, the network agency and partners have pursued and are continuing to pursue the following strategies: (1) hosting network meetings by the network partners to improve their relationships with each other and promote mutual trust; (2) representation of the interests of the care companies (e.g., through a monthly press release on care that aims to attract potential new network partners, the topics of which are selected at network meetings and subsequently written by network partners); (3) orientation of the network meetings to themes that impinge on the everyday needs of the network partners and potential future issues concerning the care sector; (4) gathering information on other networks with a similar focus with a view to cooperation and the development of new ideas and synergies; (5) formulation of the benefits of the network in financial terms and in terms of soft skills; and (6) organization of a specialist fair where the network partners each have stands.

As described earlier in this section, the structures and processes of the network strategy were agreed upon by all the actors in the network, thus creating a basis for a relationship of trust and cooperation that is conducive to interaction processes and consolidates acceptance and reliability in the network (Ritter and Pöser 2013). In addition, cooperation within the network is so dynamic and open that outside organizations are suggesting new ideas for dealing with the challenges discussed within the network. Thus, the learning cycle depicted in Fig. 2 has come full circle.

These practical network structures and processes in the *Zukunft:Pflege* network testify to the crucial coordinating role of the network agency. Over time, organized discussions have helped mutual trust to develop. At the same time, the learning processes described above have gained importance at both company and network levels. The coordinating agency takes issues out of the company context and into the network and, in its role of impartial, informed facilitator, creates a common ground for the discussion of topics such as measures for assessing guidelines. Learning together at network level is facilitated by open communication and the mediation of challenges. Moreover, the network agency has taken a step forward in reflexive learning ("are we learning the right thing properly?") together with the network partners. In late 2013, some network partners expressed ongoing doubts about the benefits, purpose, and procedures of the network. In response, a workshop was organized to discuss these points and develop appropriate strategies. Previous procedures were revised or modified. Reflections on their own work regenerated the participating organizations' trust in networking while simultaneously providing impulses for dealing with new challenges.

4 Summary and Perspectives

The prevailing operational structures in small and medium-sized long-term care companies suggest that individual companies have little potential for innovation. Most businesses are so busy coping with routine tasks in long-term care and administration that they have little capacity for developing strategies for sustainable employability. Focusing on the motivation, health and skills of their employees would, however enable ambulatory care companies to retain their qualified employees on a long-term basis and thus improve their chances of success in a booming market.

The idea behind the *Zukunft:Pflege* network is to bundle the existing resources of the individual companies together and develop them for the benefit of all network partners. The coordination of cooperation or interaction between the network actors is complex and certainly requires deliberate input. Coordination must take into account various particular features (Killich 2005) as well as criteria for efficiency and effectiveness and hence the benefits that can accrue to businesses through cooperation.

The application of human and relational capital and a strategy for organizational learning give rise to synergistic effects that promote enduring employability in long-term care, giving networked companies an advantage over businesses that operate alone. A basic prerequisite for network cooperation among a group of legally and formally independent enterprises is that they have a common objective. The common purpose of the network examined here was to facilitate sustained employability; to this end, new subjects and innovative strategies were analyzed and processed by the network agency and fed back into the network. The process can thus be defined as organized learning at a network level. This learning process begins in a company at the individual employee level and aggregates to group learning and on to organizational learning. It is the responsibility of the network agency to pass the outcomes of learning (in the case of the *Zukunft:Pflege* project, these comprise ideas for sustainable employability) on to the network level and to develop them further through coordinated discourse. For the network agency, reflexive learning means enquiring whether the chosen strategies encompass all relevant learning strategies of the network partners in terms of human and relational potential. The specific dynamics of a network, and in particular of the coordination function of the agency, lie both in the constant monitoring of its topics and in the organization and coordination of the various working groups and creative teams to determine places and processes for sharing knowledge and learning.

The human and relational potential of the involved partners is both a prerequisite and the outcome of the exchange and learning process described above. Without a certain amount of human and relational potential, no further discussion and hence no innovation is possible in a network. Trustworthy, stable network coordination can strengthen both potentials in an organization.

The partners consider the role of the network agency to be so important that the network would be unable to function without it. It is not only a question of practical coordination such as organizing meeting rooms and appointments. The network agency is regarded as a neutral party that acts as a filter for information, has a positive influence on the development of trust, conveys a sense of security when dealing with competitors, prevents rivalries or alliances from emerging and, last but not least, provides valuable information that is otherwise not easily available. Thus, the network agency is considered a crucial factor in uniting the members. It also has the important task of public relations. In addition, in future the network agency will act as the initiator of new joint projects, indicating that there are already long-term plans for the coordination agency and that the whole network has gained the necessary momentum through the coordination work. Successful coordination can be seen as a regulatory strategy that allows an exchange of knowledge that helps the participants to achieve their own objectives within the target framework of the network (Aulinger 2008, p. 18ff.).

The establishment and operation of a network in accordance with the wishes of its participants (companies, employees, supporters, and new members) can thus be the product of a joint project in which each partner's wishes are taken on board, discussed and implemented. An ongoing exchange about achievements and occasional adjustments is also important. Analyses and regular network meetings, subject-related working groups and workshops with the employees all provide a basis for a dynamic workshop. The whole network depends on a balanced measure of "give and take" on the part of every participant to operate dynamically and to be tailored to suit everyone's needs.

References

Argyris C, Schön D (1999) Organizational learning II. Theory, method, and practice. Addison-Wesley, Reading (in German: Die lernende Organisation. Grundlagen, Methode, Praxis. Klett-Cotta, Stuttgart)

Arrow KJ (1974) The limits of organization. Norton, New York

Aulinger A (2008) Unternehmensnetzwerke und Verbundnetzwerke [Network evaluation: challenges and practices for association networks]. In: Aulinger A (ed) Netzwerkevaluation. Kohlhammer, Stuttgart, pp 15–34 (in German)

Bapuji H, Crossan M (2007) Knowledge types and knowledge management strategies. In: Gibbert M, Durand T (eds) Strategic networks: learning to compete. Blackwell, Oxford, pp 8–25

Becke G, Bleses P (2013) Interaktion und Koordination: Befunde zur Arbeitssituation in ambulanten Pflegeunternehmen. In: Becke G, Behrens M, Bleses P, Jahns K, Pöser S, Ritter W (eds) Nachhaltige Beschäftigungsfähigkeit in der ambulanten Pflege. Zwischenbericht des Verbundprojekts ZUKUNFT: PFLEGE. Forschungszentrum Nachhaltigkeit, Universität Bremen, pp 33–56 (in German)

Becker T, Dammer I, Howaldt J et al (2007) Netzwerke—praktikabel und zukunftsfähig. In: Becker T, Dammer I, Howaldt J, Killich S, Loose A (eds) Netzwerkmanagement. Mit Kooperation zum Unternehmenserfolg, 2nd edn. Springer, Berlin, pp 3–11 (in German)

Crossan MM, Lane HW, White RE (1999) An organizational learning framework: from intuition to institution. Acad Manage Rev 24:522–537

Freese E (2000) Grundlagen der Organisation. Konzept–Prinzipien–Strukturen, 8th edn. Gabler, Wiesbaden (in German)

Geissler H (1998) Umrisse einer Grundlagentheorie des Organisationslernens. In: Geissler H, Lehnhoff A, Petersen J (eds) Organisationslernen im interdisziplinären Dialog, vol 5. Deutscher Studien, Weinheim, pp 162–223 (in German)
Grossmann R, Scala K (1994) Gesundheit durch Projekte fördern. Ein Projekt zur Gesundheitsförderung durch Organisationsentwicklung und Projektmanagement. Juventus, Weinheim (in German)
Ike R (2008) Performance Management—Synergiepotenziale von Wissensmanagement und Business Intelligence im Rahmen eines ganzheitlichen Ansatzes zur strategischen Unternehmenssteuerung. Selbstverleger, Göttingen
Initiative Neue Qualität der Arbeit (INQA) (2011) Gesundheitsnetzwerke—Ein Leitfaden für Klein- und Mittelunternehmen Bd 41. INQA Geschäftsstelle c/o Bundesanstalt für Arbeitsschutz und Arbeitsmedizin. Druckverlag Kettler, Bönen (in German)
Killich S (2005) Kooperationsformen. In: Becker T, Dammer I, Howaldt J, Killich S, Loose A (eds) Netzwerkmanagement—Mit Kooperation zum Unternehmenserfolg Springer. Heidelberg, Berlin, pp 13–22 (in German)
Krugman P (1991) Geography and trade. MIT Press, Cambridge
Nonaka I (1994) A dynamic theory of organizational knowledge creation. Organ Sci 5:14–37
Probst G, Büchel B (1994) Organisationales Lernen: Wettbewerbsvorteil der Zukunft. Schweizerische Gesellschaft für Organisation und Management. Gabler, Wiesbaden (in German)
Ritter W, Pöser S (2013) Gemeinsam Austauschen und Lernen für nachhaltige Beschäftigungsfähigkeit—Kooperation im Netzwerk ZUKUNFT:PFLEGE. In: Becke G, Behrens M, Bleses P, Jahns K, Pöser S, Ritter W (eds) Nachhaltige Beschäftigungsfähigkeit in der ambulanten Pflege. Zwischenbericht des Verbundprojekts ZUKUNFT: PFLEGE. Forschungszentrum Nachhaltigkeit, Bremen University, pp 83–115 (in German)
Schubert H (2008) Netzwerkkooperation: Organisation und Koordination von professionellen Vernetzungen. In: Schubert H (ed) Netzwerkmanagement: Koordination von professionellen Vernetzungen—Grundlagen und Beispiele. VS Verlag für Sozialwissenschaften/GWV Fachverlage, Wiesbaden, pp 7–105 (in German)
Siebert H (1991) Ökonomische Analyse von Unternehmensnetzwerken. In: Staehle WH, Sydow J (eds) Managementforschung. De Gruyter, Berlin, pp 291–311 (in German)
Simonin BL (1999) Ambiguity and the process of knowledge transfer in strategic alliances. Strateg Manage J20:595–623
Staehle WH (1999) Management, 8th edn. Vahlen, München
Stirzel M, Armbrüster T (2012) Innovations–Performance Measurement für Netzwerke: Leistungsmessung und Ausrichtung über Unternehmensgrenzen hinweg. Zeitschrift für Controlling & Management (ZfCM) 56(1):58–63 (in German)
Weyer J (2000) Einleitung: Zum Stand der Netzwerkforschung in den Sozialwissenschaften. In: Weyer J (ed) Soziale Netzwerke. Konzepte und Methoden der sozialwissenschaftlichen Netzwerkforschung. Oldenbourg-Wissenschaftsverlag, München, pp 1–34 (in German)
Willke H (1998) Systemisches Wissensmanagement. Lucius & Lucius, Stuttgart (in German)

Small- and Medium-Sized Enterprises' Preferences for Occupational Health Services and Willingness to Pay

Mirella Cacace, Ingrid Franz, Daniel Braun-Beustrin, and Dieter Ratz

Abstract

Small and medium-sized enterprises (SMEs) in particular benefit from occupational health services because these may help to reduce the potential costs of accidents or illnesses at work, support staff retention and recruitment, and decrease wage costs. Nevertheless, SMEs, especially microenterprises (<10 persons employed), rarely offer these services to their employees. The innovation incubator's project "Healthy at Work" offers research-based advice to private service units to support SMEs to provide occupational health services in the region of Luneburg. This chapter describes how we use an Adaptive Choice-Based Conjoint Analysis combined with a short willingness to pay (WTP) questionnaire to elicit SMEs' preferences regarding occupational health services. We found that the optimal approach is to offer a comprehensive service package tailored to the needs of the individual company on a pay-per-use basis. The private supplier benefits from cooperation with a social insurance provider: either a health insurance fund or occupational accident insurer. Further, we found that employers are willing to pay for services. Within the group that is willing to pay, WTP increases with company size. It is therefore particularly important to offer appealing and affordable occupational health services to microenterprises, preferably in cooperation with social insurance providers.

M. Cacace (✉)
SOCIUM Research Center on Inequality & Social Policy, University of Bremen, Bremen, Germany
e-mail: cacace@uni-bremen.de

I. Franz • D. Braun-Beustrin
Leuphana University Luneburg, Luneburg, Germany
e-mail: ifranz@inkubator.leuphana.de; braun-beustrin@inkubator.leuphana.de

D. Ratz
M-Forschung, Hamburg, Germany
e-mail: dieterratz@m-forschung.de

M. Wiencke et al. (eds.), *Healthy at Work*, DOI 10.1007/978-3-319-32331-2_16

1 Introduction

In Germany, the work-force is projected to decrease tremendously whereas average age is increasing (Fuchs et al. 2011). At the same time, illness-related absences from work are leading to considerable productivity losses (Bundesanstalt für Arbeitsschutz und Arbeitsmedizin 2014). In particular, small and medium-sized enterprises (SMEs) with limited personnel resources and a high demand for skilled workers will be faced with these challenges. Against this background, it is particularly important to safeguard employees' health, for example by offering occupational health services such as programs for health promotion and prevention of work-related diseases. Occupational health services may help to reduce the costs of accidents or illnesses at work, enhance the general reputation of the company, support staff retention and recruitment and, finally, decrease wage costs (Miller et al. 2002). However, SMEs, particularly microenterprises (<10 persons employed), rarely offer these services to their employees (Harrison et al. 2013). In line with the observations made by Zelfel et al. (2011), our previous studies have shown that, in addition to time requirements and financial support, a major reason for this deficit is a lack of qualified support structures (see Cacace et al. 2016).

The innovation incubator's project "Healthy at Work" supports the creation of private service units offering health education, prevention programs, and occupational rehabilitation in the region of Luneburg. The aim of our research within this project was to support these units to develop services in accordance with the preferences of SMEs. For this purpose, we conducted a conjoint analysis and, because the supplier of services would be a private for-profit provider, we also investigated whether SMEs would be willing to pay for these services. In Sect. 2, the research design and method for eliciting preferences and willingness to pay (WTP) are explained; our findings are presented in Sect. 3; and Sect. 4 discusses these against the background of the aim of the "Healthy at Work" project; namely, supporting a private service provider to supply occupational health services to SMEs, in particular to microenterprises.

2 Method and Research Design

2.1 Pre-studies

In a previous study of 72 SMEs in the region of Luneburg that was also part of the "Healthy at Work" project, 45 % of respondents agreed strongly or very strongly that a lack of qualified experts to answer questions regarding occupational health programs is a major impediment to offering these services to their employees (see Cacace et al. 2016, p. 54). Thus, SMEs need external professional expertise and support structures to offer occupational health services to their employees and these structures need to be tailored to the specific needs of each SME (Cacace et al. 2014).[1] Accordingly, the research question for the present study was how to tailor such services in accordance with the preferences of each SME, in particular microenterprises. To elicit these preferences and WTP for occupational health services in SMEs and microenterprises, we conducted a conjoint analysis followed by a short survey. These components of the study were preceded by qualitative pre-studies in which we conducted expert interviews with various actors involved in occupational health, such as accident and health insurers, trade chambers, guilds, and professional associations. Further, we organized focus group discussions with the person responsible for the purchasing decision on occupational health services. Because SMEs are by definition small, in many cases this person was also the owner of the company. To avoid omitted variable bias (see Clark et al. 2014), we considered that expert interviews and focus group discussions were essential for formulating the attributes and attribute levels for the conjoint analysis (see Sect. 2.2). Another benefit of our qualitative research was to familiarize ourselves with the particular vocabulary used by the respondents in relation to occupational health services. Finally, our analysis was preceded by an in-depth market screening of existing supply structures for occupational health services.

2.2 Conjoint Analysis

The conjoint method is based on the seminal work of Luce and Tukey (1964), who described the essential characteristics of simultaneous conjoint measurement. In market research, conjoint analysis is the preferred method for identifying consumer preferences for new products and concepts by assessing partial benefit values for single product characteristics or attributes (Green and Srinivasan 1978). Today, it is frequently used in health economics to assess consumers' preferences for health care service delivery and a range of other health applications.[2] Conjoint analysis is a decomposition method; that is, it is based on the assumption that a product or service can be described by its attributes. The value of the product or service depends on the nature and level of these attributes. Within experimental settings, conjoint analysis provides information about the importance of single attributes and the utility of attribute combinations to respondents. It "show(s) how people are willing to trade between characteristics" (Ryan and Farrar 2000, p. 1530), which is useful when deciding on the optimal way of providing a good or service when resources are limited. During conjoint interviews, through several iterations, respondents are shown different product profiles and asked to choose the favored alternative. By disaggregated evaluation of the utilities respondents assign to single

[1]These findings are in accordance with the observations made by Zelfel et al. (2011), who conducted a much larger study that surveyed 1441 SMEs all over Germany.

[2]See, for example, reviews by Clark et al. (2014), De Bekker-Grob et al. (2012), Ryan and Farrar (2000), and Ratcliffe (2000).

attribute levels, conjoint analysis estimates the overall preference or utility associated with each level of each attribute. This separation is called attribute-specific part-worth function. Overall utility scores and the relative importance of single attributes can be computed from part-worth functions (Rao 2014). In this way, conjoint analysis reveals the underlying preference function.

2.3 Research Design

For our survey, we recruited 75 SMEs in urban and rural areas in the region of Luneburg by telephone. The response rate was 100 %: all recruited SMEs participated in both the conjoint analysis and WTP sections of our study. The recruited SMEs operate in the following sectors: childcare, hairdressing, automotive and metalworking, dentistry, construction, agriculture, cleaning, long-term geriatric care, hotel/gastronomy, and food. The largest enterprise participating in the study employed 62 persons and the smallest only two. About 49 % were microenterprises (employing fewer than ten persons).

We designed the questionnaire as a computer-aided survey using the Sawtooth Software package SSI Web 8.3.2. For quality purposes, we opted for a self-administered questionnaire combined with supervision by interviewers (Bourque and Fielder 2003; Ryan and Gerard 2003). Interviewers' roles were restricted to providing explanations and assistance when respondents had questions. The field time was between 13 January and 31 March 2014 and the mean duration of interviews was 37.3 min.

Among the multiple conjoint analysis approaches available in the Sawtooth package, we opted for the Adaptive Choice-Based Conjoint Analysis (ACBC), which combines the advantage of the widely employed choice-based approaches with an adaptive element (see Orme 2013 for various conjoint methods and applications). ACBC was our preferred option because of decisions we had made regarding research design, such as sample size, complexity of attribute list, length of survey, and mode of interviewing. The ACBC interview is interactive, customizing questions to respondents' previous choices, which makes the tasks more engaging than non-adaptive versions of the choice-based approach (Sawtooth Software 2014). This factor improves the quality of evaluation of utilities, thus facilitating better prediction of real-world preferences (Cunningham et al. 2010). Moreover, ACBC incorporates elements of non-compensatory decision making, which means that there are "must-haves" in attribute levels: attributes the respondent strongly prefers and is not willing to trade off. ACBC therefore is capable of modeling simplified rules of decision, which are likely to occur within complex real-word choice tasks (Johnson 2008).

2.4 Attributes and Attribute Levels

For the construction of attributes and attribute levels, we reviewed the transcribed expert interviews and focus group discussions we conducted beforehand. We then determined the range of relevant attributes and attribute levels by considering the outcomes of our review together with the broader picture derived from market screening, thus setting up the decision framework for the conjoint interviews.

In sum, we identified ten attributes for the planned service; seven relating to the core service to be tested and another three to additional service offers. The seven attributes relating to the core service were:

1. Content of the occupational health service package (=supplied services)
2. General approach to service delivery (=approach)
3. Cooperation seeking by the private, commercial service provider, for example with a health insurance fund (=affiliation)
4. Tailoring of services to specific needs (=customization)
5. Contractual arrangements and payment form (=contract)
6. Certification of the supplier (=certification)
7. Provision of an award to the participating SME (=awarding)

Furthermore, we tested three additional services the private supplier could optionally offer to SMEs:

1. "Navigation function", including assistance with the administration of formalities related to the absence of workers and attempts to obtain (co-) funding for occupational health services (=navigation function yes/no)
2. Needs assessment tailored to the risks and requirements of the specific SME (=needs assessment yes/no)
3. Planning and conceptualization of occupational integration management[3] (=occupational integration management yes/no)

Table 1 lists the attributes together with the attribute levels, which we set up and refined after pre-testing the questionnaire. Because we presented the attribute list to all respondents in the fixed order presented in Table 1, we cannot exclude sequence effects, whereby respondents may give more attention to the attributes at the top than those at the bottom of a list (Chrzan 1994). However, we accepted this (potential) effect and refrained from rotating the attributes because their sequence—core services at the beginning and additional services towards the end—was part of our research design.

[3]According to German social legislation, the employer is obliged to offer occupational integration management (=Betriebliches Eingliederungsmanagement) to workers who have been absent because of sickness for more than 30 days/year. SMEs in particular are challenged by this obligation (Zelfel et al. 2011).

Table 1 Overview of attributes and attribute levels

Attribute	Attribute level
Supplied service	Prevention of physical illness (e.g., skin disease, back pain) Prevention of mental illness (e.g., stress reduction) Team-building Combination of physical and mental illness prevention Combination of physical illness prevention and team-building Combination of mental illness prevention and team-building Combination of physical/mental illness prevention and team-building
Approach	In-house coaching/training of employees Coaching/training of employees outside the company (e.g., in training centers) In-house coaching of the SME manager(s) (capacity building) Cross-company coaching/training to allow for exchange of ideas
Affiliation	Chamber of Trade/Chamber of Industry and Commerce Health Insurance Fund Guild/professional associations Occupational Accident Insurance Private supplier (no affiliation)
Customization	Service customized to the needs of individual employees Service customized to the needs of the individual company Branch-specific tailoring of services No customization of services
Contract	Pay per use Middle- or long-term service contract Insurance coverage of services
Supplier certification	Yes/No
Awarding	No award provided to the company Company receives the "Healthy Workplace" award Company receives TÜV[a] award
add. Service 1	Navigation function: yes/no
add. Service 2	Needs assessment: yes/no
add. Service 3	Occupational integration management: yes/no

[a]TÜV is the acronym for the German Technical Inspection Association (Technischer Überwachungsverein), a publicly renowned organization for setting and preserving technical standards in Germany. The TÜV award tested in this study, however, is fictitious

In line with the recommendations for good practice in the application of conjoint analysis in health care (Bridges et al. 2011), we also included hypothetical attribute levels in order to encompass the range salient to participants. (Currently) unavailable levels—at least according to our knowledge based on market screening—include, for example, the awarding attribute,[4] insurance-based contracting, and the navigation function.

To elicit WTP for the planned product or service, the attribute of price is frequently included in conjoint analysis (Ratcliffe 2000). However, our qualitative

[4]Both the "Healthy Workplace" and the TÜV award are fictitious.

research had shown that participants revealed little or no WTP for occupational health services in general. Two dominant motives became obvious from the focus group discussions. First, the benefits of occupational health services are rather unclear to SMEs, which is in line what Zelfel et al. (2011) found in their larger study. Second, SMEs tend to shift responsibility for paying for occupational health services to other actors, in particular to employers' liability insurance associations. Against this background, we chose to separate WTP from ACBC to avoid dominance of WTP criteria over all other attributes of the planned service. At the same time, we used the favorable time immediately after the ACBC to ask about WTP for the ideal service package, which the respondent had just constructed during the interview. This procedure had the crucial advantage that respondents had learned through the interview procedure about the service offer and its potential benefits. However, results therefore have to be interpreted with caution because they reflect WTP in relation to an individually optimal service offer.

ACBC interviews were immediately followed by short WTP interviews, in which we asked the following core questions:

1. If the "ideal" service package—that is, the one you prefer most—was available, would you in general be willing to pay for it? (answer options: yes, no)
2. If yes, how much would you be willing to pay for it? (answer options: <5 €, 5–10 €, 10–15 €, >15 € per employee and month)

The second question presented the WTP question according to the payment card technique (Ryan et al. 2004), where respondents are presented with a range of bids and asked to indicate from among them the most they would be willing to pay. The suggested monetary categories were derived from information obtained during focus group discussions. We chose to accept the ambiguity resulting from the use of ranges (Bridges et al. 2011, p. 406) because the payment card technique is more realistic than the alternative (Ryan et al. 2004).

2.5 Questionnaire and Interview Procedure

Our survey started with some socio-demographic questions, followed by the ACBC and WTP components. ACBC typically includes three core sections, namely the "build your own" (BYO) section, the screening section, and the choice tasks.[5] The ACBC is designed to cope with the complexity of a study in terms of attributes and attribute levels on the one hand and the restrictions in terms of sample size and respondent fatigue on the other. Before they started the BYO section, interviewers showed the attribute list (see Table **1**) to the respondents and explained it. This was particularly important because, as described in Sect. 2.4, we included hypothetical attribute levels such as the navigation function in our study. In the BYO section, the

[5]See Sawtooth Software (2014) for a more detailed description of the procedure.

respondents were asked to indicate their preferred attribute levels. From this information the program creates a pool of concepts that are relatively concentrated around the respondents' preferred attribute levels. The BYO task has the side benefit of also serving as a training exercise that further acquaints respondents with the attributes and levels being considered (Orme 2013, p. 4).

In the screening section, we showed respondents three different service alternatives at a time and asked them to indicate which of the three services they would consider. We repeated this procedure eight times, interspersed with phases in which we asked respondents whether one of the given attribute levels was absolutely unacceptable (five repetitions) or a "must-have" attribute (four repetitions). The "must-have" and "unacceptable" questions limit and adjust the choice sets in the last section, the choice task section. Table 2 shows the tasks performed in the screening section.

In the choice task section respondents are shown three alternative service bundles at a time composed of the chosen product concepts. Table 3 provides an example of the choice tasks.

Depending on the answers provided in previous sections, we showed a maximum of 10 different choice tasks and 20 different product concepts to the respondent. We repeated some service bundles to improve evaluation of the part-worth utilities.

We had several reasons for not using a "none" option (i.e. the respondent chooses none of the offered alternatives) in our choice section. First, because we had no status quo with which to make comparisons, a none option would not have been meaningful. Second, including a none option discourages some respondents

Table 2 Screening section: Responses to interviewer saying "Here are a few service offers you might like. For each, indicate whether or not it is a possibility"

Supplied service	Physical and mental illness prevention	Physical illness prevention	Team-building
Approach	In-house coaching/ training of employees	Cross-company coaching/training	In-house coaching/ training of employees
Affiliation	Health Insurance Fund	Occupational Accident Insurance	Guild/professional associations
Customization	Individual employees	Branch-specific	Individual employees
Contract	Pay per use	Pay per use	Insurance coverage
Certification	Supplier certification	No supplier certification	No supplier certification
Awarding	"Healthy Workplace"	No award	TÜV award
add. Service 1	Navigation function	No navigation function	No navigation function
add. Service 2	Needs assessment	No needs assessment	Needs assessment
add. Service 3	Occupational integration management	No occupational integration management	Occupational integration management
	❍ A possibility ❍ Won't work for me	❍ A possibility ❍ Won't work for me	❍ A possibility ❍ Won't work for me

Table 3 Choice task: Interviewer asks "Which of these three options would you prefer most? Attributes with identical levels are highlighted in gray to make it easier to focus on the options"

Supplied service	Physical and mental illness prevention	Physical illness prevention	Team-building
Approach	In-house coaching/ training of employees	Cross-company coaching/training	In-house coaching/ training of employees
Affiliation	Occupational Accident Insurance	Occupational Accident Insurance	Occupational Accident Insurance
Customization	Individual employees	Individual employees	Individual employees
Contract	Pay per use	Pay per use	Insurance coverage
Certification	Supplier certification	No supplier certification	No supplier certification
Awarding	"Healthy Workplace"	No award	TÜV award
add. Service 1	Navigation function	Navigation function	Navigation function
add. Service 2	Needs assessment	No needs assessment	Needs assessment
add. Service 3	Occup. integration management	No occupational integration management	Occup. integration management

from doing the cognitive work necessary to report their true opinions (Krosnick et al. 2002). However, according to Krosnick et al. (2002), it is unlikely that the quality of responses is compromised by this omission. Rather, the other way round, the quality of estimations is prejudiced by including this option. Another decision we made relates to prohibiting the offering of certain levels together. Some research has shown that these exclusions maximizes the realism of the suggested profiles. However, Lancsar and Louviere (2006) strongly advise against making such prohibitions: according to Orme (2002) this can lead to imprecise utility estimation or even to complete inability to calculate stable utilities. We therefore chose to refrain entirely from using prohibitions, partly because there were few implausible combinations of our attribute levels.

We analyzed ACBC using a multinomial logit model. We estimated part-worth utilities in the choice section using ACBC's Hierarchical Bayes program, which permits high-quality individual-level analysis (Sawtooth Software 2014, cf. also Cunningham et al. 2010). The ACBC design employed in the study offers an optimal compromise between the number of stimuli shown to respondents to produce enough data for consistent estimations and the limits imposed by the respondents' cognitive capacity. A test phase with hypothetical data showed the design was efficient, reliable, and robust.

3 Results

3.1 Adaptive Choice-Based Conjoint Analysis

The first result ACBC provides is the relative importance of each attribute, computed as an average over all respondents. Figure 1 shows the relative importance of attributes. Because we have expressed these as percentages, they add up to 100 %.

The conjoint interviews revealed that by far the most important attribute is the supplied service itself, namely prevention of physical or mental illness, team-building, or a combination thereof. This attribute is responsible for just over a quarter (26 %) of overall importance. Provision of coaching/training is also highly relevant (16 %). At 12.1 %, 11.7 %, and 11.5 %, the attributes contract, affiliation, and customization, respectively, were of almost identical importance, whereas the factor "awarding" was slightly less important (8 %). Whether the supplier is certified was of minor relevance to the respondents' choices (4 %). Also additional services, namely navigation function, needs assessment, and occupational integration management, had little relevance (3.5–4 %).

The next results we derived from ACBC were partial utility values for each single attribute level. Because partial utilities are computed across all respondents, they reflect averages, thus equalizing differences in preferences between respondents.

Figure 2 provides an overview of attribute levels and related partial utilities. When interpreting these figures, it is important to note that utilities within an attribute are adjusted to zero. Thus, partial utilities of attribute levels add up to zero within the same attribute. Consequentially, negative values do not mean negative utility. Attribute levels with negative values are beneficial, though below the average utility of that attribute. Because utilities relate to single attributes only, comparison of utilities across attributes is not useful.

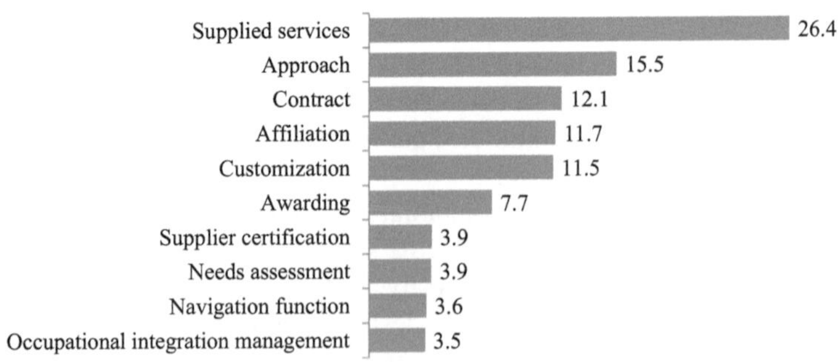

Fig. 1 Relative importance of attributes in percentages (N = 75)

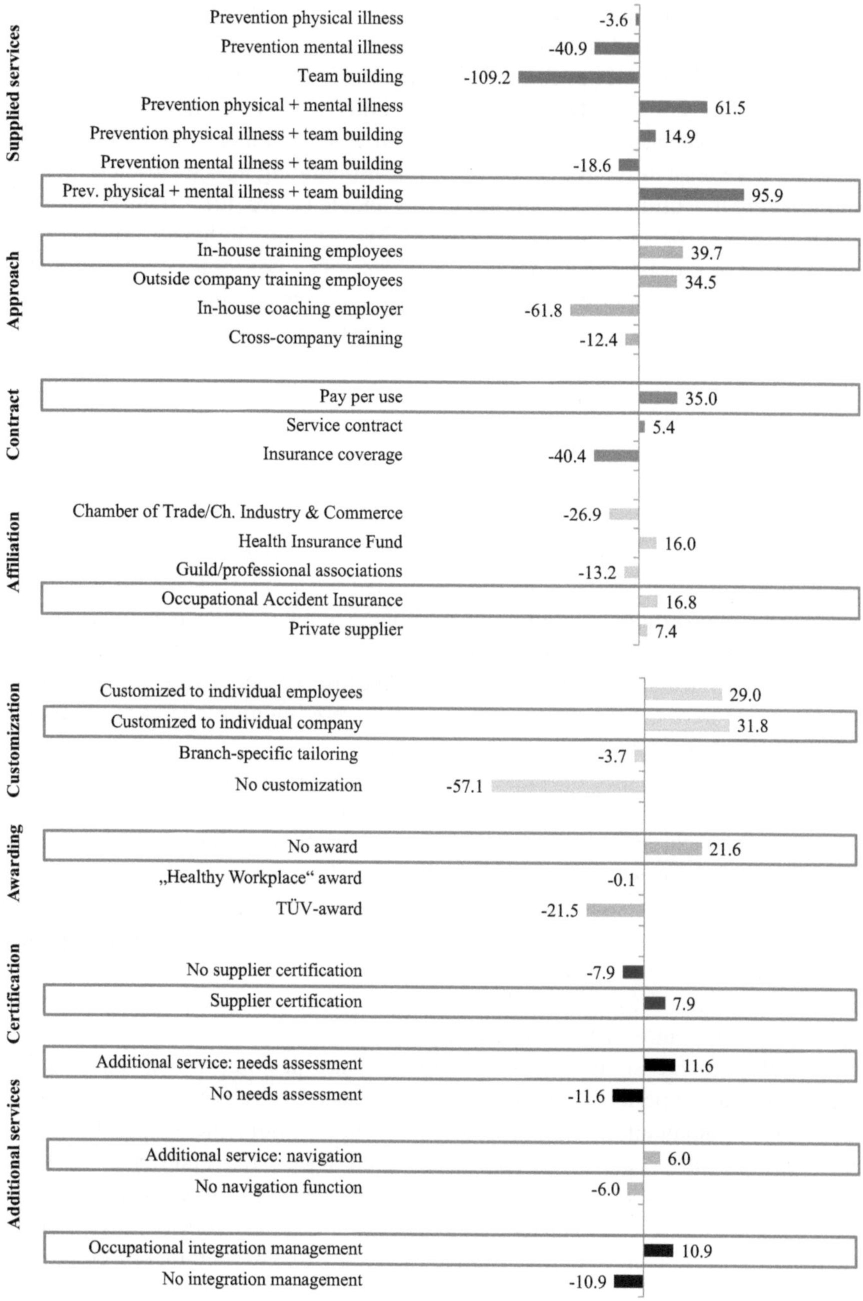

Fig. 2 Attribute levels and partial utilities (N = 75)

In Fig. 2 all attribute levels with the highest partial utility value within one attribute are surrounded by a frame; a combination of these highlighted attribute levels delivers the service package generating the highest utility.

Unsurprisingly, respondents preferred a complete service package comprising prevention of physical and mental illness and team-building efforts. Stand-alone services were least attractive; in particular, team-building as a stand-alone service showed very low utility. Trainings preferably should be provided to the employees, either in-house or outside the company. By contrast, in-house coaching for employers and cross-company training were rated below average. With respect to affiliation, health insurance funds and occupational accident insurers were particularly preferred. Private supply was also of interest: for some respondents commercial suppliers did not necessarily require cooperation. Affiliations to the chamber of trade or a guild had below-average utility values. Further, our respondents preferred customized services. Individualized services tailored to the needs of the company were sufficient; service offers matching individual employees' preferences were not perceived as of greater benefit. Generalized prevention programs were not promising alternatives. While it was considered useful for the supplier to have certification, awards to the participating SME were dispensable, especially if the award was, as we fictitiously suggested, provided by the TÜV. The supply of needs assessment, navigation function, and occupational integration management was considered beneficial, though low attribute importance needs to be taken into account. We must here re-emphasize that because choices were made without taking costs into consideration, respondents were judging the service purely according to their preferences, and not on price.

When we distinguished between number of employees/company size (not shown graphically), utilities differed significantly within the attributes of supplied services, approach, affiliation, and supplier certification. In the smallest category (up to three employees), prevention of physical illnesses was particularly highly rated as a stand-alone service. Team-building and coaching of the employer/CEO were more interesting for large companies (20+) than for smaller enterprises. Middle-sized and larger companies preferred suppliers with certification. Because company size and sector are correlated, some differences in utility values may have been caused by sector differences and vice versa.

We also found significant differences in utilities depending on whether employers had experienced major problems caused by absences of employees (not shown graphically). We identified differences within the supplied services and contract form and with respect to preferences for additional services. Employers with few or no problems caused by absences preferred services for physical illness prevention. Their preference for physical illness prevention in combination with team-building was significantly stronger than that of employers who had experienced considerable such problems. The latter tended to prefer mid- to long-term contracts or insurance contracts. Pay-per-use contracts were strongly preferred by SMEs with few problems due to absenteeism. Additional services, such as the navigation function and occupational integration management, were considered increasingly useful the more problems employers had experienced.

3.2 Willingness to Pay

Regarding the results of WTP interviews, the vast majority of SMEs (76 %) were generally willing to pay for their individual ideal service packages (see Fig. 3).

Almost 40 % were willing to pay between 5 and 10 € per month and employee. This result is surprising because our qualitative pre-studies indicated little or no WTP. A possible explanation for this discrepancy is that the prevention programs were not specified during interviews and focus group discussions whereas respondents had built up an individual service bundle that exactly met their requirements during the conjoint interviews.

Within the group of those willing to pay for services, average WTP increased with company size, as displayed in Fig. 4. Our recruitment of SMEs was guided by the desire to address sectors in which occupational health plays at least some role, either because of physical strain or stress. We recruited enterprises from four sectors: child care (sector I) and hairdressing (sector II) were both dominated by microenterprises employing fewer than ten persons. Further, to achieve relatively homogenous groups of similar size, we assembled technical-based branches into sector III and services/healthcare services into sector IV. Figure 5 shows WTP according to these sectors.

As shown in Fig. 3, about one quarter of responding SMEs had no WTP. With a share of 11 %, WTP of zero occurred least frequently in the child care sector (sector I) and most frequently in sector II, hairdressing. However, an important consideration is that financial decisions about occupational health services in the child care

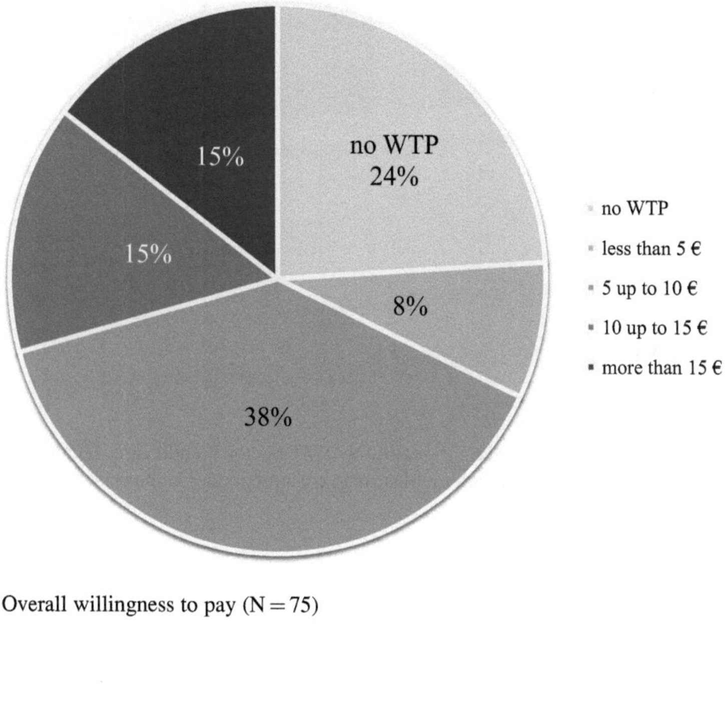

Fig. 3 Overall willingness to pay (N = 75)

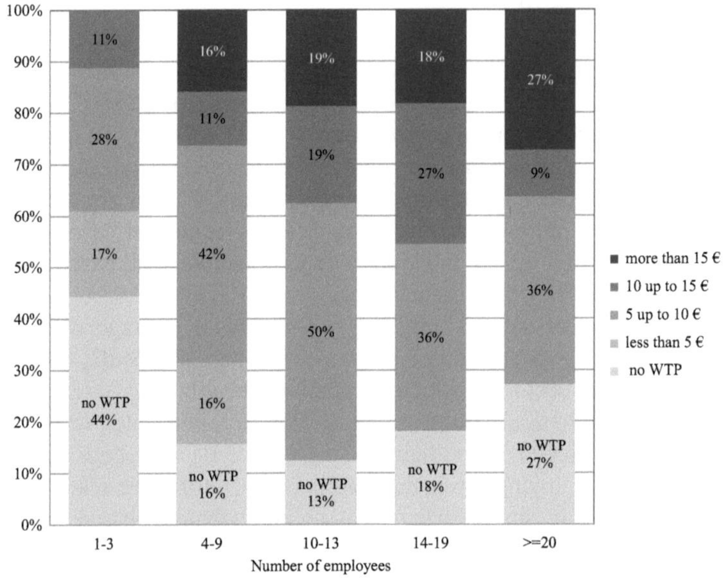

Fig. 4 Willingness to pay according to company size (number of employees) (N = 75)

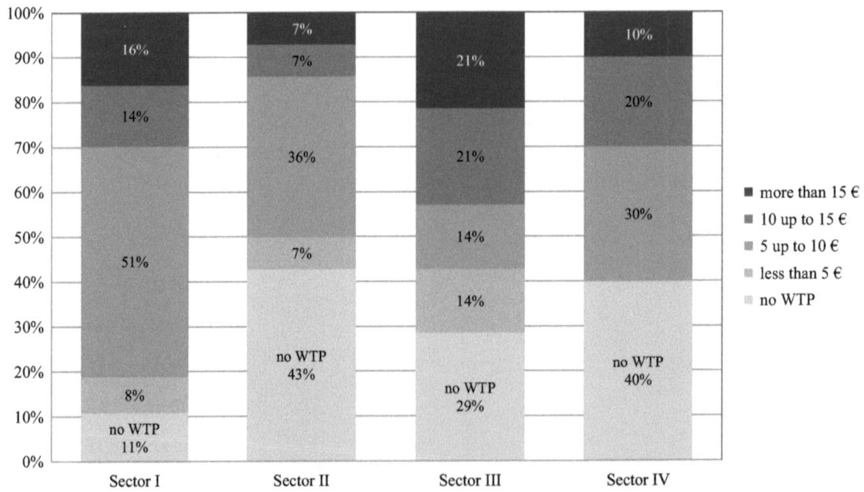

Fig. 5 Willingness to pay according to sector (N = 75). Sector I: child care; sector II: hairdressing; sector III: automotive/metalworking, dentistry, construction/agriculture; sector IV: cleaning, long-term care, hotel/gastronomy, food

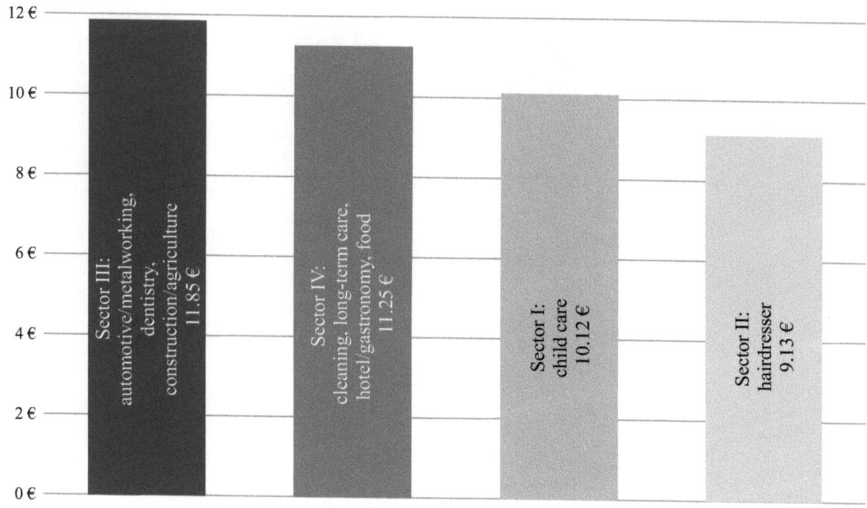

Fig. 6 Average willingness to pay according to sector (without respondents revealing zero willingness to pay) (N = 75)

sector mostly involve both the head of the center and the funding body. As a consequence, WTP in the child care sector may be over-estimated.

Figure 6 shows WTP according to sector and includes only SMEs with WTP above zero (N = 57). At 11.85 € per person and per month, the average rate for a monthly WTP was highest in the technical-based sector (III), whereas it was lowest in the hairdressing sector (II; 9.13 €).

Figure 7 shows the influence of experiences with problems due to absences caused by illness. The more severe the problems experienced, the greater was the WTP. In the group of SMEs without experience of problems caused by absences, 40 % had WTP between 5 and 10 € and another 20 % between 10 and 15 €.

Finally, we also wanted to ascertain whether there were differences between the groups' WTP in partial utilities of attribute levels, thus combining the findings of ACBC with answers from WTP interviews. These differences are shown in Figs. 8 and 9.

There were large and, in most cases, significant differences in utility values between the groups' WTP for services in all attribute categories. SMEs willing to pay find a combination of physical/mental illness prevention and team-building particularly useful; they preferred the services offered to be customized to individual employees rather than branch- or company-specific. In contrast, respondents not willing to pay did not want services offered to be customized to individual employees. A possible explanation for this difference is that the latter may suspect that costs would be higher if trainings were customized to individual employees. The group that was willing to pay for services also more strongly preferred cross-company trainings over in-house trainings than those who were not willing to pay. Mid- to long-term contracts or insurance-based contracts were preferred over

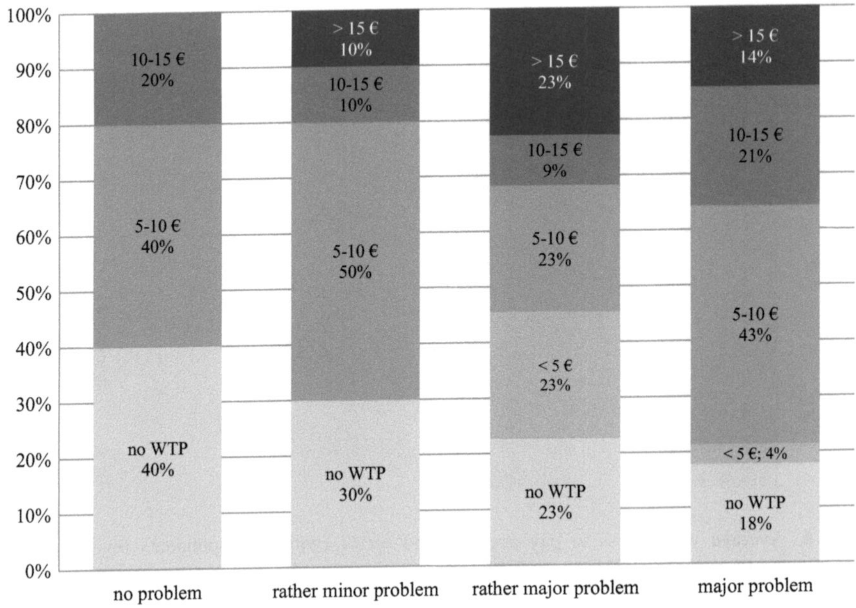

Fig. 7 Willingness to pay according to experience of problems caused by absences (N = 75)

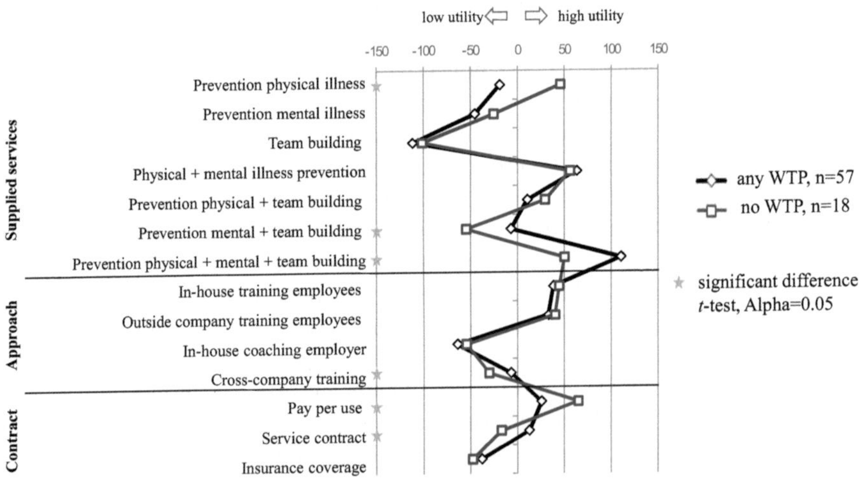

Fig. 8 Utilities by general willingness to pay (yes/no) (1/2)

contracts based on payment per use. Private supply of services was particularly attractive for SMEs willing to pay for services. SMEs that were not willing to pay preferred stand-alone services; in particular, the prevention of physical illnesses. For the latter group, affiliations with the Chamber of Trade/Chamber of Industry and Commerce were significantly more attractive than for the group that was

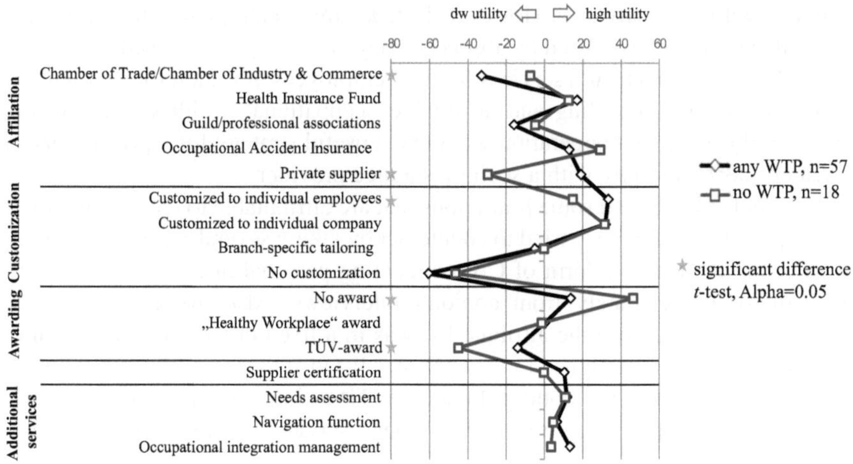

Fig. 9 Utilities by general willingness to pay (yes/no) (2/2)

willing to pay. Services offered by a Chamber may be assumed to be free of charge. For SMEs without WTP, awarding was dispensable, presumably because these respondents anticipated hidden costs associated with such awards.

4 Discussion

A major obstacle to SMEs and microenterprises providing their employees with prevention and occupational rehabilitation is a lack of qualified support structures. SMEs prefer individualized programs that take individual work-place conditions into account. Regarding occupational rehabilitation, assistance is required to coordinate the multiple players involved in the (re)integration process.

According to our findings, comprehensive service packages tailored to the needs of individual companies should be offered on a pay-per-use basis. The most preferred affiliations for private suppliers are social insurance providers, either health insurance funds or occupational health insurers. In contrast to our qualitative pre-study findings, we found that SMEs are indeed willing to pay for services. However, respondents revealed their WTP only after we had explained the services in detail and facilitated tailoring them according to their specific needs. Because the service packages are complex, it seems to be difficult for potential clients to value them. Our first recommendation to the private service unit that the "Healthy at Work" project supports is therefore to accompany market introduction with a campaign in which the facts and advantages are carefully explained. Our study further found that private supply of services is particularly attractive to employers who are willing to pay. A possible interpretation for this finding is that private suppliers supposedly provide services of a high standard. Finally, we also found that WTP within the group of microenterprises is very low and that average WTP increases with company size. However, because microenterprises are particularly vulnerable to problems caused by absences, they have the greatest need for prevention and occupational health services for their employees. It is therefore particularly important to offer appealing and affordable occupational health services to this group. In this context, cooperation would be a useful strategy for a private service provider, most advisably with a social insurance provider.

This study clearly had some limitations that are attributable to the complexity of the subject, even though we tried to counteract by taking several measures; namely, by selecting an adaptive form of CA, conducting personal interviews, and asking WTP questions separately from conjoint interviews. Also, we cannot exclude sequence effects because the attribute list was in a fixed order that likely resulted in higher utility being assigned to the attributes listed at the top of the questionnaire than to those listed at the bottom. Finally, WTP estimations must be interpreted with caution because they referred to ideal service packages for individual respondents.

References

Bourque LB, Fielder EP (2003) How to conduct self-administered and mail surveys, 2nd edn. Sage, Thousand Oaks, CA

Bridges JFP, Hauber BA, Marshall D et al (2011) Conjoint analysis applications in health—a checklist: a report of the ISPOR Good Research Practices for Conjoint Analysis Task Force. Value Health 14:403–413

Bundesanstalt für Arbeitsschutz und Arbeitsmedizin (BAuA) (2014) Volkswirtschaftliche Kosten durch Arbeitsunfähigkeit 2012 [Economic cost of absenteeism 2012]. www.baua.de. Accessed 10 July 2014 (in German)

Cacace M, Franz I, Ratz D (2014) Using conjoint analysis to elicit preferences for occupational health services in small and microenterprises. Athens J Health 1(4):237–254

Cacace M, Riegel B, Leier V (2016) Promoting workers' health in small and medium-sized enterprises: designing and evaluating a concept for preventing occupational skin diseases. In: Wiencke M, Cacace M, Fischer S (eds) Healthy at work—interdisciplinary perspectives. Springer, Berlin

Chrzan K (1994) Three kinds of order effects in choice-based conjoint analysis. Market Lett 5 (2):165–172

Clark MD, Determann D, Petrou S et al (2014) Discrete choice experiments in health economics: a review of the literature. Pharmacoeconomics 32(9):883–902

Cunningham CE, Deal K, Chen Y (2010) Adaptive choice-based conjoint analysis—a new patient-centered approach to the assessment of health service preferences. Patient 3(4):257–273

De Bekker-Grob EW, Ryan M, Gerard K (2012) Discrete choice experiments in health economics: a review of the literature. Health Econ 21:145–172

Fuchs J, Soehnlein D, Weber B (2011) Projektion des Arbeitskräfteangebots bis 2050; Rückgang und Alterung sind nicht mehr aufzuhalten [Projection of the labor force until 2050: decrease and ageing cannot be stopped]. Institut für Arbeitsmarkt- und Berufsforschung (IAB) Kurzbericht 16, IAB der Bundesagentur für Arbeit, Nürnberg (in German)

Green PE, Srinivasan V (1978) Conjoint analysis in consumer research: issues and outlook. J Consum Res 5(2):103–123

Harrison J, Woods A, Dickson K (2013) Occupational health purchasing behaviour by SMEs—a new theoretical model. Occup Med 63(73):510–512

Johnson R (2008) A perspective on adaptive CBC (What can we expect from respondents?) Sawtooth Software, Research Paper Series. http://www.sawtoothsoftware.com/. Accessed 24 Oct 2014

Krosnick JA, Holbrook AL, Beren MK et al (2002) The impact of 'no opinion' response options on data quality: non-attitude reduction or an invitation to satisfice? Public Opin Q 66:371–403

Lancsar E, Louviere J (2006) Deleting 'irrational' responses from discrete choice experiments: a case of investigating or imposing preferences? Health Econ 15(8):797–811

Luce RD, Tukey JW (1964) Simultaneous conjoint measurement: a new scale type of fundamental measurement. J Math Psychol 1(1):1–27

Miller P, Rossiter P, Nuttall D (2002) Demonstrating the economic value of occupational health services. Occup Med 52(8):477–483

Orme B (2002) Formulating attributes and levels in conjoint analysis. Sawtooth Software, Research Paper Series. http://www.sawtoothsoftware.com/. Accessed 24 Oct 2014

Orme B (2013) Which conjoint method should I use? Sawtooth Software, Research Paper Series. http://www.sawtoothsoftware.com/. Accessed 24 Oct 2014

Rao VR (2014) Applied conjoint analysis. Springer, Berlin

Ratcliffe J (2000) The use of conjoint analysis to elicit willingness to pay. Proceed with caution? Int J Technol Assess Health Care 16(1):270–290

Ryan M, Farrar S (2000) Using conjoint analysis to elicit preferences for health care. BMJ 320:1530–1533

Ryan M, Gerard K (2003) Using discrete choice experiments to value health care programmes: current practice and future research reflections. Appl Health Econ Health Policy 2(1):55–64

Ryan M, Scott DA, Donaldson C (2004) Valuing health care using willingness to pay: a comparison of the payment card and dichotomous choice methods. J Health Econ 23(2):237–258

Sawtooth Software (2014) ACBC technical paper. http://www.sawtoothsoftware.com/. Accessed 24 Oct 2014

Zelfel RC, Alles T, Weber A (2011) Gesundheitsmanagement in kleinen und mittleren Unternehmen—Ergebnisse einer repräsentativen Unternehmensbefragung [Health management in small and medium-sized enterprises—Results from a representative survey]. Das Gesundheitswesen 73(8/09):515–519 (in German)

Interpretive Patterns of Occupational Safety and Health: How Do They Affect Safety Organization and Health-Related Decisions of Enterprises? Basic Assumptions and First Impressions out of an Ongoing Project

Britta Schmitt-Howe

Implementation issues arise as a result of a range of factors including 'real world' contextual factors that are either overlooked or not captured by other research disciplines. Implementation research shines a light on those factors, providing the basis for the kind of context-specific and evidence-informed decision-making that is crucial to making what is possible in theory a reality in practice.

(Peters et al. 2013)

Abstract

In spite of a wide range of support measures for companies that should contribute to the implementation of the German occupational safety and health (OSH) law the number of companies fully complying with the core requirements of safety organization and risk assessment exceeds the number of companies running non-mandatory workplace health promotion programmes only moderately. This picture drawn by most recent findings provoked new questions in implementation research. Accordingly, it seems no longer sufficient considering the relationship between policy instruments and workplace health and safety outcome only. In fact, the understanding of employer motivation for compliance needs to be improved by asking why and how companies of different size and sector are to enhance their working environment. This is exactly what the presented research project is doing by a mixed methods approach. Since the project will be finished in 2017 this contribution is mainly confined to the description of theoretical pre-assumptions, basic objectives, research design and alignments. Nevertheless an insight is provided in preliminary conclusions based on first analyses of qualitative interviews already conducted with

B. Schmitt-Howe (✉)
Federal Institute for Occupational Safety and Health, Berlin, Germany
e-mail: schmitt.britta@baua.bund.de

M. Wiencke et al. (eds.), *Healthy at Work*, DOI 10.1007/978-3-319-32331-2_17

managers and key OSH agents in 50 companies. Interpretive patterns are to be found in the verbal material. In order to identify both implementation barriers and enablers, these patterns will be condensed into types of attitudes facilitating a subsequent quantitative validation.

1 Background

Occupational safety and health (OSH) in Germany is regulated by two legislations. These are the Occupational Safety and Health Act and the Act on Occupational Physicians, Safety Engineers and Other Occupational Safety Specialists. Companies have recourse to a wide range of supportive measures to assist them in implementing these regulations. But according to the most recent representative companies' survey by the Joint German OSH Strategy (Gemeinsame Deutsche Arbeitsschutzstrategie, GDA) conducted in 2011,[1] only 16 % of all enterprises in Germany are complying entirely with the regulations concerning workplace risk assessment. Another basic legal requirement, i.e. the provision of OSH specialists which is mandatory for German enterprises, is met only by 60 % of all companies. Comparing different sectors of OSH regulation, 60 % appears to be the maximum and 16 % the minimum compliance rate concerning aspects of mandatory safety organization. So, to explain the situation we need to identify promoting and impeding as well as other contextual factors that contribute to the question "Why get companies active in OSH management?" (cf. Parker and Lehmann Nielsen 2011, p. 2).

Compliance with the core requirements of OSH management demands basic health-related decisions, i.e. decisions mainly guaranteeing an appropriate organization and providing the necessary means to improve the safety and health protection of the employees. For example every company with at least one employee according to law has to be attended by an occupational physician as well as by an occupational safety specialist. Although there are common duties shared by both professions, a company cannot substitute the provision of the one by the provision of the other profession. So every company is obliged to sign two provision contracts: one with an occupational physician and one with an occupational safety specialist. But 40 % of all companies do not receive support from occupational safety specialists; 60 % do not provide services by occupational physicians (Sommer and Schmitt 2012).

Compliance with occupational safety regulation is particularly low when it comes to small and medium sized enterprises. As the bulk of enterprises in Germany (95 %) have less than 50 employees, provision gap is particularly salient with enterprises of this size: 43 % of them are without any supervision at all,

[1]The latest relevant data are found in the 2011 GDA companies' survey. The cross-sectional survey was part of the global evaluation of the Joint German OSH Strategy 2008–2012. A social research institute questioned 6500 enterprises by telephone about basic tasks involved in their OSH programs, for example their risk assessment processes and information for and training of employees about OSH, as well as the prevention culture prevailing in these work-places. A follow-up survey has been conducted in 2015 to evaluate the current German OSH Strategy. The results of this follow-up survey are not yet published.

i.e. they are neither attended by an occupational physician nor by an occupational safety specialist, although they are not participating in the so-called alternative provision model.[2] In contrast, there is almost no gap in provision as large companies with 250 employees and above are concerned and only a small gap (5–10 % without provision) is observed with companies with 50–249 employees (Schmitt 2014, p. 15). However, the situation is considerably worse with reference to risk assessment practice: only 51 % of all German companies do conduct risk assessments and—out of these—only 69 % pass through all mandatory steps of the risk assessment procedure. In fact, only 16 % of all German companies are able to reach the final step of efficacy testing within the risk assessment procedure (Schmitt and Hammer 2015, p. 203). Among those 51 % German companies that do conduct risk assessments the overwhelming majority, i.e. more than 90 % reflect the more obvious hazard sources like workplace layout, workplace environment and equipment. But a considerable proportion of 44 % does not incorporate the less obvious factors such as organization at the workplace and industrial organization, although reflecting work organization within the risk assessment is required by law (Nationale Arbeitsschutzkonferenz 2013, p. 79).

One might expect that companies tend to implement legally binding demands to a higher extend then voluntary standards. Consequentially, the implementation rates of work-place health promotion measures not mandated by OSH regulations should be significantly lower. But, counter-intuitively, voluntary measures can nearly compare with implementation rates of regulations by law (Ahlers and Brussig 2005; Ahlers 2011; Beck and Lehnhardt 2009; Beck et al. 2015). In the field of voluntary workplace health promotion, on average 36 % of all companies have implemented health promotion management. This share climbs up to 47 % for enterprises with more than 200 employees (Bechmann et al. 2011, p. 11). However, classifying voluntary workplace health promotion activities into different levels, the GDA data show that only "9 % of all surveyed companies exhibited the most comprehensive type of workplace health promotion including analysis, individual-directed prevention measures and participatory groups for promoting the improvement of working conditions (Beck et al. 2015, p. 1)." These findings point out that the desired outcomes by and large are not (yet) achieved, neither in the mandatory sphere of OSH nor in the non-mandatory sphere.

Legislation is usually made in anticipation of almost 100 % compliance. In the field of OSH law, however, compliance rates exceed the implementation rates of voluntary workplace health promotion activities only in a modest way. This is where implementation research comes into play. As Hasle et al. (2014) point out,

[2]The alternative provision model is available for companies up to 20 or in some cases up to 50 employees. After having attended a special seminar at the statutory accident insurance the entrepreneur him or herself can substitute both kinds of provision near-completely. However, if a company works for the first time with special hazardous substances or in case of other special inducements precisely described in the comprehensive catalogue of German Social Accident Insurances "regulation 2" the entrepreneur running the "alternative provision model" yet has to consult an occupational safety specialist or an occupational physician.

"research in state regulation is mainly aimed at compliance and efficiency of public administration, while little attention is paid to why and how public and private organizations subsequently are to improve their working environment" (p. 73). Implementation research in the field of working conditions addresses the relationship between policy instruments and workplace health and safety outcomes. Only "recently, regulation research has started to move towards an understanding of employer motivation for compliance" (Hasle et al. 2014, p. 74). As far as for example the mandatory process of risk assessment is concerned, compliance does not only mean applying solid (technical) knowledge to business processes but initiating a continuous improvement process of identifying hazards, assessing severity and probability of occurrence, deploying appropriate preventive measures and evaluating efficacy of these measures. According to law, these steps have to be repeated regularly like in a plan-do-check-act (PDCA) cycle. The legislator explicitly addresses key actors or agents within the organizations to deal with the required implementation processes. Key actors are the employer, for example, who often is represented by specialists and executive staff, as well as working councils representing employees. If implementation research wants to learn about what makes organizations take action in improving working conditions continuously, it has to put a focus firstly on certain occupational roles, predefined by law but also empirically identified as change agents in organizational behavior (Hale and Borys 2013; Elke et al. 2015) and secondly on prevention culture as a mode of organizational culture. Change agent roles are mainly taken by managers. But they are also provided by occupational physicians, occupational safety specialists, members of working councils, or by workplace health promotion managers. So the better we understand the motives of these change agents for getting active in OSH issues, "the better we can develop our knowledge about the mechanisms which will make the policy instruments work in an efficient manner" (Hasle et al. 2014, p. 76). Efficient manner in this context does not only mean to meet requirements formally. Policy instruments addressing the continuous improvement process can work in an efficient manner only if they are embedded in an appropriate occupational prevention culture.

As it is evident that core OSH policy instruments, such as regulation concerning the above mentioned mandatory risk assessment process, are not yet working truly efficiently, the Federal Institute for Occupational Safety and Health has just recently begun to explore new avenues in implementation research by focusing on the interpretive patterns of safety and health in enterprises. Accordingly, these interpretive patterns can be seen as specific modes of shared basic assumptions concerning activity, which are representing the deepest and hardly observable level of organizational culture. According to Ed Schein, concepts are needed permitting to "differentiate between organizations within a society, especially in relation to different levels of effectiveness". As Schein further states, "the concept of organizational culture has served this purpose well (Schein 1990, p. 110)." In the context of this article, effectiveness in OSH is an important dimension to differentiate between organizations. In order to understand relevant aspects of the genesis of these differences, it is necessary to look at prevention culture as a way or mode of "doing" organizational culture, just in the sense of the general culture description of "how we do things around here." Accordingly, "there is a prevention culture, where prevention is the way we do things around here, i.e. when prevention and promotion of health, safety and well-being at work are mainstreamed into business practices or macro policies and the behavior of key agents (Zwetsloot 2014, p. 30)."

Getting active in OSH issues always implies to put an emphasis on health risks. According to Niklas Luhmann, the concept of risk has emerged relatively late in human history (Luhmann 1991, p. 24) and can be defined as follows:

"For we can speak of risk only if we can identity a decision without which the loss could not have occurred (Luhmann 2008, p. 16)."[3] Only if possible damages can be ascribed to decisions, hazards can at least partly be controlled, i.e. can be changed into risks. Risk assessments usually multiply damage severity by probability of occurrence. Thus prevention is defined by Luhmann as "preparing for uncertain future losses by seeking to reduce either the probability of occurrence of losses or their extent" [4] (Luhmann 2008, p. 29). If this is "the way we do things around here", prevention culture is a mode of organizational culture tending to evaluate health risks permanently within the business process, making rational decisions and taking the possible consequences both of everyday as well as of key policy decisions and of omissions into account. Since "cultural origins and dynamics can sometimes be observed only in the power centers where elements of the culture are created and changed by founders, leaders, and powerful managers" (Schein 1990, p. 111), in the given context of OSH issues the above mentioned key agents of implementing a prevention culture are interviewed in the research project "*Variables that promote and impede implementation of OSH measures and their relationships with risk related characteristics of enterprises*" as described in detail below.

Implementation studies aimed at investigating motives and health-related intentions of key agents creating elements of organizational culture must collect data concerning how interpretive patterns of OSH interact within organizations in the process of decision making in order to observe the origin of changing procedures to improve safety and health. To understand more about the interpretive patterns that are assumed key to whether and how enterprises are engaging with OSH, the Federal Institute has conducted qualitative interviews with owners, leaders, specialists and members of working councils in enterprises of different size and sector. Starting from the basic background described above, our explorative mainly qualitative study deals with the central question concerning the relevant explicit and implicit orientations of entrepreneurs, in-plant OSH

[3]Original German version: "(...) von Risiko spricht man nur, wenn eine Entscheidung ausgemacht werden kann, ohne die es nicht zu dem Schaden gekommen wäre (Luhmann 1991, p. 25)."

[4]Original German version: "Unter Prävention soll hier ganz allgemein Vorbereitung auf unsichere künftige Schäden verstanden werden, sei es dass die Eintrittswahrscheinlichkeit, sei es dass die Höhe des Schadens verringert wird (Luhmann 1991, 38)."

professionals and working councils seen as the relevant prevention culture molders. The following chapter outlines the objectives and alignments of this project and reports some initial empirical impressions as well as an overview of the next steps.

2 Pre-assumptions and Alignments

Based on a secondary analyses of the 2011 companies' survey conducted in the context of the global evaluation of the Joint German OSH Strategy, Sommer and Schmitt (2012, p. 68ff.) derived a preliminary impact model shown in Fig. 1.

The quantitative data of the basic study raised questions about the meaning behind relevant statistical correlations. The preliminary impact model therefore expresses our first hypothesis about the directions of impact correlations. The logic model takes into account "hard" factors like sector of company, company size, OSH structures, and in-plant availability of legal knowledge as well as to some extent, "soft" factors, such as self-reported engagement with OSH. It depicts the theoretically assumed impact relations aligning our qualitative conversation guide.

Alignment was also needed for the sampling process of the project, which partly could also be provided by secondary analyses of GDA company survey data 2011. These analyses showed that differences in sector of company played a major role in OSH implementation, especially in the small (up to 49 employees) and smallest (up to 9 employees) enterprises. There were major differences in OSH engagement between enterprises with less than 50 and those with 50 or more employees; namely, on average, the OSH performance of the latter was considerably better.

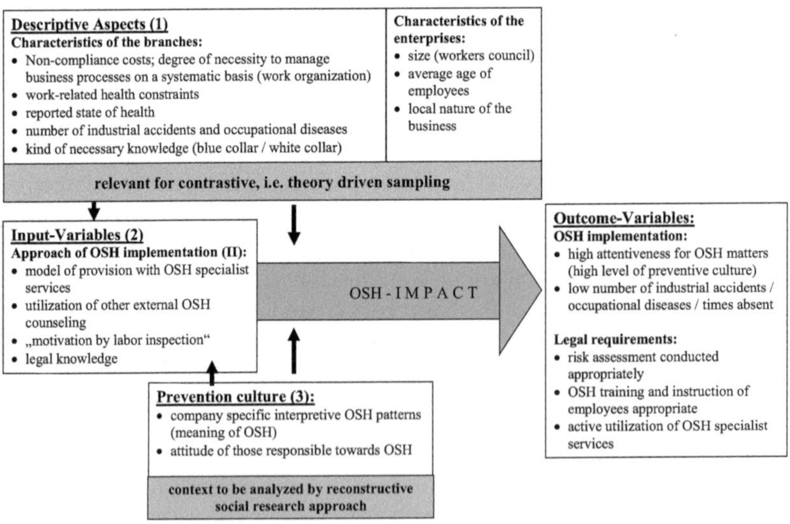

Fig. 1 "*Variables that promote and impede implementation of OSH measures and their relationships with risk related characteristics of enterprises*": Preliminary impact model

Among these larger enterprises, there were smaller differences in OSH performance between economic sectors than between larger and smaller companies. Nevertheless, some differences between sectors in outcome variables such as assessment of risk were identified. The lowest rates of performing risk assessments were found in GDA sector group[5]containing real estate broking and letting, lawyer services, public relations, engineering and other services predominantly for enterprises (36 %), in communication, finance and similar services (37 %) and in consumables, retail trade and hospitality industries (49 %), whereas the highest rates occurred in GDA sector group nutrition industry and farming (72 %), in production of capital and investment goods (68 %), and in health care together with social services (64 %). Data were similar regarding intermediate OSH variables like provision of occupational safety specialists' and occupational physicians' services or of OSH information and instruction.

The research team of the Federal Institute suggested two possible explanations for these differences between sector groups: First that every group faces different hazards and exposures, and second, that different sector groups are subject to different degrees of systematic surveillance of their business processes by authorities. As a consequence, companies' ability to avoid penalties for non-compliance is tentatively higher in sectors that also have to comply with consumer safety and environmental law such as for example waste industry and food industry.

The findings of these secondary analyses thus gave rise to the following two hypotheses: (1) the more obvious the hazards in the work-place, the more intense the OSH engagement of the company; and (2) the higher the potential costs of non-compliance in spheres that impinge on OSH such as consumer safety and environmental law, the higher the probability of intense OSH engagement.

These hypotheses were used to construct a two-dimensional coordinate system for clustering sector groups with a similar level of safety risks to enable sampling of contrasting businesses for the qualitative part of the project. As shown in Fig. 2, sector groups—regardless whether they are highly, mediocrely and poorly performing with reference to OSH outcomes—were clustered into the four quadrants of the coordinate system generated by the horizontal dimension of obviousness of hazards and the vertical dimension of potential costs of non-compliance.[6]

---

[5]The GDA company survey 2011 is based on a pooling of NACE sector codes into risk-oriented sector groups. NACE is an abbreviation for the French "Nomenclature statistique des activités économiques dans la Communauté européenne". The NACE system is known as the European Statistical Classification of Economic Activities developed by the European Commission to harmonize official business statistics within the European Community.

[6]Potential costs result from non-compliance with OSH regulation itself or in fields impinging on OSH, i.e. with environmental law or with consumer rights. Service sectors for example—especially knowledge based services and services predominantly for enterprises—occur a small risk of attracting sanctions, whereas they face huge potential costs for non-compliance with money laundering or anti-corruption law. The insignificance of the risk of non-compliance with OSH

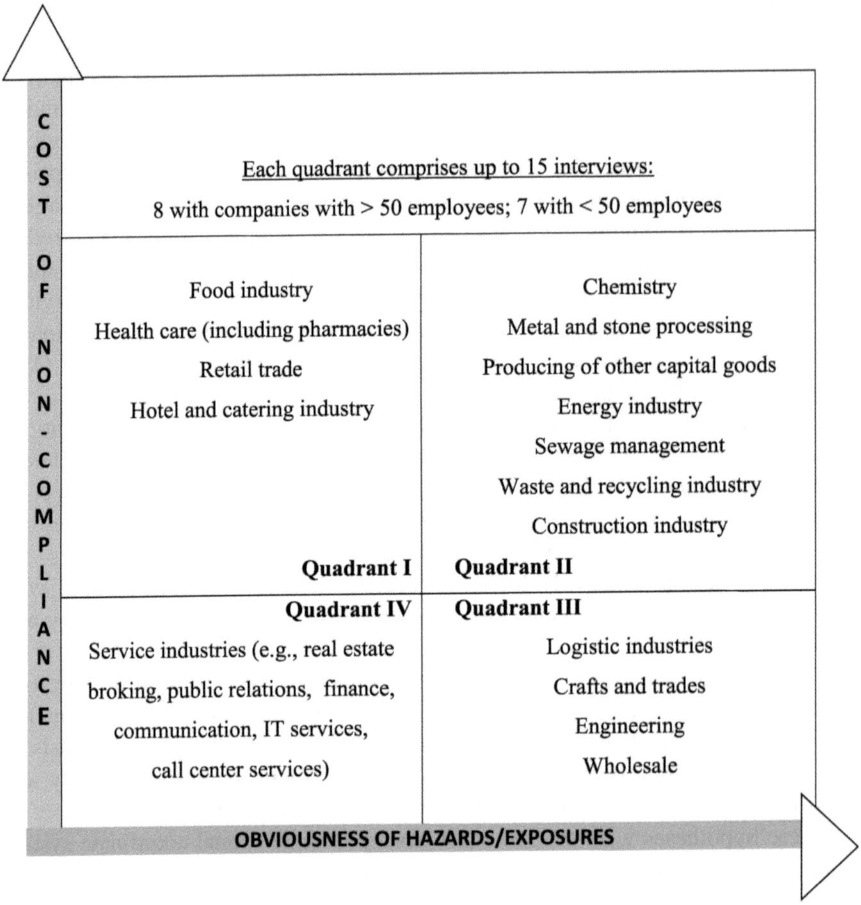

Fig. 2 Sampling technique

How sector groups were allocated to this coordinate system will be illustrated quadrant by quadrant in the next few paragraphs. To assist the decision on allocation to quadrant I, II, III or IV, among others the regulations of the mandatory OSH provisions in Germany were checked: The number of working hours occupational safety professionals and physicians must contribute to provide basic supervision is determined by the number of employees and to whether the company has high, medium or low risk as defined by the German Social Accident Insurance (Liersch

---

and environmental consumer protection law is partially attributable to the increasing time in-between inspections: there are less labor inspectors in the Federal States of Germany than in the past and labor inspectors for the statutory accident insurance institutions do not have enough time to visit small and medium enterprises in low-risk sectors of businesses, such as those in the service sector, on a regularly basis.

2014; Sommer and Schmitt 2012; Priester 2005; Marquardt et al. 2010; DGUV Regulation 2 2012). So the German Social Accident Insurance's high, medium or low risk categorization of economic sectors was one source of allocation to quadrants within the project, since this categorization is connected with our concept of obviousness of hazards.[7] Other sources of allocation in this dimension where (a) the GDA companies survey 2011, (b) the supplementary OSH survey as part of the German micro-census 2009 and 2013, (c) the Report—The State of Occupational Safety and Health in the European Union—Pilot Study of the European OSH agency (EU-OSHA 2000) and (d) the risk-oriented concept of companies' control developed by the OSH supervisory authorities of the Federal states (LASI 2014, p. 22ff.).

Allocation in the dimension of possible costs for non-compliance was both guided by estimation based as far as ascertainable on the number of OSH surveillance visits, regulatory actions, scandals and criminal investigations per sector (BAuA 2014; DGUV-Statistiken für die Praxis, 82) and by findings of the second European Survey of Enterprises on New and Emerging Risks (EU-OSHA 2015).

Quadrant I: Retail trade, hotel and catering industry, food industry and health care (GDA sector groups I, VII and XI) were characterized by low risk of accidents and insufficient obviousness of physical hazards but high risk of non-compliance costs, since companies of these sectors face sanctions if they for example do not observe basic hygiene requirements. Non-compliance can be costly for such businesses either indirectly via loss of clients or patients because of poor hygiene or directly via official penalties. These sector groups are affected less by obvious hazards and more by costs associated with non-compliance concerning consumer or patient safety requirements. Therefore they were allocated to quadrant I. Accordingly quadrant I contains all sectors with higher potential costs for non-compliance than obviousness of hazards.

Quadrant II: By contrast, chemistry, metal and stone processing and producing other capital goods as well as sewage management, energy and construction industries (GDA sector groups II and III) are characterized both by relatively highly obvious hazards and high possible costs associated with non-compliance, especially in the field of environmental law. Therefore they are allocated to quadrant II, which is the high-risk quadrant in both dimensions since it is most distant from the *imagined* neutral point[8] in the coordinate system.

Quadrant III: Logistic industries, crafts and trades, engineering and wholesale (GDA sector groups IV, V and VI) were allocated to quadrant III because the

[7]Even though the level of emotional stress presumably is in many economic sectors at least as salient as the danger of falling, stumbling, or developing musculoskeletal disorders, we classify emotional stress to be less obvious. Therefore emotional stress is not integrated in the concept of high obviousness of hazards.

[8]Since there are no metric variables depicting conspicuousness of hazards or possible costs for non-compliance there is no "neutral point" of the described coordinate system in the proper meaning of the word. Data depicting the relevant features at best are ordinal scaled. Therefore it is only valid to speak of an *imagined* neutral point.

obviousness of hazards is greater for enterprises belonging to these economic sectors than the risk of sanction.

Quadrant IV: The already mentioned service sector was allocated to quadrant IV, which is closest to the *imagined* neutral point. Quadrant IV contains all sectors with low level in both obviousness of hazards and possible costs for non-compliance with OSH, environmental or customer safety law. These sectors are namely real estate broking and letting, lawyer services, public relations and other services predominantly for enterprises as well as communication, finance and similar services (GDA sector groups VIII and IX).

It has to be added, that businesses of the sector groups belonging to quadrant IV are likely to have a highly restrictive approach to OSH, since the OSH outcome rates are the lowest in these sectors. In contrast, businesses of the sector groups belonging to quadrant II are likely to have a highly conducive approach to OSH, since their OSH outcome rates are among the highest. The described pooling of economic sectors in quadrants enables the selection of widely contrasting companies because it separates them by characteristics and contexts that are expected to influence OSH performance systematically. The biggest differences were expected between companies in diagonally opposite quadrants, namely II versus IV, and I versus III.

It was expected that the main OSH topics would differ between the four quadrants, i.e. between the sector groups they contain. Accordingly, the way company representatives thought and spoke about OSH would likely differ between economic sectors and sizes of companies. The clustering produced by the coordinate system (Fig. 2) thus allows a theory-driven sampling of contrasting businesses for qualitative interviews.

3 Project Objectives, Approach and Research Design

Objectives

The basic objective of the ongoing project is to identify how interpretive patterns of OSH act as enablers of or impediments to companies' OSH initiatives and engagement in various contexts. Thus, qualitative interviews will clarify company-specific approaches to and daily routine in implementing OSH management tasks. It is assumed that describing the company-specific approach in "doing OSH" is always driven by specific interpretive patterns sometimes explicitly, more often implicitly expressed by interview partners while answering for example to the question: "Can you please describe a typical decision process in OSH matters in your company?"—Hermeneutic interpretation of the given answers to this as well as to other questions can reveal for example what are seen as components of this management task and what are not, what are seen or implicitly acting as action drivers and impediments in the specific framework of shared experience in one company's case.

Methodology of data collection

For data collection two types of qualitative interviews are used: In enterprises with 50 or more employees as a rule a group interview is conducted with 3–4 representatives of relevant in-plant function groups, i.e. (1) one chief executive as the addressee of all OSH acts and regulations; (2) one of the two OSH experts whose provision is mandatory for each company employer in Germany, namely either an occupational safety specialist or occupational physician who works in-plant or as an external consultant for the company; and (3) one trade union representative if there was one. Thus, each interview would be performed with one manager, one employee representative, and one internal or external health and safety expert. In enterprises with less than 50 employees the semi-structured interview is conducted with the working proprietor only. These two interview options both are conducted in a problem-centered style (Witzel 2000; Witzel and Reiter 2012).

The assumption informing this approach was that the four above-mentioned function groups are those primarily negotiating the actual configuration of OSH structures and decisions within a company; i.e. for example how risk assessments are conducted as well as the design of special OSH and work-place health promotion measures that influence the occupational prevention culture. In other words: the four function groups are seen as drivers of change or “powerful managers” (Schein 1990) in OSH.

Because companies—and not individuals—are the focus of our study, small group interviews were further considered an appropriate instrument for observing parts of the negotiation processes between the relevant occupational function groups, including both the spoken word and the kinds of interactions that occurred between the respondents during the interviews. Negotiation processes lead to decisions. Since the decisions adopted reflect specific organizational values and actions resulting from interpretive patterns of safe and healthy work, attention to these interpretive patterns while analyzing text passages from the transcribed interviews would tell us a lot about various aspects of organizational culture.

This would also be true of the one-respondent interviews with small business owners or chief executives because they would also make decisions according to their own values and interpretive patterns. In these cases, the chief executives’ or (working) owner’s interpretive patterns of OSH could be assumed to be decisive for the company’s tendency to getting more active in managing OSH issues or not. Thus, it was considered that working proprietors would provide sufficient insight into the OSH relevant elements of the organizational culture of their enterprises. However, if the proprietor wanted to bring in a second respondent, for example in the case of two equal-ranking chief executives, this was accepted.

The qualitative interviews conducted within our project attempt to capture different views, descriptions and stories by raising the following four main topics: (1) anchoring and roles in OSH; (2) change processes and cooperation, including looking back at the companies’ histories; (3) in-plant communication about OSH and learning; and (4) corporate OSH identity and motivation. All interviews are to take place in the company’s premises.

Sampling

The companies selected for interviews could be located in any part of Germany. The interviewees were recruited through intermediaries such as chambers of crafts and trade, trade unions, employers' associations, regional business networks, science networks, and sometimes also by directly contacting companies cited in the business sections of newspapers. Being aware of the unavoidable tendency toward positive selection of companies willing to be interviewed about OSH matters, we anticipated that companies that were less concerned about the safety and health of their employees were unlikely to be among our respondents. Nevertheless, some of the company representatives interviewed expressed considerable skepticism about OSH regulations or work-place health promotion and made no attempt to hide their reservations. To ensure that the contacted enterprise representatives expressed themselves freely, there was one invariable sampling rule set by the Federal Institute; namely, that recruitment is never organized by labor inspectors. This ensures that the interviews are not constrained and provides the strongest standard of data privacy.

To assist the decision on the cut-off for the size of companies in our study, both the regulations of the mandatory OSH provisions and the percentage of actually provisioned small businesses in Germany were checked again resulting in the following consideration. Since the alternative supervision model is available for small enterprises and since the lack of OSH provision is biggest among them, it can be assumed that occupational safety specialists are rarely employed by small enterprises for more than a couple of hours a year. These companies also rarely have trade union representatives. Thus, these functional groups are hardly available as interviewees in small businesses. Furthermore proprietors working in their own enterprise determine OSH culture of their companies all the more, as they are often the drivers within their own alternative OSH supervision model. Thus, this was an additional pragmatic reason for conducting qualitative interviews with the working proprietor or chief executive only in small enterprises and for setting the dividing line between small and large companies at the turn from 49 to 50 employees.

Our ambitious initial target was to conduct maximally 60 qualitative interviews, which means up to approximately 30 interviews with small and up to approximately 30 interviews with medium or bigger companies. However, the number of interviews added up only to 50, since a point of theoretical saturation was reached then. Theoretical saturation as a methodological principle means that point of time additional interviews will no longer yield any new concepts, categories, or variables (Strauss and Corbin 1996, p. 159). At the point of theoretical saturation additional interviews neither enable further enrichment of already developed categories any more. Since the enrichment of the categories is done not only by desk work consideration but also by ongoing theoretical sampling, this was be the point of closing the sampling. Till the point of saturation was reached theoretical sampling in case of our project meant searching within the four quadrants for new companies' cases that are probably different from the already analyzed cases especially in specifications of one or several already developed categories.

Data triangulation

During the interviews, inquiries are also made about barriers to acceptance and implementation, and attempts made to notice specific features of knowledge or consciousness, and circumstances facilitating implementation. In both cases, i.e. in small as well as in bigger enterprises, the interviews were additionally accompanied by sighting of OSH documentation in order to check if documentation is up to date and safety measures might be carried out like reported in the interview. Sighting of OSH documents if when provided is also helpful to learn about the various approaches companies choose to conduct risk assessments in practice. Thus, the interviewer will take notes of every document sighting as well as of his or her general impressions of the company.

Altogether the chosen approach of data collection generated three kinds of data: (1) verbal material of group interviews, (2) verbal material of one-respondent interviews and (3) research notes. By diversification of data sources and settings we pursue the methodological strategy of triangulation (Flick 2011).[9] This is in line with the following recommendation of Glaser/Strauss:

> Different kinds of data give the analyst different views or vantage points from which to understand a category and to develop its properties; these different views we have called slices of data. While the sociologist may use one technique of data collection primarily, theoretical sampling for saturation of a category allows a multi-faceted investigation, in which there are no limits to the techniques of data collection (Glaser and Strauss 1967, p. 65).

According to Grounded Theory, ongoing comparisons are central to recognizing specifications of categories (Strauss and Corbin 1996, p. 44ff.). These specifications are most readily perceived when there is maximal contrast, even when the contrast initially seems too great to contribute to any understanding of the phenomenon under scrutiny. However, the actual sampling strategy had to take into account that searches for valid patterns of interpretation comprise only searches for similarities within differences or differences within similarities. Thus, it was considered that type building was easier if there were always at least two companies of the same sector and size in each quadrant, which also facilitate masking of data by mixing sector details of at least two cases; therefore singular instances in the majority of cases had been avoided.

Data analyzing and triangulation of methods

The group interviews as well as the one-respondent interviews seek to generate narratives about examples of occupational practice and OSH-relevant decisions as well as verbal statements of company specific approach, vision and opinion. Thus, the transcribed interview material reflect a wide variety of perspectives about what is seen by German companies as successful or adequate prevention of work-related

[9]By triangulation we understand a research strategy applying different methods and approaches to the same phenomenon or using various data to explore one phenomenon. By doing so, this strategy intends to compensate the weaknesses of the one approach by the strengths of the other.

health and safety issues. The whole verbal material is to be analyzed according to the methodology of Grounded Theory, in which specific context-driven types of decision-making as well as negotiation modes are searched for, the latter being especially important in larger enterprises. Grounded Theory starts analyzing of every verbal material by labeling phenomena with so called concepts. Subsequently concepts are grouped to form categories. Categories, which are seen as "higher level, more abstract concepts", are enriched by "developing them in terms of their properties and dimensions" (Corbin and Strauss 1990, p. 420), and finally they are interrelated to each other. Clearly defined contents of manifest verbal material, i.e. phenomena like concrete actions within a negotiation process are necessary for all these steps of theory building. Interpretive patterns however are often articulated implicitly, i.e. between the lines and express rather an attitude or an approach. They often come into view only if the whole case, the whole story of a specific enterprise is considered. In order to find an easier access to this kind of interpretive patterns our research team decided to complement Grounded Theory methodology with the hermeneutic approach of documentary method which helped us not only to do initial case studies but to integrate indicators for only implicitly articulated interpretive patterns into our code system.

For example, reconstruction of crucial turning points in the company's history is vital in ascertaining how internal or external impulses are perceived and how they affect organizational behavior by inducing change processes in OSH. Thus, if crucial turning points were reported in an interview or if a vivid interaction between interviewees is observed, this case might be seen as a cornerstone case that shall be initially analyzed by the more complex documentary method which is based on sociological theory of knowledge (Bohnsack 2010, p. 22).

As a contribution to the triangulation of methods, a proportion of appropriate interviews has thus been interpreted with the help of Bohnsack's documentary method to further inform the developing Grounded Theory code system. Since the documentary method draws on the aims of sociology of knowledge regarding the generation of social types and already has the required tools to do so, this currently helps to identify types of interpretive patterns in enterprises and validate such identification in combination with Grounded Theory codings. However not every interviewed company reports change processes and not every interview transcript documents a vivid story. These are the cases that we use for comparison but as a general rule not for hermeneutic analyzing in greater depth.

Hermeneutic in-depth analyses according to documentary method were developing core cases' interpretation by a research team in two basic steps—the so called formulating interpretation and the so called reflecting interpretation leading to a case study. These hermeneutic case studies help to develop interpretive types of enterprises as well as to enrich the Grounded Theory code system for analyzing all the material. Currently twelve case studies are already completed (effective: June 2016). It was intended to conduct hermeneutic in-depth analyzes of core cases for both small and bigger companies in every quadrant.

Thus we will show specifically how enterprises of different sector and size, and perhaps also of similar sector and size, deal with OSH tasks and why the identified approaches are adopted; that is, in what social context. The field work began in June 2014 and continued until December 2015.

4 Preliminary Conclusions, First Impressions and Prospects

Altogether, 28 interviews with larger and 22 interviews with small companies have been conducted (effective June 2016). Although these interviews have been spread over all four quadrants, identification of meaningful differences and similarities has so far only been accomplished for pharmacies, bakeries, joineries, sheltered workshops, chemical companies, sewage companies, and IT services. Obviously, these comparisons are only feasible with two or more cases per sector. Hermeneutic interpretations of core cases and detailed Grounded Theory coding exercises of the available interview material have so far shown the following three possible interpretive patterns of OSH:

(i) The working proprietors of small businesses in particular, frequently perceive OSH as a dimension of professional training and believe that they and their employees have undergone more or less the same vocational education. However, OSH is one very small component of vocational training. Thus, these respondents believe that if workers or craftsmen want to be professionals, they only need to comply with the OSH rules they learnt in their job training and nothing other than their professionalism is required. Because flat hierarchies often prevail in small businesses, this interpretive pattern leads to a laissez-faire attitude toward OSH.

(ii) In sharp contrast with this first pattern of interpretation, we can identify a social type of larger enterprises in which managers, employee representatives, and OSH specialists strongly believe that OSH matters are means of self-protection. In this context, which occurs especially in high-risk types of business like the sewage industry, self-protection is understood by those responsible as protection against shock and feelings of guilt rather than protection against hazards itself. When they think about OSH tasks in their company, the central concern of authorities is the avoidance of fatal occupational accidents that would force them to notify the bad news to the relatives of the accident victim.

(iii) The third pattern of interpretation of OSH that has been generated from the material obtained thus far concerns allegiance to a desire to avoid investigation or imposition of penalties by authorities while simultaneously believing that the requirements of the OSH regulations are excessively formal. For this type of enterprise, which we have found in both small and larger companies, the only motivation to attend to matters of OSH is extrinsic.

With ongoing type building, these preliminary findings will be refined, contextualized, completed, and supplemented. Only then will the findings be sufficiently mature to be validated by quantitative methods.

On the basis of the qualitative research findings, namely a flexible OSH impact model according to different sectors of enterprises, the Federal Institute for Occupational Safety and Health plans a quantitative validation by a structured survey of about 600–1000 companies after completion of the qualitative research part in autumn 2016. The theory thus developed in two steps—firstly qualitative, secondly quantitative—would likely assist governmental and non-governmental OSH players to determine how to activate companies with different preventive cultures under different professional and economic conditions. Predominantly OSH professionals and supervision authorities will benefit from the project outcomes since they will be enabled to identify the specific OSH type of each enterprise and thus can express their input in ways that make sense not to themselves, but to the companies.

References

Ahlers E (2011) Wachsender Arbeitsdruck in den Betrieben. In: Kratzer N, Dunkel W, Becker K et al (eds) Arbeit und Gesundheit im Konflikt. Analysen und Ansätze für ein partizipatives Gesundheitsmanagement. edition sigma, Berlin, pp 35–60 (in German)

Ahlers E, Brussig M (2005) Gefährdungsbeurteilungen in der betrieblichen Praxis. WSI-Mitteilungen 58(5):517–523, available via: http://www.boeckler.de/wsi-mitteilungen_53826.htm (in German)

BAuA (ed) (2014) Sicherheit und Gesundheit bei der Arbeit 2013. Unfallverhütungsbericht Arbeit. Bundesanstalt für Arbeitsschutz und Arbeitsmedizin, Dortmund, available via: http://www.baua.de/dok/5746626 (in German)

Bechmann S, Jäckle R, Lück P et al (2011) iga Report 20. Motive und Hemmnisse für Betriebliches Gesundheitsmanagement (BGM). Umfrage und Empfehlungen, 2nd edn. BKK Bundesverband, Essen, available via: http://www.iga-info.de/veroeffentlichungen/iga-reporte/iga-report-20.html (in German)

Beck D, Lehnhardt U (2009) Verbreitung der Gefährdungsbeurteilung in Deutschland. Prävention und Gesundheitsförderung 4(1):71–76 (in German)

Beck D, Lenhardt U, Schmitt B et al (2015) Patterns and predictors of workplace health promotion: cross-sectional findings from a company survey in Germany. BMC Public Health. doi:10.1186/s12889-015-1647-z

Bohnsack R (2010) Rekonstruktive Sozialforschung. Einführung in qualitative Methoden, 8th edn. Verlag Barbara Budrich/UTB, Opladen, in German

Corbin J, Strauss A (1990) Grounded theory research: procedures, canons and evaluative criteria. Zeitschrift für Soziologie 19(6):418–427

DGUV (2012) DGUV Vorschrift 2. Anlage 2 Abschnitt 4 des Mustertextes (Vollständige Liste der Zuordnung der Betriebsarten zu den Betreuungsgruppen mit den Angaben aller Unfallversicherungsträger gemäß Anlage 2 Abschnitt 4 des Mustertextes der DGUV Vorschrift 2 in der Fassung vom 1.1.2012). Available via: http://www.dguv.de/de/Prävention/Vorschriften-Regeln-und-Informationen/DGUV-Vorschrift-2/Downloads/index.jsp (in German)

DGUV (2013) DGUV-Statistiken für die Praxis. Available via: http://www.dguv.de/de/Zahlen-und-Fakten/Broschüren/index.jsp (in German)

Elke G, Gurt J, Möltner H, Externbrink K (2015) Arbeitsschutz und betriebliche Gesundheitsförderung—vergleichende Analyse der Prädiktoren und Moderatoren guter Praxis [Occupational safety and health and workplace health promotion—comparative analysis of predictors

and moderators of good practice]. Bundesanstalt für Arbeitsschutz und Arbeitsme-dizin, Dortmund (in German)
EU-OSHA (2000) Report: the state of occupational safety and health in the European Union—pilot Study. Available via: https://osha.europa.eu/en/tools-and-publications/publications/reports/402/view
EU-OSHA (2015) Zweite Europäische Unternehmensbefragung über neue und aufkommende Risiken—ESENER-2. Available via: https://osha.europa.eu/de/surveys-and-statistics-osh/esener (in German)
Flick U (2011) Triangulation, 2nd edn. VS Verlag für Sozialwissenschaften, Wiesbaden
Glaser B, Strauss A (1967) The discovery of Grounded Theory: strategies for qualitative research. Aldine, Chicago
Hale A, Borys D (2013) Working to rule, or working safely? Part 1: A state of the art review. Saf Sci 55:207–221
Hasle P, Limborg HJ, Nielsen KT (2014) Working environment interventions: bridging the gap between policy instruments and practice. Saf Sci 68:73–80
LASI (ed) (2014) Überwachungs- und Beratungstätigkeit der Arbeitsschutzbehörden der Länder—Grundsätze und Standards. Länderausschuss für Arbeitsschutz und Sicherheitstechnik. Available via: http://lasi-info.com/publikationen/lasi-veroeffentlichungen (in German)
Liersch A (2014) Arbeitsunfälle und arbeitsbedingte Gesundheitsprobleme. Ergebnisse einer Zusatzerhebung im Rahmen des Mikrozensus 2013. In: Statistisches Bundesamt (ed) Wirtschaft und Statistik, Wiesbaden (in German)
Luhmann N (2008) Risk: a sociological theory. Aldine Transaction, New Brunswick
Luhmann N (2012 [1991]) Soziologie des Risikos. de Gruyter, Berlin (in German)
Marquardt N, Gardes R, Robelski S et al (2010) Determinanten menschlicher Fehler in Risikoindustrien. In: Trimpop R, Gericke G, Lau J (eds) Psychologie der Arbeitssicherheit und Gesundheit. Sicher bei der Arbeit und unterwegs—wirksame Ansätze und neue Wege. 16. Workshop 2010. Asanger, Kröning (in German)
Nationale Arbeitsschutzkonferenz (2013) Zwischenbericht zur Dachevaluation der Gemeinsamen Deutschen Arbeitsschutzstrategie. Berlin. Available via: http://www.gda-portal.de/de/Evaluation/Evaluation2008-12.html (in German)
Parker C, Lehmann Nielsen V (eds) (2011) Explaining compliance. Business responses to regulation. Edward Elgar, Cheltenham, UK
Peters DH, Tran NT, Adam T (2013) Implementation research in health: a practical guide. WHO Library Cataloguing-in-Publication Data, Geneva, http://www.who.int
Priester K (2005) Mindestens jede fünfte Krankheit ist arbeitsbedingt—Risikobranchen Bau-, Metall-, Agrar- und Verkehrsberufe. Gute Arbeit 17(8/9):53–55 (in German)
Schein E (1990) Organizational culture. Am Psychol 45(2):109–119
Schmitt B (2014) Statutory and operational framework conditions for best possible efficacy of external OSH services in Germany: problems and possible solutions as reflected by current research and recent legislation. In: Documentation of TAIEX—Workshop on the evaluation and review of external occupational health and safety services, 16–17 Jan 2014, Ankara. Available via: http://www.ec.europa.eu/enlargement/taiex/dyn/taiex-events/library/detail_en.jsp?EventID=54396
Schmitt B, Hammer A (2015) Für welche betrieblichen Kontexte ist der Prozess der Gefährdungsbeurteilung anschlussfähig? WSI-Mitteilungen 68(3):202–211 (in German)
Sommer S, Schmitt B (2012) Company and employee survey 2011 in framework of the evaluation of the Joint German Occupational Safety and Health Strategy (GDA). GESIS Data Archive, Cologne. ZA5634 Data file Version 1.0.0, doi:10.4232/1.11483
Strauss A, Corbin J (1996) Grounded theory: Grundlagen qualitativer Sozialforschung. Weinheim, Beltz (in German)

Witzel A (2000) Das problemzentrierte Interview [25 Absätze]. Forum Qualitative Sozialforschung/Forum Qual Soc Res 1(1):Art. 22, http://nbn-resolving.de/urn:nbn:de:0114-fqs0001228

Witzel A, Reiter H (2012) The problem-centred interview. Sage, London

Zwetsloot GIJM (2014) Evidence of the benefits of a culture of prevention. From risk to vision zero. In: International symposium on culture of prevention—future strategies, Helsinki, Volume: Proceeding, pp 30–35. Available via: http://www.researchgate.net/publication/268510584_Evidence_of_the_benefits_of_a_culture_of_prevention

Working Parents: Challenged Employees, Challenged Health Promotion?

Benjamin Fuchs

Abstract

This chapter examines whether parenthood is causally related to an individual's health, specifically for the workers in the USA. Although society values employment and parenthood individually, there is widespread public concern that a combination of the two may be too much of a good thing, especially in the case of mothers of young children. Against this background, this chapter outlines a sociological framework for analyzing the health impact of parenthood. The hypothesis that children have a detrimental impact on employees' health is derived. The exceedingly modest public and private policies for supporting working parents in the USA are described briefly and it is argued that neither are likely to mitigate the theoretical relationship. Some quantitative research has indeed shown a negative impact of parenthood. However, many studies have found no relationship or even positive impacts. Possible reasons for these apparent contradictions are discussed. Despite the unclear state of research, the following three clear recommendations are made. First, occupational health management should, for now, desist from introducing specific health promotion for employees with children. Second, public policy should expand existing measures to support working parents. Third, more rigorous research that utilizes more appropriate methods for causal analysis than have been implemented previously is needed in this field.

B. Fuchs (✉)
Institute for Employment Research of the Federal Employment Agency, Regensburger Strasse 104, 90478 Nuremberg, Germany
e-mail: benjamin.fuchs@iab.de

M. Wiencke et al. (eds.), *Healthy at Work*, DOI 10.1007/978-3-319-32331-2_18

1 Introduction

Two main societal trends that have occurred within the last four decades in Western societies have gained sociologists' attention: increasing participation of women in the labor market in general, and maternal employment in particular. As in other OECD countries, the participation rate of women has been increasing in the USA since the 1960 and 1970s. Whereas 43 % of women aged over 16 years were in the labor force in 1975, this share had increased to 60 % by the year 2000 and has remained that high. Maternal employment has also been on the rise. Strong growth has even occurred in the subgroup of mothers with young children. Whereas only 34 % of mothers of children aged under 3 years were in the labor force in 1975, this share had grown by more than 20 percentage points by 2012 (US Bureau of Labor Statistics 2014).

Of course, these trends have also been noticed by researchers in other scientific disciplines, occupational health management, and the public; the last seems somewhat worried. Although society values employment and parenthood individually, it seems to believe that both together may be too much of a good thing, especially for women. Various folk theories dealing with the stressfulness of parenthood have arisen. They have offered a range of explanations for this perceived stressfulness and the media has been addressing such issues since the outlined societal trends became evident (Verbrugge 1983). This rather negative perspective on motherhood may now have spread to fatherhood. With egalitarian couple arrangements slowly becoming more widespread, some fathers have been maximizing participation in both work and family. Thus, the everyday stresses of combining extensive engagement in both work and family may now be threatening not only the well-known "supermoms" (De Meis and Perkins 1996) but also the new "superdads" (Kaufmann 2013). Critical thinkers may be asking themselves whether these concerns are being exaggerated, while scholars may be asking themselves whether there is hard scientific evidence to support these perceptions. Occupational health management that is aware of societal trends may also be interested, because if these perceptions are correct, approaches to specifically promoting the health of such challenged employees would be important. This chapter aims to address these questions simultaneously. We have focused on the USA as the biggest and most influential economy in the investigated time period (from the 1970s to today). Compared to other OECD countries, its labor market is more dynamic and less regulated, possibly implying tougher conditions for working parents. Its heterogeneous population and economic structure make it possible to address the issue of the generalizability of results while keeping the focus on one country and considering the specific institutional contexts of working parents.

This chapter is structured as follows. First, we briefly describe existing public and private policies for supporting working parents in the USA. Second, we outline a sociological framework for analyzing health outcomes of employment and parenthood. We use a conceptual model based on role theory to derive a hypothesis about the impact of parenthood on the health of employees. We then relate this hypothesis to a wide range of empirical studies and summarize the state of research.

The chapter concludes with recommendations for health management, public policy, and science.

2 Policies for Supporting Working Parents in the USA

2.1 Main Instruments of Existing Public Policies

In the USA, there are minimal measures specifically designed to support working parents. Currently, Ruhm (2011) considers there are only two measures on the federal level that are both (indirectly) related to health and significant aids for working parents; namely, the Family and Medical Leave Act of 1993 (FMLA) and the provision of early care and education (ECEC). We will not be discussing other public measures, such as financial support, public schools, parent education policies and so on, that support parents in this chapter (see, e.g., Bogenschneider and Corbett 2010).

The explicit rationale of the FMLA has been to balance the demands of the workplace and the needs of families. Before introduction of the FMLA, there was no right to unpaid leave on the federal level. Eligible employees now have the right to take up to 12 weeks of job-protected leave during any 12-month period to take care of children. Such leave can also be used to take care of one's own or other family members' health problems. The employee must be placed in the same or a similar position when resuming work after FMLA leave. It is obvious that this measure offers potential health benefits for working parents (Chatterji and Markowitz 2004; Gjerdingen et al. 1995). The importance of this measure is magnified by the fact that there is no maternity protection at all in the USA. Anticipated negative effects, such as on women's employment or wages, did not occur (Waldfogel 1999); thus, the FMLA appears to be a effective instrument for supporting working parents in the view of many scientists. Bailey (2008) states that "the FMLA gave millions of Americans the opportunity to care for themselves and their family without sacrificing job security and continues to protect the American family unit today". However, it has numerous limitations, the first one being eligibility. Except for the public sector, it applies only to workers in businesses with 50 or more employees within a 75-mile radius. The employee must have worked for the employer for at least 12 months and at least 1250 h during that 12 months. For "key" employees, job reinstatement is not mandatory. Thus, only about half of the workers in the private economy are eligible for FMLA leave (Ruhm 2011). The second limitation is that, in contrast to parental leave regulations in all other OECD countries, FMLA is unpaid. Third, it is very short, its 12 weeks being far fewer than the OECD average of 58.5 weeks in 2013 (OECD 2015a). It can be seen more as a baseline rather than definitive form of parental leave, accordingly the provisions are sometimes expanded on the firm or state level (Bailey 2008). Some states have dropped the threshold of 50 employees, expanded the definition of family, or otherwise relaxed the eligibility criteria. Additionally, some states have introduced paid family leave programs, California being the first one in 2004; however, such leave is rather short and poorly-paid (Baum and Ruhm 2013). Fourth, it is difficult to amend FMLA by choosing part-time employment on resumption of work. There is no general legal entitlement to part-time employment in the USA and, because of the 1250 work hours necessary for eligibility for FMLA leave, employees might subsequently be ineligible for further FMLA leave. Thus, it is not surprising that part-time employment is not widespread in the USA: in 2013, only 12.3 % of employed persons worked part-time (less than 30 h a week) in that country, which is below the OECD average (OECD 2015b).

Provision of ECEC is another way of supporting working parents; this has a more subtle link to their health than the FMLA. In the USA, childcare is mainly seen as the private responsibility of parents; consequently, there are no federal or state universal child care systems. Nevertheless, there is some public funding for child care. The situation can be characterized as a mix of programs and services that vary by state and involve various combinations of public and private funding and public and private delivery (Allhusen et al. 2006). The *Child Care and Development Block Grant*, which consolidated various established child care programs in 1996, is the largest contemporary federal program. It is complemented by state-level funding and provides childcare subsidies for children aged up to 13 years, the majority being received for children under the age of six. Only families with incomes equal to or less than 85 % of the median incomes of their states are eligible. However, only 20 % of those whose incomes make them eligible use these subsidies (Ruhm 2011). In the fiscal year 2013, US$8601 billion was spent on 874,200 families, mostly in the form of vouchers or cash (Office of Child Care 2015a, b). Families can choose between center-based care, family daycare, and in-home care. Another federal program targeted at low-income families (and those with disabled children) is *Head Start*, which since 1965 has provided not only childcare and education, but also health, nutrition and mental health services for children aged between 3 and 4 years of age. However, promoting (women's) employment is not the underlying intention of Head Start, as evidenced by the fact that few Head Start programs offer full day care (Allhusen et al. 2006). This measure has been complemented by the *Early Head Start* program, which since 1993 has provided Head Start services for children under the age of three, mostly in center-based settings. This has led to an increase in the number of families utilizing Head Start. It provides services for an estimated 47 % of income-eligible 3 and 4 year olds and 3 % of income-eligible children under the age of three. For the most part, the funds are paid directly to local public and private service providers (Ruhm 2011). In the fiscal year 2014, US$8598 billion was spent on Head Start and Early Head Start, which together provided services for 1076 million children (Head Start 2015).

Childcare is also subsidized for working parents in middle- and upper-income families via the tax system, which has two major programs. Introduced in 1976, *the Child and Dependent Care Tax Credit* was designed to ensure that caring for children does not force individuals to leave the labor market. Today, depending on their adjusted gross incomes, income-eligible working parents can claim up to US$3000 for one child or US$6000 for two or more children to compensate for childcare expenses (IRS 2015). The second major program is the *Dependent Care*

Assistance Plan, which is an alternative to the Child and Dependent Care Tax Credit. Under this plan, employees in firms with flexible spending accounts can tax-shelter up to US$5000, which offers advantages specifically for high-income families (IRS 2015). Both programs allow a free choice of type of childcare; the childcare provider does not even have to be registered (Allhusen et al. 2006).

Both public childcare and consequently public funding have inconsistent stances regarding working parents' health. On the one hand, in many cases childcare is a prerequisite of working and thereby enhancing household income, both of which are positively related to health. On the other hand, if childcare is utilized to enable a return to work very soon after childbirth, which is a common practice among working parents in the USA, several studies have shown that this is negatively related to maternal health compared with parental leave (e.g., Chatterji and Markowitz 2004; Gjerdingen et al. 1995). The public policy of effectively non-existent maternity leave, short parental leave entitlements and subsidized out-of-home daycare in the USA leads to many mothers returning quickly to full-time work and not breastfeeding. They thus miss out on the associated health benefits such as reduced risk of postpartum depression, cancer and type 2 diabetes (for an overview see Ip et al. 2007). Furthermore, they and their partners (if present) do not enjoy the indirect effect of enhanced child health (Gartner et al. 2005). Therefore, if the focus is on health outcomes, an expansion of public childcare as recommended by a large proportion of the scientific community, does not appear to be promising. The overall effects of ECEC policies are much more uncertain than those of parental and maternity leave and are likely to vary according to the interval between childbirth and mother's return to work (Ruhm 2011).

2.2 Private Policies

In contrast with other OECD countries, the USA does not rely heavily on government spending in the field of social policy; rather, it encourages employers and other private actors to provide social services and benefits (Hacker 2002). Thus, the lack of support for working parents on the federal and state level, as described in Sect. 2.1, is not surprising. So the question is to what extent is this gap filled by private actors. Concerning parental leave, some employers support working parents by continuing to pay employees who have taken unpaid FMLA leave. However, this is true for only a minority of employees: 12 % of USA employees received such employer-paid leave for care of any family members in 2012 (Hara and Hegewisch 2013). There is minimal support by employers for childcare. The most common form is information delivery, for example, by phone. As described in Sect. 2.1, some medium and large businesses offer access to the Dependent Care Assistance Plans to enable parents to make tax deductions for childcare expenses. Rarely, they also offer on- or off-site childcare facilities: only about 5 % of employees working full-time in private industry have access to such facilities. This small proportion is not surprising because most US-American employees work for small businesses, which are unlikely to have the resources to provide this (Allhusen et al. 2006).

Some charitable organizations also provide childcare for USA families; however, this is not usually specifically targeted at working parents but to members of other groups, such as parents of disadvantaged children or members of church communities. Examples of such providers are religious organizations, the United Way, and philanthropic groups (Allhusen et al. 2006).

All in all, it can be concluded that the private policies to support working parents cannot compensate the gap which exists in the field of public policy, neither with respect to parental or maternity leave, nor with respect to childcare. As Gornick and Meyers (2004, p. 107) put it: "The results of the American experiment with market-based solutions (to the needs of dual-earner families) have been calamitous for many American parents and children".

3 Theoretical Considerations

3.1 A Sociological Framework for Analyzing Health Outcomes of Employment and Parenthood

Sociological perspectives can yield fruitful insights into health issues. The public and private policies for working parents in the USA described in Sects. 2.1 and 2.2 are so limited that they are unlikely to significantly ameliorate or compensate for the outlined theoretical relationships. From a sociological perspective, working and caring for children are not just physical processes, but rather have meanings that are ascribed by society. This means that both are *social roles* that individuals can adopt.[1] Roles are "socially defined expectations that a person in a given status, or social position, follows" (Giddens 2009, p. 264). There are different types of social roles. Occupational (e.g., worker) and family roles (e.g., parent, spouse), which are the crucial ones for the topic of this chapter, are categorized as position or status roles. This type of role is linked with positions in organizations or other formally organized groups (Turner 2006, p. 234). The link to health issues lies in the theoretical construct of *role strain*, which involves feelings of being under pressure, anxiety, and worry and is believed to prejudice health (Goode 1960). Role strain can result, among other things, from *inter-role conflict* or *role overload*. Both of these causes can occur only if one individual simultaneously has multiple roles. Inter-role conflict results when behavior in one role violates values in another role. Role overload describes situations in which individuals have more roles than they have time, energy, or resources for (Turner 2006, p. 249). Obviously, both considerations are relevant in the case of working parents.

[1] For a criticism of role theory see, for example, Barbalet (2006, p. 583).

3.2 Combining Work and Family: A Multiple Role Perspective on Health

A multiple role perspective has been carried over, sometimes consciously, sometimes subconsciously, into health-related disciplines such as epidemiology, medicine, and psychology. It is reflected in one of the two basic theoretical models that underpin the many published studies exploring the influence of social circumstances on health. The *role overload model* proposes that individuals who have to balance the demands and obligations of different social roles are in danger of "overloads" that prejudice their health (Bartley et al. 1992). Analogously, the *multiple burden hypothesis* (Gove 1984) suggests that the role strain associated with simultaneously having a paid job and dependent children leads to deterioration in women's health, particularly when they have full-time work. This role strain is caused by the role overload and inter-role conflict involved in simultaneously fulfilling two major social roles: the roles of parent and employee.

Role overload can develop quickly in such cases because (continuously) employed mothers who are also mainly responsible for childcare generally have less time for recuperation after childbirth than non-employed mothers, especially if they are single mothers. The difference in values between the occupational role, which is associated with tough competition, and the parental role, where empathy and love is expected, contributes to the development of inter-role conflict. This theoretical perspective originally referred primarily to women. However, given the slight but significant increase in numbers of egalitarian couple arrangements within the last decades, it seems worth investigating whether this perspective can also be applied to men.

Before we proceed, we must mention that there is also a diametrically opposite position to the perspective outlined above. The *multiple attachment hypothesis*, the corresponding theoretical model of which is the *role expansion model*, predicts that simultaneously fulfilling parental and occupational roles would benefit health (Nathanson 1975). More recently, more complex theoretical models that try to integrate different perspectives have been formulated; for example by Arber (2005). We will not describe these here. We have chosen to focus on the multiple burden hypothesis because the scientific mainstream in the humanities seems to have adopted a rather negative view of parenthood, as we will now show.

4 Employment, Parenthood and Associated Health Outcomes

Decisions by health management to consider promoting the health of working parents should be grounded in clear scientific evidence for parenthood having a negative causal effect on employees' health (or on the health of all parents). Failure to invalidate the multiple burden hypothesis under various circumstances could provide such evidence, as we will now explore.

We have selectively reviewed existing published material with a focus not on similarities between the studies, as has been done by previous reviewers, but on differences in outcomes. From a scientific perspective, engagement of occupational health management to specifically promote working parent's health can only be recommended if these differences are rather small or non-existent. To ensure that these differences do not result from differences in quality of the cited studies, we have mainly focused on peer-reviewed publications in scientific journals.

4.1 Descriptive Patterns

Before accepting a causal perspective, it is reasonable to investigate health disparities between adults with and without children. A strong negative causal effect would be expected to result in parents having worse health than adults without children, whether employed or not. Gove's studies (1977, 1979, 1984) have been particularly influential in creating a negative view of parenthood. Analyzing survey data of 18–60-year-old married Chicago citizens collected in 1974, Gove and Geerken (1977) found that mothers of young children (0–4 years) had more psychiatric symptoms than those without children or with older children, regardless of their employment status. For men, psychiatric symptoms increased in parallel with the age of the youngest child. Additionally, parents with young children had more psychiatric symptoms than those without children in the household. Despite a rather small sample size and limited geographic area, the authors drew the strong conclusion that "having children in the household generally contributes to poor mental health". They argued that this relationship is caused by the detrimental effects of role strain associated with having children, which later became known as the multiple burden hypothesis. Subsequent studies suggested that these patterns also hold true for non-married men and women (Gove and Hughes 1979) and that this association was not restricted to mental health but also applied to physical health: married women with children in the household more often felt physically impaired than those without children. However, this relationship did not hold true for men (Gove 1984). These findings were subsequently at least partly replicated for other health measures or health-related outcomes (Goldsteen and Ross 1989; Ross and Van Willigen 1996).

Conversely, some studies failed to show that parents are generally in worse health than their childfree counterparts. Using data from a prospective cohort study that examined the health of older adults aged 50–71 years in the USA, Eisenberg et al. (2011) examined the relationship between fertility biographies and men's health. They found that men with a moderate number of children (two or three) had better self-reported health, lower rates of diabetes and smaller age-adjusted mortality rates than men without children or those with many children. However, this study (and most other studies on parenthood and men's health) did not consider a possible interaction between parenthood and employment status. Furthermore, some published studies have reported descriptive patterns such as poorer health only in specific subgroups of parents. They have also reported that these patterns differ in their health-related outcomes (Deaton and Stone 2014; Evenson and Simon 2005). In conclusion, there is no clear evidence that parents generally or specifically employed parents have worse health than their counterparts without children. Nor is there a clear descriptive pattern in the way that health decreases in parallel with number of children. However, when controlling for related life circumstances, parenthood may nevertheless turn out to be detrimental, a possibility that is explored in the next section.

4.2 Studies of Associations Between Parenthood and Health

The fairly negative perspective on parenthood suggested by Gove's studies (1977, 1979, 1984) was quickly challenged by associative studies, which tried to attribute the poorer health of employed or non-employed parents reported by Gove to different characteristics of adults with and without children, mostly by using regression analysis. Using daily health records of Detroit citizens collected in 1978, Verbrugge (1983) analyzed the effects of different social roles on various health indicators. Controlling for sociodemographic characteristics, Verbrugge found that being a parent had a positive, but insignificant effect on various health (-related) outcomes such as subjective health, number of chronic problems, job limitations, number of days of restricted activity, and drugs taken. The negative perspective has also been challenged by studies that showed that the association of parenthood with health depends on the specific form of being a parent (age of children, stepchildren vs. biological children etc.) and that it depends on which health-related outcomes are examined (Hughes 1989; McLanahan and Adams 1989). However, some studies have replicated Gove's findings (Goldsteen and Ross 1989; Kandel et al. 1985).

In sum, a negative perspective has more or less prevailed within the scientific community. In an extensive review of reports published in a highly reputable journal, Ross et al. (1990) summarized the then current state of research on the impact of parenthood on health and well-being. They argued that it was only in the best and relatively uncommon circumstances that children did not decrease parental health or well-being and concluded that "overall, children at home decrease adult well-being". Interestingly, in the meantime, Gove had already retracted his previously trenchant position (Umberson and Gove 1989); however, in their review Ross et al. (1990) did not mention the study in which this retraction was made.

From the 1990s onwards, the geographic focus, which had mostly been limited to the USA, expanded to other countries (e.g., see Artazcoz et al. 2004; Bartley et al. 1999; Lahelma et al. 2002; Steptoe et al. 2000). The confusing findings reported in the 1980s also led to more sophisticated theoretical models which, for example, included issues of social class (Arber 2005). Longitudinal data, which had rarely been used to investigate this topic, had begun to be analyzed. For example, Barnett and Marshall (1991) analyzed longitudinal data from a study in which adult women were interviewed in 1985 and 1986 in Boston. They found that parent-role occupancy was not negatively associated with psychological distress.

Ten years after the negative verdict of Ross et al. (1990), Arendell (2000) summarized the next decade's research on the impact of motherhood in a review of published studies; again, a negative view more or less prevailed. Concerning the influence of social roles, the author identified child care as a primary strain for women and listed a range of studies that showed negative associations between mothering and psychological well-being and distress, stating that "many mothers (and fathers) pay a high personal price trying to balance work and family demands". It is noteworthy that, although a negative view of the effects on health of motherhood (and sometimes, but clearly increasingly less so, of fatherhood) has prevailed in many more recent publications (Deaton and Stone 2014; Evenson and Simon 2005; Glavin et al. 2011), studies have regularly been published refuting this view (e.g., Sneed et al. 2012). For men, the mainstream position seems to have changed within recent years. Smith and Zick (1994) and Eisenberg et al. (2011) reported positive associations between fatherhood and health for the USA. An extensive review of published studies on the impact of fatherhood on men's health drew the conclusion that "in the long term, the weight of evidence indicates that fatherhood is beneficial to a man's health" (Bartlett 2004).

Despite this verdict, there are so many contradictions between studies so that it must be concluded that no clear pattern is evident concerning the overall effect of parenthood on health.

4.3 Parenthood, Employment and Health

Some studies have looked at potential interactions between parenthood and health specifically for working individuals. The multiple burden hypothesis suggests that motherhood has a negative impact on employed women; as described in Sect. 4.2, this may also hold true for men. Verbrugge (1983) investigated this interaction by analyzing the effects of interactions between different social roles on various health indicators. Controlling for sociodemographic characteristics, no significant interaction between the social roles of employee, parent, and spouse was found for men or for women. These findings clearly contradict the concept that health-impairing role strain occurs when the parental role is combined with the employee role. Using representative survey data collected in the USA in 1976, McLanahan and Adams (1986) investigated the relationship between parenthood and experienced worries. They found that having preschool children in the household had a negative effect on working single mothers, but not on working married mothers or men. This suggests a three-way interaction rather than the simple two-way interaction proposed by the multiple burden hypothesis. Aneshensel et al. (1981) analyzed epidemiological data from Los Angeles County collected in 1979. They found that for employed women living with a partner, having children in the household was associated with fewer depressive symptoms. The same was true for men, which is the opposite of what the multiple burden hypothesis suggests. Their findings suggest that the marital rather than the occupational role is a crucial determinant of the impact of parenthood:

single women with children in the household were significantly worse off, whereas those with a partner were not.

In the meantime, studies that support the multiple burden hypothesis have been published. Cleary and Mechanic (1983) analyzed a small sample of Wisconsin citizens and found that minors in the household increased the likelihood of depression for employed women, but not for homemakers. Employed married men were not affected by having children and, as in many other studies, there were too few unemployed men to investigate the effects of children on them. In a sample of Chicago citizens interviewed in 1978 and controlling for other factors, Goldsteen and Ross (1989) found that mothers' perceived burdens increased in parallel with the number of children. However, this effect varied with age but not with employment status. In a study of a small sample of women living in the New York Metropolitan area, Kandel et al. (1985) found that, although descriptive results suggested having children in the household had a positive effect, regression analysis showed that motherhood exacerbated occupational stress.

However, just as is true of studies investigating the overall effect of parenthood, there are many contradictions between published studies concerning interactive effects of parenthood and occupational role. For example, Barnett and Marshall (1991) found in their longitudinal study on women in Boston that parent-role occupancy was not negatively associated with psychological distress. Furthermore, the parental role buffered the distressing effects on employees of having little authority to make decisions at work. Other studies have provided a different perspective and argued that employment has a beneficial effect on mothers (e.g., see Frech and Damaske 2012).

In sum, it must be concluded here that there is no clear pattern in the effect of interactions between parenthood and occupational roles: as is true of the overall effect of parenthood, the contradictions between the studies outweigh the similarities. Thus, there is no clear evidence for detrimental impact of parenthood on employees.

4.4 Do Children Causally Affect Health?

Inconclusive research findings make it difficult to identify any causal relationship between parenthood and health for employees or for other groups such as homemakers or the unemployed. Various empirical studies have shown that parenthood has different impacts at different time points, in different geographical areas and on different health indicators; correspondingly, the health status of (working) parents has sometimes been reported to differ from that of other groups. The multiple burden hypothesis had been invalidated under some circumstances but not others. No pattern that could account for the inconsistencies in findings has thus far been identified.

Given that a contextual explanation of variations in findings is elusive, there may be a methodological one. Because most studies dealing with this topic are based on certain types of simple cross-sectional regression models and not on experimental designs, endogeneity may explain the inconsistencies. The findings analyzed in this chapter are drawn from an era of legalized and widely realized abortion, during which the transition to parenthood was more a decision made by individuals than in earlier eras. Such decisions are likely to depend on a range of circumstances that are associated with health and are not fully controlled for by the variables included in these models.

One technique that can be used in such situations is the instrumental variable estimator. The main principle behind this econometric modeling technique is to use only exogenous variation within an independent variable to estimate an instrumental variable regression (e.g., see Wooldridge 2010). In theory, this reveals the unbiased effect and therefore allows causal claims even in cross-sectional studies. Thus far, only two studies have used this methodology to investigate the causal effect of parenthood on health. Cáceres-Delpiano and Simonsen (2012) analyzed the impact of parenthood on women's blood pressure, weight and Medicaid receipt using repeated cross-sectional data for the period 1982–2003, data that were representative of the USA as a nation over that period. Using multiple births as a source of exogenous variation in the number of children a mother has, they showed a rather negative impact of parenthood: it significantly increased the likelihood of mothers having high blood pressure and obesity and using Medicaid. Although the other study in this field that has used the instrumental variable technique did not involve the USA, we nevertheless mention it because of its specific methodological approach. Using a large panel dataset from interviews conducted in 2004 and 2006 in 13 European countries, Kruk and Reinhold (2014) analyzed the effects of the number of biological children on depression in older persons (50–90 years). They were able to differentiate between the impact of unexpected and desired children by using multiple births and sex proportions of the children as sources of exogenous variation in the number of children. These researchers found considerable heterogeneity in the effect of children. For women, each unexpected child increased the likelihood of depression by 10–20 percentage points, whereas desired children did not. Men were not affected at all.

So, do children endanger one's health? Although both studies that make causal claims suggest that they do, the evidence is far too limited and inconclusive to reliably answer this question. If we consider the many studies that contradict each other, that only two studies with causal claims exist, and that their methodology has recently been questioned (Wagner 2013), we must conclude that although employed parents may face serious challenges because of their children, the multiple burden hypothesis can neither be seen as confirmed nor as refuted. *There is as yet no clear scientific evidence that children harm (working) parents' health.* The only pattern that has prevailed in these studies is that having children has a negative impact on single women. Thus, parenthood does not seem to interact with the occupational role, whereas it may, but not definitely, interact with the marital role. Table 1 summarizes the empirical studies reviewed in this contribution.

Table 1 Overview of studies on the impact of parenthood on health in the USA

Geographical area	Authors (year)	Sample	Methods	Health measure(s)	Findings
Detroit metropolitan area	Verbrugge (1983)	White adults, interviewed in 1978 (n = 589)	Ordinary least squares regression models	Subjective health, number of chronic problems, job limitations, number of days of restricted activity, drugs taken	Positive, but non-significant association with most health measures; no interaction with employment status
Federal territory	McLanahan and Adams (1989)	Adults (≥21 years), interviewed in 1957 and 1976 (n = 4664)	Binary logistic regression models	Physical health, anxiety and worries	No association with physical health or anxiety; negative association with worries
Los Angeles County	Aneshensel et al. (1981)	Adults, interviewed in 1979 (n = 1000)	Two-way analysis of variance, analysis of covariance	Depression	Negative association for single women but not for all other investigated groups, no interaction with employment status
Three communities in Chicago metropolitan area	Goldsteen and Ross (1989)	Women (aged 17–59 years) with children in the household, interviewed in 1978 (n = 549)	Ordinary least squares regression models	Perceived burden	Negative association
New York metropolitan area	Kandel et al. (1985)	Adult women, interviewed in 1979 (n = 197)	Two-way analysis of variance, analysis of covariance, ordinary least squares regression models	Depression	No effect of parenthood as such, but exacerbating effects of occupational stress
Federal territory	Umberson and Gove (1989)	Adults who have been married at least once, interviewed in 1974–75 (n = 1753)	Ordinary least squares regression models	Depression	No effect of children at home regardless of age; positive effect of children no longer living in the household

(continued)

Table 1 (continued)

Geographical area	Authors (year)	Sample	Methods	Health measure(s)	Findings
Boston, MA (including surrounding area)	Barnett and Marshall (1991)	Women (aged 25–55 years) working in licensed practical nursing or social work, interviewed three times between 1985 and 1986 (n = 403)	Pooled ordinary least squares regression models	Psychological distress	No association; parental role-occupancy buffered the distressing effects on employees of having little decision authority at work
Federal territory	Deaton and Stone (2014)	34–46 year old adults, interviewed between 2008 and 2012 (sample size not reported)	Ordinary least squares regression models	Anger, worry, stress, physical pain	Negative association with health measures, but positive association with other indicators relating to life quality
Federal territory	Evenson and Simon (2005)	Adults, interviewed in 1987–1988 (n = 11,473)	Ordinary least squares regression models	Depression	All investigated forms of parenthood negatively associated with health; only some of the associations significant
Federal territory	Glavin et al. (2011)	Adults in the paid labor force, interviewed in 2006–2007 (n = 1042)	Ordinary least squares regression models	Distress	Negative, but non-significant association
Federal territory	McLanahan and Adams (1986)	Adults (aged $\geq$ 21 years), interviewed in 1957 and 1976 (n = 2806)	Binary logistic regression models	Worries	Negative effect of preschool children on working single mothers, but not on other groups
Pittsburgh metropolitan area	Sneed et al. (2012)	Healthy adults (aged 18–55 years), experimentally exposed to a common cold virus between 1993 and 2004 (n = 795)	Binary logistic regression models	Getting a cold	Positive association with host resistance

Federal territory	Smith and Zick (1994)	Individuals who were married during 1968 and 1969, interviewed between 1970 and 1987 (n = 2604)	Paired hazard rate models	Mortality	Positive association
California, Florida, Louisiana, New Jersey, North Carolina and Pennsylvania	Eisenberg et al. (2011)	Male members of the American Association of Retired Persons (aged 51–71 years), interviewed between 1995 and 2005 (n = 137,903)	Cox proportional hazard models	Cardiovascular mortality	Positive association
Wisconsin	Cleary and Mechanic (1983)	Adults, interview dates not mentioned (n = 1026)	Ordinary least squares regression models	Depression	Minors in the household negatively associated with depression, but only for employed women
Federal territory	Cáceres-Delpiano and Simonsen (2012)	Women (20–45 years), interviewed between 1980 and 2003 (sample size not mentioned)	Instrumental variable regression	Medicaid usage, high blood pressure, obesity	Negative impact on all investigated health measures
Federal territory	Hughes (1989)	Adults, interviewed in 1957 and 1976 (n = 4724)	Multiple classification analysis	Anxiety, worries	Negative impact on worries, but only for men; Negative impact on anxiety, but only for unmarried men

5 Implications

The inconclusive state of research leads to a clear recommendation for occupational health management. Since there is considerable doubt that parenthood has a negative effect on employees, *specific health promotion for working parents should not yet be introduced by occupational health management*. Whereas costs of such health promotion would be certain, returns would be uncertain. The single study with enhanced methodological quality that reported partial evidence for a negative effect of parenthood in the USA is insufficient to justify specific health promotion, especially given that it did not consider a possible interaction between parenthood and employment and only investigated women. Employed parents are certainly specifically challenged employees; however, whether this is an issue for occupational health management is as yet not conclusively known. Therefore, an engagement would be premature.

Health promotion already exists for parents, whether working or not. Given that prenatal classes, gym courses for pregnant women, and so on are partly offered or funded by certain public or private institutions, occupational health management should desist from engaging in this field. A substitution effect could occur if employers started to offer measures already provided by other public or private institutions. The latter would then have an incentive to retreat from this field and allocate resources otherwise. A similar recommendation can be made regarding parental leave and maternity protection. It has been shown that these are measures with high possible health benefits for working parents. Again, this is a matter for public policy rather than occupational health management. Not individual employers, but the USA as a nation must recognize that it is one of only four countries in the world which offer no paid maternity leave, the other three being Liberia, Papua New Guinea and Swaziland (Heymann et al. 2007). Furthermore, question of both pay and time need to be addressed. The FMLA needs to be expanded, at least up to 6 months, which would then cover the 6 months of breastfeeding recommended by the World Health Organization. California's expansion of parental leave can be seen a first step in the right direction.

ECEC policies consequently need to focus more on quality than quantity. An expansion of FMLA would reduce the necessity for childcare subsidies for infants; the resulting saved resources could then be invested in the quality of childcare. Nevertheless, employers should be aware of the needs of their employees who have become parents, particularly those in vulnerable groups. As shown in Sect. 2, single mothers form one such group. This should also be considered by policy makers. Single working parents, who face substantial financial and time pressures, might benefit most from the recommended expansion and wage continuation of FMLA. Although their poor health status may be related to selection into single parenthood, social policy seems to be a crucial factor for supporting the health of this growing social group (Lahelma et al. 2002).

There are also clear recommendations for science. In this chapter, only studies from the USA were reviewed. The question of generalizability of their findings to other countries must be addressed in future research, preferably by means of meta-analysis. Given that the USA is a large and heterogeneous country, it is not implausible that the findings may hold true for other countries, at least those in which there are similarly restricted social policies for supporting working parents. Furthermore, better research that avoids the two main mistakes of previous studies is needed in this field. First, empirical research has focused on cross-sectional associations. However, to reduce bias and draw valid causal claims, the focus needs to be changed to analysis of longitudinal data. Because parenthood is a major biographical event and biographies take place within a temporal continuum, the aspect of time needs to be given greater consideration: this is crucial for the concept of causality. Given the more numerous and longer-lasting panel data sets that have become available within recent years, this is an achievable task for scholars. The second main mistake may be the most difficult to avoid. We have shown that major reviews of published reports in this field have drawn premature conclusions. In the future, we must be much more cautious and focus more on studies that contradict rather than confirm our theories, an approach that is not new but simply the core of good science.

References

Allhusen VD, Clarke-Stewart KA, Miner JL (2006) Childcare in the United States: characteristics and consequences. In: Melhuish E, Petrogiannis K (eds) Early childhood care and education. International perspectives. Routledge, New York, pp 7–26

Aneshensel CS, Frerichs RR, Clark VA (1981) Family roles and sex differences in depression. J Health Soc Behav 22:379–393

Arber S (2005) Opening the black box, inequalities in women's health. In: Abbot P, Payne G (eds) New directions in the sociology of health. Falmer, Hampshire, pp 35–54

Arendell T (2000) Conceiving and investigating motherhood: the decade's scholarship. J Marriage Fam 62:1192–1207

Artazcoz L, Benach J, Borell C et al (2004) Unemployment and mental health: understanding the interactions among gender, family roles, and social class. Am J Public Health 94:82–88

Bailey NJ (2008) Family and Medical Leave Act. In: Loue S, Sajatovic M (eds) Encyclopedia of aging and public health. Springer, New York, pp 351–352

Barbalet J (2006) Social roles. In: Turner BS (ed) The Cambridge dictionary of sociology. Cambridge University Press, Cambridge, pp 583–584

Barnett RC, Marshall NL (1991) The relationship between women's work and family roles and their subjective well-being and psychological distress. In: Frankenhaeuser M, Lundberg U, Chesney M (eds) Women, work and health: stress and opportunities. Springer, New York, pp 111–138

Bartlett EE (2004) The effects of fatherhood on the health of men: a review of the literature. J Men Health Gender 1:159–169

Bartley M, Popay J, Plewis I (1992) Domestic conditions, paid employment and women's experience of ill-health. Sociol Health Illness 14:313–343

Bartley M, Sacker A, Firth D et al (1999) Social position, social roles and women's health in England: changing relationships 1984–1993. Soc Sci Med 48:99–115

Baum CL, Ruhm CJ (2013) The effects of paid family leave in California on labor market outcomes. NBERWP 19741

Bogenschneider K, Corbett TJ (2010) Family policy: becoming a field of inquiry and subfield of social policy. J Marriage Fam 72:783–803

Cáceres-Delpiano J, Simonsen M (2012) The toll of fertility on mother's wellbeing. J Health Econ 31:752–766
Chatterji P, Markowitz S (2004) Does the length of maternity leave affect maternal health? National Bureau of Economic Research, working paper 10206
Cleary PD, Mechanic D (1983) Sex differences in psychological distress among married people. J Health Soc Behav 24:111–121
Deaton A, Stone AA (2014) Evaluative and hedonic wellbeing among those with and without children at home. Proc Natl Acad Sci USA 111:1–6
De Meis DK, Perkins HW (1996) "Supermoms" of the nineties: homemaker and employed mothers' performance and perceptions of the motherhood role. J Fam Issues 17:777–792
Eisenberg ML, Park Y, Hollenbeck AR et al (2011) Fatherhood and the risk of cardiovascular mortality in the NIH-AARP Diet and Health Study. Hum Reprod 26:3479–3485
Evenson RJ, Simon RW (2005) Clarifying the relationship between parenthood and depression. J Health Soc Behav 46:341–358
Frech A, Damaske S (2012) The relationships between mothers' work pathways and physical and mental Health. J Health Soc Behav 53:396–412
Gartner LM, Morton J, Lawrence RA et al (2005) Breastfeeding and the use of human milk. Pediatrics 115:496–506
Giddens A (2009) Sociology. Polity, Cambridge
Gjerdingen DK, McGovern PM, Chaloner KM et al (1995) Women's postpartum maternity benefits and work experience. Fam Med 27:592–598
Glavin P, Schiemann S, Reid S (2011) Boundary-spanning work demands and their consequences for guilt and psychological distress. J Health Soc Behav 52:43–57
Goldsteen K, Ross CE (1989) The perceived burden of children. J Fam Issues 10:504–526
Goode WJ (1960) A theory of role strain. Am Sociol Rev 25:483–496
Gornick JC, Meyers MK (2004) Families that work. Sage, New York
Gove WR (1984) Gender differences in mental and physical illness: the effects of fixed roles and nurturant roles. Soc Sci Med 19:77–91
Gove WR, Geerken MR (1977) The effect of children and employment on the mental health of married men and women. Soc Forces 56:66–76
Gove WR, Hughes M (1979) Possible causes of the apparent sex differences in physical health: an empirical investigation. Am Sociol Rev 44:126–146
Hacker JS (2002) The divided welfare state. Cambridge University Press, Cambridge
Hara Y, Hegewisch A (2013) Maternity, paternity, and adoption leave in the United States. Institute for Women's Policy Research Briefing Paper A143
Head Start (2015) Head Start Program Facts Fiscal Year 2014. Available from http://eclkc.ohs.acf.hhs.gov/hslc/data/factsheets/docs/hs-program-fact-sheet-2014.pdf Accessed 3 Apr 2015
Heymann J, Earle A, Hayes J (2007) The work, family, and equity index. How does the United States measure up? Project on Global Working Families, Harvard School of Public Health and Institute for Health and Social Policy, McGill University
Hughes M (1989) Parenthood and psychological well-being among the formerly married: are children the primary source of psychological distress? J Fam Issues 10:463–479
Kruk KE, Reinhold S (2014) The effect of children on depression in old age. Soc Sci Med 100:1–11
Ip S, Chung M, Raman G et al (2007) Breastfeeding and maternal and infant health outcomes in developed countries. Evid Rep Technol Assess 153:1–186
IRS (2015) Child and dependent care expenses. Available from http://www.irs.gov/pub/irs-pdf/p503.pdf. Accessed 4 Apr 2015
Kandel DB, Davies M, Raveis VH (1985) The stressfulness of daily social roles for women: marital, occupational and household roles. J Health Soc Behav 26:64–78
Kaufmann G (2013) Superdads: how fathers balance work and family in the 21st century. New York University Press, New York
Lahelma E, Arber S, Kivelä K et al (2002) Multiple roles and health among British and Finnish women: the influence of socioeconomic circumstances. Soc Sci Med 54:727–740

McLanahan S, Adams J (1986) Explaining the decline in parent's psychological well-being: the role of employment, marital disruption and social integration. Center for Demography and Ecology, Working Paper No. 85-25. University of Wisconsin, Madison
McLanahan S, Adams J (1989) The effects of children on adults' psychological well-being: 1957–1976. Soc Forces 68:124–146
Nathanson CA (1975) Illness and the feminine role: a theoretical review. Soc Sci Med 9:57–62
OECD (2015a) Length of maternity, paternity and parental leave. Available from http://www.oecd.org/gender/data/lengthofmaternitypaternityparentalleave.htm Accessed 4 Apr 2015
OECD (2015b) Part-time employment rate. Available from http://www.oecd-ilibrary.org/employment/part-time-employment-rate/indicator/english_f2ad596c-en. Accessed 2 Apr 2015
Office of Child Care (2015a) CCDF fiscal year 2013 State spending from all appropriation years. Available from http://www.acf.hhs.gov/programs/occ/resource/ccdf-fiscal-year-2013-state-spending-from-all-appropriation-years. Accessed 2 Apr 2015
Office of Child Care (2015b) FY 2013 Preliminary data Table 1—average monthly adjusted number of families and children served. Available from http://www.acf.hhs.gov/programs/occ/resource/fy-2013-ccdf-data-tables-preliminary-table-1. Accessed 4 Apr 2015
Ross CE, Van Willigen M (1996) Gender, parenthood and anger. J Marriage Fam 58:572–584
Ross CE, Mirowsky J, Goldsteen K (1990) The impact of the family on health: the decade in review. J Marriage Fam 52:1059–1078
Ruhm CJ (2011) Policies to assist parents with young children. Future Child 21:37–68
Smith KR, Zick CD (1994) Linked lives, dependent demise? Survival analysis of husbands and wives. Demography 31:81–93
Sneed RS, Cohen S, Turner RB et al (2012) Parenthood and host resistance to the common cold. Psychosom Med 74:567–573
Steptoe A, Lundwall K, Cropley M (2000) Gender, family structure and cardiovascular activity during the working day and evening. Soc Sci Med 50:531–539
Turner RH (2006) Role theory. In: Turner JH (ed) Handbook of sociological theory. Springer, New York, pp 233–254
Umberson D, Gove W (1989) Parenthood and psychological well-being: theory, measurement, and stage in the life course. J Fam Issues 10:440–462
US Bureau of Labor Statistics (2014) Women in the labor force: a databook. BLS Report 1049. Washington, DC
Verbrugge LM (1983) Multiple roles and physical health of women and men. J Health Soc Behav 24:16–30
Wagner S (2013) Rusty instruments? Revisiting the twin approach to estimating the relationship between fertility and maternal labour market outcomes. Paper presented at the annual meeting of the Society of Labor Economists, Madison, 2–3 May 2014
Waldfogel J (1999) The impact of the family and medical leave act. JPAM 18:281–302
Wooldridge GM (2010) Econometric analysis of cross section and panel data. MIT Press, Cambridge, MA

Part IV

Evidence-Based Studies and Interventions Promoting Occupational Health

Occupational e-Mental Health: Current Approaches and Promising Perspectives for Promoting Mental Health in Workers

Dirk Lehr, Anna Geraedts, Robert Persson Asplund, Zarnie Khadjesari, Elena Heber, Jessica de Bloom, David Daniel Ebert, Peter Angerer, and Burkhardt Funk

D. Lehr (✉)
Institute of Psychology, Leuphana University Lueneburg, Lueneburg, Germany
e-mail: lehr@leuphana.de

A. Geraedts
ArboNed, Utrecht, The Netherlands
e-mail: Anna.geraedts@arboned.nl

R. Persson Asplund
Department of Behavioral Sciences and Learning, Linköping University, Linköping, Sweden
e-mail: robert.persson.asplund@liu.se

Z. Khadjesari
Addictions Department, Institute of Psychiatry, Psychology and Neuroscience (IoPPN), King's College London, London, UK
e-mail: zarnie.khadjesari@kcl.ac.uk

E. Heber
Division of Health Training Online, Leuphana University Lueneburg, Lueneburg, Germany
e-mail: elena.heber@inkubator.leuphana.de

J. de Bloom
School of Social Sciences and Humanities, University of Tampere, Tampere, Finland
e-mail: Jessica.de.Bloom@uta.fi

D.D. Ebert
Department of Clinical Psychology and Psychotherapy, Friedrich-Alexander-University Erlangen, Erlangen, Germany
e-mail: david.ebert@fau.de

P. Angerer
Institute of Occupational and Social Medicine, Heinrich-Heine-University Düsseldorf, Düsseldorf, Germany
e-mail: Peter.Angerer@uni-duesseldorf.de

B. Funk
Department of Information Systems Research, Leuphana University Lueneburg, Lueneburg, Germany
e-mail: funk@uni.leuphana.de

M. Wiencke et al. (eds.), *Healthy at Work*, DOI 10.1007/978-3-319-32331-2_19

Abstract

During the past few years, the Internet has started to change lifestyles and affect all life domains, including working life. It is also increasingly used for targeting mental health issues. The "application of information technology in mental and behavioral health" (Andersson G, Riper H, Carlbring P (2014) Editorial: Introducing Internet interventions—a new open access journal. Internet Intervent 1:1–2) is becoming common in health-care; interventions have already been incorporated into routine care in countries such as the Netherlands, Sweden, the UK, Australia, and the USA. As a next step, Internet interventions in the area of occupational health are progressively emerging. They may offer an evidence-based, cost-effective, and convenient way of promoting workers' mental health on a large scale. Currently, Internet interventions for workers are the most promising approach in the field of occupational e-mental health. The evolution of occupational e-mental health is embedded in interdisciplinary research, practice, and policy. In the first section of this chapter, the origins of occupational e-mental health will be outlined and a definition proposed. Following this, different approaches to occupational e-mental health will be described and their potentials elucidated. A comparison between Internet interventions and traditional stress-management trainings will provide further insights into the design and characteristics of the most elaborated approach in occupational e-mental health. Subsequently, various Internet training programs will be introduced and the evidence for their efficacy summarized. Finally, important topics for further research and implementation will be outlined.

1 Introduction

In 1950, the International Labor Organization and World Health Organization defined the aims of occupational health as "the promotion and maintenance of the highest degree of physical, mental and social well-being of workers". Later, in the 1970s, the psychosocial aspects of work increasingly gained interest, attention being paid to the consequences of occupational stress regarding safety issues, health, and mental health. In the late 1980s and the 1990s, efforts were made to bridge the gap between the disciplines of occupational health and psychology. The potential of psychology to contribute to occupational health was described as "industrial/organizational psychology provides expertise in work organization and job design, and the fields of health psychology and clinical and counseling psychology provide expertise on stress, health and mental health" (Sauter et al. 1999). As a result, a new discipline was established: occupational health psychology. Cox et al. (2000, p. 101) stated that occupational health psychology is "the contribution of applied psychology to occupational health". Similarly, Sauter et al. (1999) referred to the definition of occupational health psychology proposed by the American National Institute for Occupational Safety and Health as "the application of psychology to improving the quality of worklife, and to protecting and promoting the safety, health and well-being of workers" (National Institute for Occupational Safety and Health 2005). Houdmont and Leka (2010) emphasized the applied, evidence-based, and multidisciplinary character of occupational health psychology and its focus on interventions as particularly desirable characteristics.

The term e-health had come into use by the year 2000 and has since become widely prevalent (Pagliari et al. 2005). Although there is no consensus about its definition (Oh et al. 2005; Pagliari et al. 2005), e-health can be summarized as "the use of emerging information and communications technology, especially the Internet, to improve or enable health and healthcare" (Eng 2001, p. 8) and "in a broader sense, the term characterizes not only a technical development, but also a new way of working, an attitude, and a commitment for networked, global thinking, to improve health-care locally, regionally, and worldwide by using information and communication technology" (Eysenbach 2001). E-mental health was defined by Christensen et al. (2002, p. 17) as a "form of e-health which deals with mental health and mental health disorders".

The term occupational e-mental health combines the trajectories of occupational health psychology and e-mental health. In general, occupational e-mental health can be described as the application of e-mental health in the specific life domain of work; its aims include improving the quality of working life and protecting and promoting the safety, health, and well-being of workers. Adopting a more detailed definition of e-mental health proposed by Riper et al. (2010) and adapting it to the working-life context, we define occupational e-mental health as a generic term that describes the use of information and communication technology to deliver psycho-education, health risk assessment, work-place health promotion, preventive interventions (universal, selected, or indicated), treatment, relapse prevention, and return-to-work assistance for the mental health of workers as well as to improve occupational health-care delivery, professional education (e-learning), and online research in the field of occupational mental health. Thus, occupational e-mental health covers person- and organization-focused approaches. Beyond mental health, the more generic term "occupational e-health" refers to the promotion and maintenance of the highest degree of physical, mental, and social well-being of workers using information and communications technology.

Currently, major challenges for occupational medicine and occupational psychology concerning mental health include: (1) the prevention of ill health and sick leave (physical and mental) caused by adverse psychosocial work conditions ("work stress"); (2) provision of fast and adequate health service to employees who develop mental health problems that become apparent in the work-place; and (3) providing support for those who return to work after prolonged absences because of mental disorders. For prevention, the central tools are mental health risk assessment and subsequent interventions such as redesign of work organization and tasks or stress-management training; however, successfully achieving these requires considerable effort. Provision of fast and adequate health services to persons with mental health problems is a challenge to the entire health system and especially to responsible stakeholders in enterprises; prolonged illness without appropriate diagnosis and treatment may cause presenteeism, difficulties at work, unnecessarily long sick leave, and job losses, all of which can be heavy burdens for employers and employees. As for early treatment, successful return to work helps to prevent job loss and early retirement because of mental disorder, which are one of the most frequent reasons for prematurely quitting work life. The emerging resource of occupational e-mental health may provide solutions to these challenges.

2 Current Approaches and Upcoming Trends in Occupational e-Mental Health

The proposed definition of occupational e-mental health covers a wide variety of intervention approaches including Internet-based interventions; m-health or mobile health technologies; social media; serious gaming and gamification; v-health or virtual reality; and providing care via videoconferencing, telephone, or instant messaging (cf. Mohr et al. 2013). We will now describe these approaches in further detail (cf. Fig. 1).

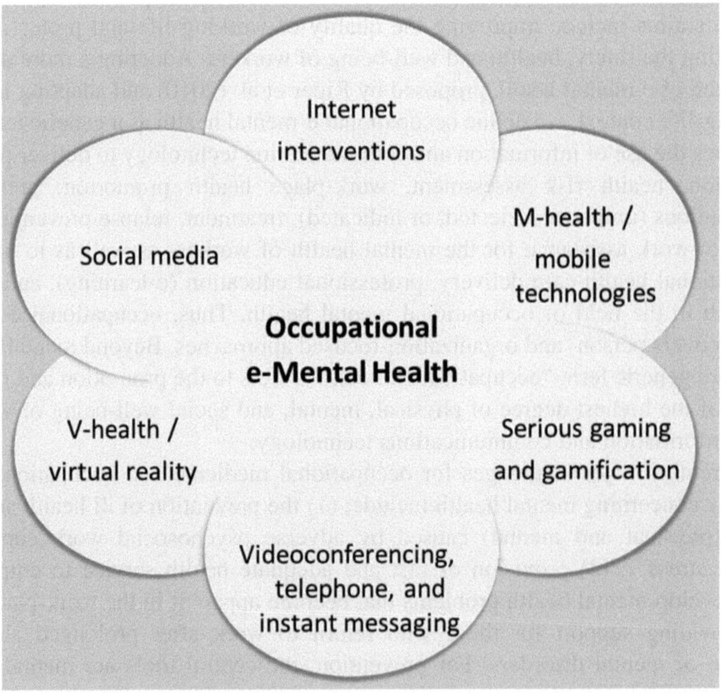

Fig. 1 Overview of the current approaches and upcoming trends in Occupational E-mental Health

2.1 Internet Interventions

So far, most research has been conducted in the field of Internet interventions targeting mental health outcomes. Ritterband and Thorndike (2006) described Internet interventions as "typically behaviorally or cognitive-behaviorally-based treatments that have been operationalized and transformed for delivery via the Internet. Usually, they are highly structured; self- or semi-self-guided; based on effective face-to-face interventions; personalized to the user; interactive; enhanced by graphics, animations, audio, and possibly video; and tailored to provide follow-up and feedback". This definition provides an insight into the typical features employed in Internet interventions and differentiates Internet interventions from websites providing information and education. Guidance is usually provided by a health-care professional. Most Internet interventions are designed for desktop computers.

2.2 Mobile Technologies

Mobile technologies have the potential to introduce interventions into everyday life. Because most people carry mobile or smartphones with them and these devices are widely available, mobile technologies can deliver "ecological momentary interventions" under real-world and real-time conditions to individuals on a very large scale (Heron and Smyth 2010). Interventions using mobile technologies can be delivered as stand-alone programs or in combination with desktop-optimized Internet interventions. Combined or hybrid formats are particularly useful for the components of interventions that focus on ecological moment-to-moment assessment and interventions such as regular mood or behavior assessments, diaries, frequent exercises, or messages (e.g., reminding or encouraging messages). Moreover, built-in devices (e.g., cameras), sensors (e.g., GPS), or wearables (e.g., glasses, watches) connected to smartphones may also play roles in such interventions. Finally, traditional devices such as accelerometers, actigraphy devices, and blood pressure or heart rate monitors may also be connected indirectly to the Internet via smartphones. Moreover, such devices can also have direct Internet access, a phenomenon that has been described as the Internet of Things. While the dynamic in the development of mental health smartphone applications has been extraordinary, empirical evidence for its efficacy is very limited, although promising (Fiordelli et al. 2013). Notably, the distinction between Internet-based and mobile health interventions is not clear-cut and should be regarded as in a heuristic stage. We believe that interventions delivered via mobile devices that are connected to the Internet such as tablet computers or smartphones may also be labeled Internet-based interventions.

2.3 Social Media

Social media focus on the exchange of content created by its users in the context of social networks, discussions, or support groups. Social media can be a component of Internet-based interventions that enables participants to give each other feedback and provide encouragement. Although there is some evidence that support groups moderated by a health-care professional can be beneficial (Winzelberg et al. 2003), it is possible for non-moderated groups to have no or even negative effects because of social contagion (Takahashi et al. 2009). In addition to their use as a component of Internet-based interventions, social media (namely, Facebook, Twitter, WhatsApp) may also be useful tools for health communication and to increase the reach and efficiency of (occupational) health services (Capurro et al. 2014; Moorhead et al. 2013).

2.4 Serious Gaming and Gamification

Games for serious purposes or serious games related to e-health are "digital games with the purpose to improve an individual's knowledge, skills, or attitudes in the 'real' world" (Graafland et al. 2014). This approach is based on the idea of using "entertaining games with non-entertainment-related goals" such as the promotion of mental health (Mohr et al. 2013). A related approach is to identify the ingredients that make games attractive and use these features to increase the enjoyableness, reach, and efficacy of Internet interventions or smartphone applications. Generally, gamification has been defined as "the use of game design elements in non-game contexts" (Deterding et al. 2011). Some of these design elements are rewards, badges, leaderboards, points and levels-systems, challenges and quests, specific goals, social engagement loops and onboarding, and the provision of cover-stories (Zichermann and Cunningham 2011).

2.5 V-Health or Virtual Reality

The potential of v-health approaches for mental health was first successfully investigated in virtual reality exposure therapy for anxiety disorders (e.g., social phobia, fear of flying) (Opriş et al. 2012). Virtual reality exposure has been defined by Parsons and Rizzo (2008) as a method "in which users are immersed within a computer-generated simulation or virtual environment (VE) that updates in a natural way to the user's head and/or body motion." When a user is immersed in a VE, they can be systematically exposed to specific "feared" situations and learn to apply new behavioral skills. Regarding occupational health, an avatar-based virtual environment "Family of Heroes" has been found to be an effective tool for helping family members to motivate their veterans to seek help from Veterans Administration when those veterans are exhibiting evidence of post-traumatic stress disorder, traumatic brain injury, depression, or suicidal ideation (Albright et al. 2012).

2.6 Videoconferencing, Telephone, and Instant Messaging

Real-time or synchronous communication is a shared characteristic of videoconferencing (e.g., Skype, Google Hangout), telephoning, or instant messaging. Compared with in-person or face-to-face settings, videoconferencing eliminates some elements of nonverbal communication (e.g., olfactory or haptic cues); however, it does provide restricted but similar visual information. Talking to someone on the telephone removes the visual channel from communication and instant messaging is restricted to written material. There is a growing body of evidence that videoconferencing is an effective mode of delivering psychotherapeutic care to patients in their own homes, which is particularly relevant for persons staying or living in remote areas (including military personnel and people in civil war zones or developing countries) or for providing support and expertise from specialized health-care professionals to remote clinics (Hilty et al. 2013; Richardson et al. 2009). Occupational health videoconferencing has successfully been used to treat mental disorders in veterans (Hilty et al. 2013). Telephone-delivered psychotherapy for treating veterans reportedly yields mixed results (Mohr et al. 2011); however, telephone-delivered psychotherapy is generally effective in other settings (Mohr et al. 2008).

Most research in the field of occupational e-mental health has been conducted with Internet-based trainings as defined by Ritterband and Thorndike (2006). Accordingly, the following sections will focus on this approach. As to the other approaches just described, the immediacy and wide reach of m-health and social media make their potential obvious; developing applications in this area is a popular topic and researchers have begun to employ stronger research designs. Likewise, the potential benefits of serious gaming and v-health are apparent; however, the high costs of developing virtual reality or high-quality games are currently a barrier to widespread research and implementation that may be overcome in the next few years with the development of more affordable solutions. Videoconferencing and telephone-delivered health-care have already proven their potential. However, for technical reasons it has only been in the last few years that video- and telephone conferencing has begun to be Internet-based and therefore to qualify as a form of e-mental health. In the future, the aforementioned approaches are expected to be increasingly combined and integrated. For example, incorporation of smartphone applications, social media features, or videoconferencing into Internet-based trainings is feasible and promising. Likewise, boundaries become blurred as responsive web design allows delivery of the same interventions via mobile devices such as smartphones and via desktop computers (e.g., Dagöö et al. 2014).

3 Characteristics of In-Person and Internet Interventions

In this section, we compare the characteristics of traditional stress-management trainings with those of Internet-based trainings for promoting mental health in workers (c.f. Lehr et al. 2014).

Stress-management trainings are often based on cognitive behavioral therapy, which is the most common intervention used in occupational health to promote psychological well-being. Several meta-analyses have demonstrated the effectiveness of stress-management trainings with regard to stress, depression, and anxiety (Martin et al. 2009; Richardson and Rothstein 2008; van der Klink et al. 2001; Bhui et al. 2012). As outlined in Sect. 2, most research in occupational e-mental health has targeted Internet-based interventions as opposed to other approaches. Accordingly, Internet-based and face-to-face training formats can be considered the standard approaches in the online and offline worlds, respectively.

Over the past decade, well-established training concepts have been adopted and converted into Internet-based formats. Internet-based trainings are usually based on the same theoretical background and offer the same content as traditional trainings. Moreover, traditional and Internet-based trainings both require investment of time and challenge working people to change their health behavior or stress-related cognitions in spite of their daily, sometimes heavy, workload. In addition to these shared fundamentals, both training formats have their own specific characteristics.

Traditional trainings are typically face-to-face or in-person interventions (Table 1) with groups of 10–20 participants led by one or two health-care professionals.

The trainings usually comprise between four and ten sessions. Although weekly sessions are usually recommended, trainings are often conducted in block courses as single events. Participation in traditional trainings requires participants to be willing to work in groups on their personal health issues. In groups aimed at addressing personal stress management, participants are expected to disclose personal information to other group members to enable them to work on their individual strengths and weaknesses. A group atmosphere can discourage some individuals, particularly those with social anxiety, from opening up or participating. However, groups can offer social support, which may increase motivation for participation, adherence, and effectiveness. The group setting provides opportunities for interactive training techniques for promoting social skills, such as role-playing. Additionally, group members can provide direct feedback. With emotionally demanding exercises such as imagination exercises, trainers can provide participants with immediate support. Verbal sharing of personal experiences in group settings also requires a certain degree of verbal competence and affinity. Regarding organizational factors, participation in group trainings requires fixed dates and times over several weeks or block seminars. Additionally, extra time for traveling by private or public transportation to the venue has to be taken into account. The availability of such trainings may be limited, especially in rural areas. Furthermore, trainings often take place in closed groups, which means that groups cannot be joined at any time and potential participants must wait for the next training course to begin.

In contrast with the group setting of traditional trainings, Internet trainings are usually conducted individually. The interventions characteristically comprise five to ten training modules that can be completed weekly. This procedure underlines the necessity of performing exercises repeatedly and regularly. Each module is

Table 1 Typical characteristics of current Internet trainings and traditional stress-management trainings

Characteristic	Guided Internet training	Traditional training
Location	• Connected via Internet	• In person/face-to-face
Social setting	• Individual training	• Group setting
Role of health-care professional	• Provides individual guidance to participants	• Coach works with 10–20 participants at the same time
Access	• Relatively low threshold, depending on intensity of guidance	• Depends on the availability of courses nearby
Start	• Flexible training start, low latency from participation intention to start	• Start of training is time-bound, waiting times are likely to occur
Venue	• Anywhere with Internet access	• Fixed venue, means of transportation, and travel time necessary
Appointments	• Time flexibility	• Determined by coach or group
Pace of training	• According to individual preferences	• Pace depends on schedule, coach or group
Choice of topics	• Depends largely on a priori choices made by training developers	• Newly relevant topics can be included in training flexibly
Deepening of themes	• Working on individually important topics, can be performed in depth and repeatedly	• Individual focus on personal relevant topics only in accordance with coach and group
Avoidance of topics	• Easily possible	• Depending on coach and group, soft pressure for participants to confront difficult topics
Self-disclosure	• Easy to adjust degree of self-disclosure according to personal norms	• Coach and group sets social norm for degree of self-disclosure
Feedback	• E-coach and/or automated feedback, or alternatively no feedback	• Receive feedback from coach and other participants
Social support	• E-coach • Integration of social media features can enable social support from other participants	• Receive support from coach and other participants • Providing helpful support to other participants may promote self-confidence • Being confronted with problems of other participants can be an additional burden
Crises	• Delayed, asynchronous crises interventions	• Immediate crises interventions
Self-management skills	• Highly flexible training times require setting priorities and protecting the training time against urgent daily duties or tendencies to procrastinate	• Fixed training times require longer term time management skills to organize free time slots

(continued)

Table 1 (continued)

Characteristic	Guided Internet training	Traditional training
Communication skills	• Reading and writing skills	• Talking and listening skills
Anonymity	• Perceived anonymity ranges between complete and total transparency	• Varies with group composition, low in intra-corporate setting, higher with external participants
Specific anxieties	• Surveillance and publication of personal data on the Internet	• Social anxieties in groups, disclosure of confidential information by group members
Training dropout	• Low threshold to dropping out of training	• High threshold depending on (perceived) cohesiveness of group

usually designed to be completed in approximately 30–60 min; the exercises within a module may all be performed at once or distributed over some time. Some interventions are pure self-help trainings without any support from a health-care professional (unguided self-help training). Self-help trainings in particular have the potential to be disseminated on a large scale and to be available at any time. Hence, training can be started when desired and no waiting time is necessary. The training can be conducted anywhere that is convenient, with Internet access being the only precondition. Pure self-help trainings most appropriately compared with traditional self-help books whereas guided trainings provide the possibility of contacting a health-care professional, who is sometimes referred to as an e-coach, via email or a messaging system. The e-coach guides the participants through the training, answers questions, and provides support with the exercises. Typically, this support is provided for 15–30 min after every session. Alternatively, with coaching-on-demand formats, participants receive support only when requested. Formats that require a higher coaching intensity, similar to the amount of time needed in psychotherapy, may also be used. The greater the guidance provided, the more personnel resources are needed, which limits the reach and immediacy of access to Internet training.

Interactive elements such as audio and video clips are often included. Furthermore, online diaries offer participants the opportunity to monitor their own health and training progress. Despite the increased use of multimedia elements, Internet-based interventions are often characterized as being largely text-based. Written exercises require substantial reading and writing skills and also an affinity with these communication channels. The lack of nonverbal signals may cause misunderstandings in asynchronous communications that can only be dealt with after a time delay. Likewise, only delayed rather than immediate support may be available for emotional crises that are, for example, triggered by certain exercises. Unless monitoring is part of the training, the participants must initiate sharing of information on, for example, symptom deterioration; alternatively, the e-coach can ask for such information at regular intervals.

Another specific characteristic of Internet-based interventions is that participants can work at their own pace and review materials and relevant topics as often as they wish. Because their experiences with exercises are in written form, the participants can also easily access their previous responses. Individuals can decide according to their personal preferences or habits on the extent to which they will self-disclose and self-reflect during the training process. However, this may also lead to avoidance of certain topics.

In general, there is a low access threshold for participating in Internet-based trainings. However, easy-in also means easy-out; the threshold for dropping out is also low. Hence, specific measures for fostering adherence, such as gamification, may be required. Anonymity is an important characteristic of Internet-based trainings, especially regarding their low threshold for access. Anonymity is often regarded as an advantage of Internet trainings because it facilitates engagement in topics associated with feelings of shame or guilt (e.g., alcohol consumption for managing stress). Although Internet trainings may appear to be completely anonymous, the degree of anonymity actually depends on several factors. These include individual habits in handling personal data and ensuring data security, the provision of data security by the Internet or training provider or both, and legal or illegal activities by third parties who may read, save, or pass on data. These possibilities may elicit fear and discourage some individuals from participating.

Table 1 summarizes the characteristics of Internet and traditional trainings. However, because of the rapidity of technological developments, these characteristics are constantly changing and only reflect present circumstances. They should be regarded as an orientation to determining the best match between the preferences, needs, and skills of participants and training characteristics.

4 Internet Interventions for Workers: A Review of Concepts and Evidence of Efficacy

The efficacy of Internet-based interventions has so far mainly been investigated in the field of psychotherapy. Internet interventions have proven to be effective in the reduction of depression (Richards and Richardson 2012) and anxiety (Cuijpers et al. 2009; Haug et al. 2012), and have also demonstrated substantial effects in reducing sleep disorders (Cheng and Dizon 2012).

Several Internet interventions targeting mental health in workers have also been investigated in randomized controlled trials, which are regarded as the gold standard for investigating efficacy. We will now describe the content of those trainings and their basic results.

Ruwaard et al. (2007) evaluated an email-based training for stress reduction in the work-place in a sample of 342 individuals. The training consisted of the following seven modules: self-monitoring and perception of stress symptoms; relaxation; rumination, worrying, and coping with negative thoughts; positive self-verbalization; social competence training; time management; and relapse prevention. On average, each participant received 5 h of coaching. Compared with the wait-list control subjects, participants in the intervention group showed significantly greater improvements in stress, depression, anxiety, and burnout.

Hasson et al. (2005) evaluated a stress toolbox in 303 employees in the information technology and media sector. Participants in the control group had access to an online stress diary with feedback and received information concerning stress and health whereas the intervention group underwent a program that offered exercises covering relaxation, sleep enhancement, cognitive restructuring, time management, regulation of emotion, and self-esteem. Moreover, a chat was available to participants of the intervention group. These investigators used the following biological stress markers as indicators of efficacy: chromogranin A, dehydroepiandrosterone, neuropeptide Y, and tumor necrosis factor-α. The intervention group showed significantly better results for all biological markers.

Cook et al. (2007) investigated a comprehensive health promotion program that included stress management, nutrition/weight management, and fitness/physical activity. The participants were 419 employees of a human resources company. Information and guidance was provided on a website for 3 months. The web-based program was completely self-guided, highly interactive, and used videos, audio, and graphics. Participants in the web-condition were compared on a variety of outcome measures with an active control group that received high-quality print material on the same topics. The subjects in the web-condition improved more in the areas of diet and nutrition than did those in the print-condition; however, there were no significant differences between the groups in stress reduction and improvements related to physical activity. Adherence to the intervention was generally poor, particularly for stress management and physical activity. However, the authors identified a dose-response relationship: there was a significant relationship between the number of times participants accessed the web program and the strength of effect across the dependent measures.

A prevention program for stress, depression, anxiety, and substance abuse was examined by Billings et al. (2008) in a sample of 309 employees. The intervention group had 3 months' access to a self-help program that was audio-narrated and incorporated videos and graphics. The training was based on different modules, such as goal setting, problem-solving, identification and restructuring of negative thoughts, and time management. Compared with a wait-list control group, the investigators found a small reduction in stress but no effects on depression or anxiety. However, the intervention group showed a more positive attitude toward seeking psychological help for crises.

Yamagishi et al. (2008) evaluated a training course for nursing staff based on the assumption that unfavorable occupational identity is a key source of stress. Its effectiveness was examined in nursing staff ($N = 60$). The training consisted of the following four modules: definition of occupational identity; assumptions regarding participants' own occupational identity, characteristics of nursing staff, and alternative careers; and management and planning of occupational identity. They found no differences between the intervention and control groups regarding health benefits.

Abbott et al. (2009) developed a 10-week online resilience training and evaluated the intervention in 53 sales managers. The goal of the training was to learn the following seven core competences with the aim of improving resilience: regulation of emotion, impulse control, optimism, analysis of causality of problems, empathy, self-efficacy, and improvement in positive aspects of life such as close contact with other people or facing challenges. Only seven people completed the training, which may explain the absence of significant intervention effects.

Geraedts et al. (2014a, b) investigated the effects of the Happy@Work training in employees with symptoms of depression. The intervention consisted of six sessions and comprised a problem-solving training and additional features adopted from cognitive therapy. The study included 231 employees who received support from an e-coach. Although the training group showed considerable improvements in depressive symptoms, the investigators identified neither short- nor long-term effects (after 12 months) compared with care as usual (Geraedts et al. 2014a, b). Notably, an economic evaluation of this clinic trial showed that the majority of costs were attributable to presenteeism; however, the intervention was not considered to be cost-effective or cost-saving for the employer (Geraedts et al. 2015).

Ebert, Lehr et al. (2014) evaluated a similar problem-solving training in a study of 150 schoolteachers with symptoms of depression. The intervention comprised five sessions of problem-solving training and participants received guidance from an e-coach. Compared with a wait-list control group, participants showed a moderately greater reduction in depressive symptoms and this improvement remained stable over 6 months. Similar effects were observed in general self-efficacy and work-specific self-efficacy. Several predictors of training efficacy such as age, sex, weekly working hours, experience with therapy or training, emotional exhaustion, and general self-efficacy were tested; however, these variables did not predict change in severity of depression (Junge et al. 2015).

A recreation training designed to improve restorative sleep, mental detachment from work, and recreational activities was developed by Thiart, Lehr et al. (2015). The training consisted of six lessons and employed established methods from cognitive behavioral therapy for managing insomnia such as sleep restriction, stimulus control, and hygiene interventions, as well as techniques targeted at reducing rumination such as a gratitude diary and the promotion of recreational activities. In a study of 128 teachers receiving personal support from an e-coach, the investigators found that the training substantially reduced sleep problems and fostered recreational behavior and mental detachment from work. The beneficial effects were maintained over 6 months. Only 10 % of the participants had had prior experiences with other occupational mental health trainings, indicating that this training reached workers who had not been reached by traditional training formats. In a replication study Ebert et al. (2015) found similar effects for the unguided version of the recreation training. Moreover, Thiart et al. (in press) conducted the first economic evaluation of an internet intervention from an employer's perspective and found that the focusing on sleep improvement using the recreation training may be a cost-effective strategy in occupational health care.

Heber et al. (2016) evaluated a web-based and mobile stress-management training course for employees. The training consisted of seven sessions and was based on problem-solving and emotion-regulation techniques. The 256 participants were supported by an e-coach and could receive text messages on their mobile phones alongside a web-based intervention that introduced training exercises and motivational support in real life and real time. A considerable difference between the intervention and a wait-list control group was found for reduction of perceived stress post-test and at a 6-month follow-up as there is new evidence: Significant medium to large between-group effects were also found for relevant secondary outcomes concerning mental health (e.g., depression, anxiety), work-related health (e.g., emotional exhaustion), and stress-related skills (e.g., emotion-regulation competencies).

Wolever et al. (2012) developed a mindfulness-based training for employees that focused on work-related stress, work-life balance, and self-care. "Similar effects for primary and secondary outcomes were found in a replication study investigating the unguided self-help version of the same intervention (Ebert et al. 2016)." The training comprised 1-h sessions over 12 weeks and daily exercises. The same training was offered in two different formats; namely, a traditional local group training and a virtual group training over the Internet with real-time bidirectional communication. Ninety-six individuals took part in the study: the investigators found no differences between the virtual and in-person training formats. When they compared the Internet training group with a wait-list control group, they found a medium to large effect on stress reduction.

Feicht et al. (2013) evaluated a 7-week happiness intervention in 147 employees of a health insurance company. The intervention was based on principles of positive psychology and mindfulness and included components such as improving social relationships, mindfulness exercises, identification and use of personal strengths, and a happiness and gratitude diary. Participants received a weekly email at work explaining the intervention topic (10–15 min) and were asked to perform the exercises at home. Compared with a wait-list control group, the investigators found a moderate effect on reduction of subjective experience of stress post-test. They also found large between-group effects on happiness, satisfaction, and quality of life in favor of the intervention group at post-test and 4 weeks after completion of training. They found no differences between the groups on objective physiological measures (serum cortisol and alpha-amylase concentrations).

Bolier et al. (2014) examined the effects of a workers' health surveillance module including tailored feedback and an online intervention in 366 nurses and allied health professionals. After an initial screening, they offered participants tailored advice about their mental health and an online intervention in the following areas: well-being and mental fitness, depression, stress, panic, and alcohol. Compared with a wait-list control group, positive mental health and psychological well-being were significantly enhanced in the intervention group at post-test and follow-up. No significant effects were found for work engagement. However, because the uptake of the interventions was lower than expected, the authors suggested that the positive effects may have been attributable to the increased awareness caused by the screening process and tailored, personalized advice.

Ly et al. (2014) evaluated the effectiveness of a smartphone-based training in a sample of 73 mid-level managers. The 6-week training was based on the principles of Acceptance Commitment Therapy and incorporated mindfulness exercises to help participants focus on the present and reduce multitasking, avoid equating thoughts about a situation with reality, learn to accept unpleasant thoughts and emotions without avoidance, and identify their own values and translate them into action goals. The investigators found moderate positive differences between the intervention and wait-list control groups.

Kawakami et al. (2006) developed an unguided online training for supervisors. The training comprised the following nine sessions: a general introduction; the importance of mental health in the work-place; official guidelines for promoting mental health in the work-place and the role of the supervisor; listening to and advising subordinate workers and the use of mental health services; support for workers who return to work after sick leave; improving the work environment regarding stress prevention; awareness of and coping with stress; knowledge about mental health; and a final summary. The investigators randomly assigned 46 supervisors to either training or non-training work-places. With regard to job stressors, social support received by distressed workers did not differ between trained (N = 81) and untrained (N = 108) supervisors. However, there were some small favorable effects for work autonomy and friendly atmosphere in the work-place in favor of work-places with trained supervisors.

Overall, Internet trainings for workers have been investigated using a variety of health outcomes ranging from biological stress markers to self-reported stress, depression, or expert-diagnosed insomnia. Participants have been recruited from within companies and from the general population. Some studies have applied almost none and others rigorous inclusion criteria; some have used a universal and others a selected or indicated approach to prevention. Some studies have focused on specific occupational groups or settings whereas others have been designed for a broader range of participants. Trainings have also differed in length, theoretical background, and the number of different skills or coping strategies addressed. Finally, support from a personal coach was provided in some trainings, whereas others were designed as pure self-help trainings.

In summary, it can be concluded that Internet trainings for workers are feasible. However, findings concerning efficacy are heterogeneous, varying from reportedly having small or non-significant effects through to significant effects. The availability of support and guidance from health-care experts appears to be one important explanation for differences in efficacy. However, there are currently far too few studies to identify the key factors associated with effective Internet trainings.

5 Agendas for Future Research

Occupational e-mental health is still in its infancy. This section outlines topics that might be important for accelerating its growth. Some of these refer to general challenges in e-mental health research (cf. Mohr et al. 2013), whereas others are specific with regard to the occupational field.

5.1 Creating a Critical Mass of Efficacy Research

Compared with the amount of research into, and experiences with, integration of psychotherapeutic interventions into routine health-care, particularly regarding the treatment of depression and anxiety, there is scant evidence for the efficacy of occupational e-mental health. Few randomized controlled trials investigating the efficacy of Internet interventions in this field have been conducted. Because a critical mass of research is needed to ascertain the potential of occupational e-mental health and identify blind alleys and fruitful routes, the number of studies employing strong research designs must increase substantially. The research described in Sect. 4 indicates that Internet interventions in particular have the potential to be effective; however, there is insufficient current evidence to draw general conclusions. In addition to investigating general efficacy, it is also importance to determine how these interventions work, and for whom. For example, research could focus on the identification of subgroups for which training is more successful (i.e. Junge et al. 2015). Comparison of Internet-based and traditional trainings to identify the characteristics that determine why traditional trainings work better for some people and Internet-based trainings work better for others would also be useful (Wolever et al. 2012).

5.2 Investigating Reach, Adoption, and Adherence

The potential reach and availability of Internet and m-health interventions is huge because increasing numbers of people own computers with Internet access or use smartphones, or both. This is often seen as a major advantage of e-mental health over traditional health-care. However, evidence concerning the actual reach of public health interventions via the Internet is very limited and less optimistic. Research is needed to better characterize the actual reach and investigate better strategies for increasing acceptance and utilization rates (Bennett and Glasgow 2009). Greater effort and more creative methods are needed to engage males and less educated people, these groups being characteristically under-represented. Especially regarding occupational health, it is important to recommend that companies adopt e-mental health and to investigate factors that influence its adoption. Moreover, even when workers have been reached, Internet interventions are often plagued by high attrition rates. The advantage of easy-in, particularly in self-help trainings, is counter-balanced by the problem of easy-drop-out. Several methods for improving adherence have been proposed, including managing expectations prior to commencing an intervention, tailoring the content to the users' situations, enhancing usability, providing positive feedback, adding personal support (via email, phone, or other media), offering incentives, and applying gaming elements to make the intervention more engaging (Bennett and Glasgow 2009; Mohr et al. 2013).

5.3 Applying Theoretical Frameworks for Occupational Stress in Training Design

Traditional and third-wave cognitive behavioral therapies are often used as the theoretical background for the trainings described here, reflecting the clinical and therapeutic origins of practice and research in the field of Internet interventions. However, with few exceptions, these trainings have not been designed based on a theoretical framework for occupational stress. The recreation training (Thiart, Lehr et al. 2015) described in Sect. 4 is one example of a training whose content is based on a theoretical model of recreation proposed by Lehr et al. (2012). This model claims that recreational activities, mental detachment from work, and restorative sleep are the three core components of recovery from work-related stress. Designing interventions that incorporate the most prominent models of occupational stress would necessitate that such a training cover the topics of coping with effort–reward imbalance situations (Siegrist 2002) or dealing with work-places characterized by low decision latitude (i.e., autonomy and control) in combination with high work demands (Karasek and Theorell 1990). Applying other fruitful frameworks such as the "stress as offense to self" approach (Semmer et al. 2007) to the design of a training could result in trainings that focus on strengthening self-esteem (i.e. through success experiences) or protecting the self against offenses in the work-place such as disrespectful social interactions, illegitimate tasks (i.e., unreasonable or unnecessary tasks), and illegitimate stressors (i.e., pressure to use inadequate tools to accomplish work tasks). However, even traditional stress-management interventions have rarely been designed according to a well-established theory of occupational stress such as the effort–reward imbalance model (i.e., Limm et al. 2011). Applying theoretical frameworks of occupational stress can help to design trainings that specifically target problems that are relevant to participants, thus making the trainings more tailored and sensitive to their work situations. Moreover, application of evidence-based frameworks ensures that trainings focus on topics that have been proven to be associated with mental health.

5.4 Considering Environmental and Organizational Factors

Research on delivering psychotherapy using the Internet has focused almost exclusively on individual factors. However, occupational health has a much broader arena and should take into account the creation of healthy working conditions and the responsibility of employers to ensure workers' health. As Semmer (2006, p. 515) has emphasized, "the promotion of health and the prevention of health problems should predominantly focus on creating a work environment that does not induce an undue amount of stress and that compensates for unavoidable stresses by characteristics such as high control and high rewards". The above cited review of Richardson and Rothstein (2008) illustrates that this widely accepted claim often fails to be acknowledged by the economic world; whereas individual stress management interventions that target the individual factors of work life are effective, interventions targeting organizations have been found to have little or no effects. However, as Lamontagne et al. (2007) have shown in a comprehensive review, if changes in the work-place are actually implemented, their effects on workers' mental health are positive. This finding has been corroborated in later trials (Bourbonnais et al. 2011; Weigl et al. 2013). Kawakami et al. (2006) were among the first to design an Internet intervention for worksite mental health promotion. More research is needed to clarify how to design effective Internet trainings aimed at improving the work environment and organizational factors. Such interventions could either focus on fostering individual skills in "job-crafting" or enable those who are responsible for the organization of work (i.e., supervisors, managers, personnel officers) to create a healthy work environment, or both. Social media in particular has the potential to increase participation, which is necessary for achieving accepted and meaningful organizational changes (Semmer 2006).

5.5 Combining Assessment and Intervention

In many countries, health risk assessment is demanded by the occupational safety and health authorities. The mental health risk component of health risk assessment has only recently received adequate attention and is now increasingly being implemented in the work-place. However, because both mental health and psychosocial work-place factors contributing to physical and mental health are much more difficult to measure precisely, interpret, and change than physical or chemical factors, substantial new expertise is required. Moreover, not only what is done but also how it is done, that is, the implementation process, is extremely important because many power groups within an enterprise must work together to achieve a health risk assessment (Weigl et al. 2015). Recently, the European Agency for Safety and Health at Work launched the OiRA: an online interactive risk assessment project. Internet-based assessment tools provide advantages such as being accessible on a large scale, easy to update, and providing the opportunity to link further information and interventions to an assessment. These assessments could focus on the work environment and organization of work (e.g., identifying hazards, stressors) or on the health status of the individual worker (e.g., level of perceived stress, health behavior), or both. Online health assessments were found to be preferred and well accepted by employees across six different public and private sector organizations in England, especially when highly stigmatized mental health problems such as unhealthy alcohol consumption are targeted (Khadjesari et al. 2015). First experiences from linking Internet-based assessments with personalized feedback on health behaviors has demonstrated the feasibility of delivering screening in combination with brief interventions in the work-place (Khadjesari et al. 2015). Similarly, Bolier et al. (2014) combined screening, tailored advice, and tailored choice of Internet interventions for promoting mental health in nurses in the Netherlands. However, more work is needed to identify successful strategies for engaging a larger proportion of employees in online health assessments (Khadjesari et al. 2014), and in doing so to attract those workers with unhealthy behaviors who would most benefit from the interventions (Khadjesari et al. 2015). Nevertheless, a very promising route seems to be the investigation of delivering Internet-based health assessments that provide valid individual feedback on yellow or red flags, thus helping workers prioritize their health issues and enhancing their motivation for change. Such feedback could be combined with immediate access to tailored interventions that are either Internet-based or in-person. Moreover, information and communications technology easily allows combining data from such worker-directed health assessments and work-place-directed assessments. This could enable the initiation of comprehensive and coordinated actions that are again both worker- and work-place-directed.

5.6 Fostering Implementation

Investigation of sound strategies for integrating occupational e-mental health into existing occupational care is required. Lessons could be learnt from the successful implementation of Internet-based treatments into routine psychiatric and psychotherapeutic care. Development of blended trainings (combining in-person and Internet-based training sessions) and integration of occupational e-mental health into a stepped-care approach are promising avenues. The mainstays of successful implementation are transparent ongoing communication and the participation of all groups in an enterprise (e.g., employer's representatives, employees, health experts, staff representatives) from the very beginning. Moreover, embedding the interventions into existing structures, for example, within the occupational safety and health departments and their existing health promotion activities, is key to success.

Blending in-person and Internet-based training may be especially promising because they combine the virtues of both approaches and facilitate implementation of training in the complex setting of a work-place. Social support and immediate feedback as experienced in traditional interventions (so called health circles; Aust and Ducki 2004), or stress management interventions (Limm et al. 2011; Bourbonnais et al. 2011) are probably among the main factors that promote long-term benefits. In health circles that aim to identify stressors at work and at redesigning the work environment for better work design, exchanges in such groups and an atmosphere of mutual support contribute to creative solutions and broad acceptance of changes. The emotional exchanges that occur in stress-management groups create social support, mutual understanding, and the recognition that others also experience work-related stress. Regarding implementation, trust is of outmost importance. A combination of face-to-face meetings in the work-place with Internet-based interventions and its obvious advantages seems optimal for creating a framework of trust and social support.

Anonymity is often seen as an advantage of Internet interventions, especially with regard to stigmatized mental health issues. However, concerns about data security may pose a major threat for the broad incorporation of occupational e-mental health into routine occupational health-care. Illegal tracking of personal information is of concern, as is the employment by legal authorities of barely legal methods of surveillance. Khadjesari et al. (2015) found 27 % of workers to be concerned about the safety of health-related personal information on the Internet. Trust in data security is especially important because mental health problems are often stigmatized and accompanied by feelings of insufficiency, shame, and guilt.

Finally, from an ethical point of view, workers should be free to choose either a traditional or Internet-based intervention. Unfortunately, technological changes tend to enter a field as an additional opportunity and become obligatory with the passage of time. It is therefore important to emphasize freedom of choice as a paramount value from the birth of occupational e-mental health.

5.7 Capitalizing on Interdisciplinary Research

Research in e-mental health has so far been driven predominantly by research in clinical or health psychology and psychiatry. Especially with occupational e-mental health, there is a need for stronger cooperation with experts from work and organizational psychology and occupational medicine. Moreover, research in occupational e-mental health should go beyond health sciences such as medicine and psychology and capitalize on interdisciplinary research, namely computer science and economics.

Computer science and information systems research can contribute to occupational e-mental health in multiple ways: First, through Internet interventions and health risk assessments, a wealth of data (e.g. sensor data from smartphones, data from ecological assessments in the moment, and simple questionnaires) can be collected that can help to identify people at risk and improve the quality of interventions (Mohr et al. 2013). Application of techniques from machine learning to these data can not only derive informative features for enhancing and personalizing interventions but also contribute to a general understanding of workers' behavior and mental health (cf. Both et al. 2008). Second, occupational e-mental health can learn from software engineering how to bridge the gap between the pilot studies currently being carried out in many research settings and their full implementation in practice. This may involve various aspects, including considering basic principles of software architecture, balancing the conflicting requirements of different stakeholders, and leveraging approaches like "privacy by design" (Gürses et al. 2011). Third, information systems research can shed some light on user acceptance of, and engagement in, occupational e-mental health interventions (cf. Kleine Stegemann et al. 2012).

Economic evaluations can provide valuable information on the economic costs and benefits of occupational e-mental health interventions (van Dongen et al. 2014). So far, little is known about the effects of Internet trainings on the costs of absenteeism and presenteeism (Geraedts et al. 2015). Additionally, future research should also consider the effects on costs of work-place accidents, staff turnover, work engagement, and organizational citizenship behavior. However, whether economic evaluations demonstrating occupational e-mental health interventions to be cost-beneficial from an employer's perspective would lead to greater investments is an unanswered question. In his review, Semmer (2006) found little evidence that even sound cost–benefit analyses have an important impact on companies' decisions regarding their investment in health-oriented interventions. Other factors appear to be more relevant to decision-making regarding investments in health promotion. For example, it seems particularly important for decision makers to perceive health as a distinct value, independently of financial interests. Moreover, trusting relationships in those who deliver such interventions appear decisive. Further research is needed to investigate the impact of economic evaluations and other factors on the willingness to invest in occupational e-mental health.

References

Abbott J, Klein B, Hamilton C et al (2009) The impact of online resilience training for sales managers on wellbeing and performance. E J Appl Psychol 5(1):89–95

Albright G, Goldman R, Shockley KM et al (2012) Using an avatar-based simulation to train families to motivate veterans with post-deployment stress to seek help at the VA. Game Health J 1(1):21–28

Andersson G, Riper H, Carlbring P (2014) Editorial: Introducing Internet interventions—a new open access journal. Internet Intervent 1:1–2

Aust B, Ducki A (2004) Comprehensive health promotion interventions at the workplace: experiences with health circles in Germany. J Occup Health Psychol 9(3):258–270

Bennett GG, Glasgow RE (2009) The delivery of public health interventions via the Internet: actualizing their potential. Annu Rev Public Health 30:273–292

Bhui KS, Dinos S, Stansfeld SA et al (2012) A synthesis of the evidence for managing stress at work: a review of the reviews reporting on anxiety, depression, and absenteeism. J Environ Public Health. doi:10.1155/2012/515874

Billings DW, Cook RF, Hendrickson A et al (2008) A web-based approach to managing stress and mood disorders in the workforce. J Occup Environ Med 50(8):960–968

Bolier L, Ketelaar SM, Nieuwenhuijsen K et al (2014) Workplace mental health promotion online to enhance well-being of nurses and allied health professionals: a cluster-randomized controlled trial. Internet Intervent 1:196–204

Both F, Hoogendoorn M, Klein M et al (2008) Modeling the dynamics of mood and depression. In: Proceedings of the 18th European conference on artificial intelligence, Amsterdam, pp 266–270

Bourbonnais R, Brisson C, Vézina M (2011) Long-term effects of an intervention on psychosocial work factors among healthcare professionals in a hospital setting. Occup Environ Med 68(7):479–486

Capurro D, Cole K, Echavarría MI et al (2014) The use of social networking sites for public health practice and research: a systematic review. J Med Internet Res 16(3):e79

Cheng SK, Dizon J (2012) Computerised cognitive behavioural therapy for insomnia: a systematic review and meta-analysis. Psychother Psychosom 81(4):206–216

Christensen H, Griffiths KM, Evans K (2002) E-mental health in Australia: implications of the internet and related technologies for policy. Information Strategy Committee discussion paper, vol 3. Commonwealth Department of Health and Ageing, Canberra

Cook RF, Billings DW, Hersch RK et al (2007) A field test of a web-based workplace health promotion program to improve dietary practices, reduce stress, and increase physical activity: randomized controlled trial. J Med Internet Res 9(2):e17

Cox T, Baldursson E, Rial-Gonzalez E (2000) Occupational health psychology. Work Stress 14(2):101–104
Cuijpers P, Marks IM, van Straten A et al (2009) Computer-aided psychotherapy for anxiety disorders: a meta-analytic review. Cogn Behav Ther 38(2):66–82
Dagöö J, Persson Asplund R, Andersson Bsenko H et al (2014) Cognitive behavior therapy versus interpersonal psychotherapy for social anxiety disorder delivered via smartphone and computer: a randomized controlled trial. J Anxiety Disord 28:410–417
Deterding S, Dixon D, Khaled R et al (2011) From game design elements to gamefulness: defining 'gamification'. In: Proceedings of the 15th international academic MindTrek conference: envisioning future media environments, New York. ACM, New York, pp 9–15
Ebert DD, Lehr D, Boss L et al (2014) Efficacy of an internet-based problem-solving training for teachers: results of a randomized controlled trial. Scand J Work Environ Health 40(6):582–596
Ebert DD, Berking M, Thiart H et al (2015) Restoring depleted resources: efficacy and mechanisms of change of an internet-based unguided recovery training for better sleep and psychological detachment from work. Health Psychol 34:1240–1251
Ebert DD, Heber E, Berking M et al (2016) Self-guided internet-based and mobile-based stress management for employees: results of a randomised controlled trial. Occup Environ Med 73(5):315–323
Eng TR (2001) The eHealth landscape: a terrain map of emerging information and communication technologies in health and heath care. Robert Wood Johnson Foundation, Princeton
Eysenbach G (2001) What is e-health? J Med Internet Res 3(2):e20
Feicht T, Wittmann M, Jose G et al (2013) Evaluation of a seven-week web-based happiness training to improve psychological well-being, reduce stress, and enhance mindfulness and flourishing: a randomized controlled occupational health study. Evid Based Complement Alternat Med 2013:676953
Fiordelli M, Diviani N, Schulz PJ (2013) Mapping mHealth research: a decade of evolution. J Med Internet Res 15(5):e95
Geraedts AS, Kleiboer AM, Twisk J et al (2014a) Long-term results of a web-based guided self-help intervention for employees with depressive symptoms: randomized controlled trial. J Med Internet Res 16(7):e168
Geraedts AS, Kleiboer AM, Wiezer NM et al (2014b) Short-term effects of a web-based guided self-help intervention for employees with depressive symptoms: randomized controlled trial. J Med Internet Res 16(5):e121
Geraedts AS, van Dongen JM, Kleiboer AM et al (2015) Economic evaluation of a web-based guided self-help intervention for employees with depressive symptoms; results of a randomised controlled trial. J Occup Environ Med 57(6):666–675
Graafland M, Dankbaar M, Mert A et al (2014) How to systematically assess serious games applied to health care. JMIR Serious Games 2(2):e11
Gürses S, Troncoso C, Diaz C (2011) Engineering privacy by design. Available from https://www.cosic.esat.kuleuven.be/publications/article-1542.pdf. Accessed 7 Apr 2015
Hasson D, Anderberg U, Theorell T et al (2005) Psychophysiological effects of a web-based stress management system: a prospective, randomized controlled intervention study of IT and media workers. BMC Public Health 5(1):78
Haug T, Nordgreen T, Öst LG et al (2012) Self-help treatment of anxiety disorders: a meta-analysis and meta-regression of effects and potential moderators. Clin Psychol Rev 32(5): 425–445
Heber E, Lehr D, Ebert DD et al (2016) Web-based and mobile stress management intervention for employees: a randomized controlled trial. J Med Internet Res 18(1), e21
Heron KE, Smyth JM (2010) Ecological momentary interventions: incorporating mobile technology into psychosocial and health behaviour treatments. Br J Health Psychol 15(1):1–39
Hilty DM, Ferrer DC, Parish MB et al (2013) The effectiveness of telemental health: a 2013 review. Telemed J E Health 19(6):444–454

Houdmont J, Leka S (2010) An introduction to occupational health psychology. In: Leka S, Houdmont J (eds) Occupational health psychology. Wiley-Blackwell, Chichester, pp 1–30
Junge M, Lehr D, Bockting CLH et al (2015) For whom are occupational e-mental health interventions effective? Predictors of internet-based problem-solving training outcome. Internet Intervent 2(1):39–47
Karasek R, Theorell T (1990) Healthy work: stress, productivity, and the reconstruction of working life. Basic Books, New York
Kawakami N, Takao S, Kobayashi Y et al (2006) Effects of web-based supervisor training on job stressors and psychological distress among workers: a workplace-based randomized controlled trial. J Occup Health 48(1):28–34
Khadjesari Z, Freemantle N, Linke S et al (2014) Health on the web: randomised controlled trial of online screening and brief alcohol intervention delivered in a workplace setting. PLoS One 9(11):e112553
Khadjesari Z, Newbury-Birch D, Murray E et al (2015) Online health check for reducing alcohol intake among employees: a feasibility study in six workplaces across England. PLoS One 10(3):e0121174
Kleine Stegemann S, Ebenfeld L, Thiart H et al (2012) Towards measuring user engagement in internet interventions for common mental disorders. In: Proceedings of HCI 2012, The 26th BCS conference on human computer interaction. Available from http://ewic.bcs.org/upload/pdf/ewic_hci12_pcp_paper7.pdf. Accessed 7 Apr 2015
Lamontagne AD, Keegel T, Louie AM et al (2007) A systematic review of the job-stress intervention evaluation literature, 1990–2005. Int J Occup Environ Health 13(3):268–280
Lehr D, Heber E, Thiart H (2012) Regeneration als Ressource: Erholungsverhalten als Antwort auf berufliche Herausforderungen [Regeneration as an major resource. Coping with occupation stress by fostering recreational behaviour]. PADUA 7:182–187 (in German)
Lehr D, Eckert M, Baum K et al (2014) Online-Trainings zur Stressbewältigung—eine neue Chance zur Gesundheitsförderung im Lehrerberuf? [Online stress-management-interventions—an effective approach to fostering mental health in school teachers?]. Lehrerbildung auf dem Prüfstand 1:191–212 (in German)
Limm H, Gündel H, Heinmüller M et al (2011) Stress management interventions in the workplace improve stress reactivity: a randomised controlled trial. Occup Environ Med 68(2):126–133
Ly KH, Asplund K, Andersson G (2014) Stress management for middle managers via an acceptance and commitment-based smartphone application: a randomized controlled trial. Internet Intervent 1(3):95–101
Martin A, Sanderson K, Cocker F (2009) Meta-analysis of the effects of health promotion intervention in the work place on depression and anxiety symptoms. Scand J Work Environ Health 35(1):7–18
Mohr DC, Vella L, Hart S et al (2008) The effect of telephone-administered psychotherapy on symptoms of depression and attrition: a meta-analysis. Clin Psychol Sci Pract 15(3):243–253
Mohr DC, Carmody T, Erickson L et al (2011) Telephone-administered cognitive behavioral therapy for veterans served by community-based outpatient clinics. J Consult Clin Psychol 79(2):261–265
Mohr DC, Burns MN, Schueller SM et al (2013) Behavioral intervention technologies: evidence review and recommendations for future research in mental health. Gen Hosp Psychiatry 35: 332–338
Moorhead SA, Hazlett DE, Harrison L et al (2013) A new dimension of health care: systematic review of the uses, benefits, and limitations of social media for health communication. J Med Internet Res 15(4):e85
National Institute for Occupational Safety and Health (2005) Occupational health psychology. Available from http://www.cdc.gov/niosh/topics/ohp/. Accessed 18 Feb 2015
Oh H, Rizo C, Enkin M et al (2005) What is eHealth (3): a systematic review of published definitions. J Med Internet Res 7(1):e1

Opriş D, Pintea S, García-Palacios A et al (2012) Virtual reality exposure therapy in anxiety disorders: a quantitative meta-analysis. Depress Anxiety 29(2):85–93

Pagliari C, Sloan D, Gregor P et al (2005) What is eHealth (4): a scoping exercise to map the field. J Med Internet Res 7(1):e9

Parsons TD, Rizzo AA (2008) Affective outcomes of virtual reality exposure therapy for anxiety and specific phobias: a meta-analysis. J Behav Ther Exp Psychiatry 39(3):250–261

Richards D, Richardson T (2012) Computer-based psychological treatments for depression: a systematic review and meta-analysis. Clin Psychol Rev 32:329–342

Richardson KM, Rothstein HR (2008) Effects of occupational stress management intervention programs: a meta-analysis. J Occup Health Psychol 13:69–93

Richardson LK, Frueh BC, Grubaugh AL et al (2009) Current directions in videoconferencing tele-mental health research. Clin Psychol Sci Pract 16(3):323–338

Riper H, Andersson G, Christensen H et al (2010) Theme issue on e-mental health: a growing field in internet research. J Med Internet Res 12(5):e74

Ritterband LM, Thorndike F (2006) Internet interventions or patient education websites? J Med Internet Res 8(3):e18

Ruwaard J, Lange A, Bouwman M et al (2007) E-mailed standardized cognitive behavioural treatment of work-related stress: a randomized controlled trial. Cogn Behav Ther 36:179–192

Sauter SL, Hurrel JJ, Fox HR et al (1999) Occupational health psychology: an emerging discipline. Ind Health 37(2):199–211

Semmer NK (2006) Job stress interventions and the organization of work. Scand J Work Environ Health 32(6):515–527

Semmer NK, Jacobshagen N, Meier LL et al (2007) Occupational stress research: the "stress-as-offense-to-self" perspective. In: Houdmont J, McIntyre S (eds) Occupational health psychology: European perspectives on research, education and practice, vol 2. Nottingham University Press, Nottingham, pp 43–60

Siegrist J (2002) Effort-reward imbalance at work and health. In: Perrewe P, Ganster D (eds) Research in occupational stress and well-being, vol 2, Historical and current perspectives on stress and health. JAI Elsevier, New York, pp 261–291

Takahashi Y, Uchida C, Miyaki K et al (2009) Potential benefits and harms of a peer support social network service on the internet for people with depressive tendencies: qualitative content analysis and social network analysis. J Med Internet Res 11(3):e29

Thiart H, Ebert DD, Lehr D et al. (in press) Internet-based cognitive behavioral therapy for insomnia: a health economic evaluation. Sleep.

Thiart H, Lehr D, Ebert DD et al (2015) Log in and breathe out: internet-based recovery training for sleepless employees with work-related strain: results of a randomized controlled trial. Scand J Work Environ Health pii:3478. doi:10.5271/sjweh.3478 [Epub ahead of print]

van der Klink JJ, Blonk RW, Schene AH et al (2001) The benefits of interventions for work-related stress. Am J Public Health 91(2):270–276

van Dongen JM, van Wier MF, Tompa E et al (2014) Trial-based economic evaluations in occupational health: principles, methods, and recommendations. J Occup Environ Med 56(6): 563–572

Weigl M, Hornung S, Angerer P et al (2013) The effects of improving hospital physicians working conditions on patient care: a prospective, controlled intervention study. BMC Health Serv Res 13:401

Weigl M, Herbig B, Bahemann A et al (2015) Empfehlungen zur Durchführung einer Gefährdungsbeurteilung psychischer Belastungen: Positionspapier der Deutschen Gesellschaft für Arbeitsmedizin und Umweltmedizin, erstellt von der AG Psychische Gesundheit in der Arbeit der DGAUM [Recommendations on developing and carrying out psychosocial risk evaluations at the workplace]. Arbmed Sozialmed Umweltmed 50:660–665 (in German)

Winzelberg AJ, Classen C, Alpers GW et al (2003) Evaluation of an internet support group for women with primary breast cancer. Cancer 97(5):1164–1173

Wolever RQ, Bobinet KJ, McCabe K et al (2012) Effective and viable mind-body stress reduction in the workplace: a randomized controlled trial. J Occup Health Psychol 17(2):246–258

Yamagishi M, Kobayashi T, Nakamura Y (2008) Effects of web-based career identity training for stress management among Japanese nurses: a randomized control trial. J Occup Health 50(2): 191–193

Zichermann G, Cunningham C (2011) Gamification by design: implementing game mechanics in web and mobile apps. O'Reilly Media, Sebastopol, pp 35–67

Monitoring Mental Stressors at Work with the Work Health Audit Instrument Factors: Results of Validation Studies

Sebastian Fischer, Kai-Michael Kleinlercher, and Wulf Rössler

Abstract

To monitor the mental health status of their employees, businesses rely on ready-made tools that are easy to use, effective and not only assess their employees' mental health, but also provide clues about the origins of any difficulties, thus facilitating choice of appropriate interventions. In this chapter, we describe the development and validation of an instrument for measuring mental health at work that we call Work Health Audit Instrument (WHAI) and provide an overview of its reliability and validity. We propose 18 WHAI factors that measure distinct causes for mental health difficulties at work and show that these factors relate to various relevant outcome criteria within businesses, such as work satisfaction, work engagement, irritation, and mental health. Thus, this chapter aims to give employers reasons for using this tool to measure the mental health of their employees. Additionally, we suggest that the WHAI factors can be used in further scientific inquiry, for example as a screening tool for studies on intervention effectiveness.

1 Introduction

Mental disorders account for more than a quarter of all illnesses worldwide (Menken et al. 2000). In Germany, mental factors are having an increasing impact on productivity in the work-place and accounting for an increasing number of sick days and medical treatments (Jacobi 2009). Mental disorders have severe effects on

This chapter is an abbreviated and translated version of a report (Kleinlercher et al. 2015). For the sake of brevity, we have eliminated some of the statistical section and focused on the rationale for the study and its results. Additional statistical information can be obtained from the authors.

S. Fischer (✉) • K.-M. Kleinlercher • W. Rössler
Leuphana University of Lüneburg, Lüneburg, Germany
e-mail: sebastian.fsc@gmail.com

M. Wiencke et al. (eds.), *Healthy at Work*, DOI 10.1007/978-3-319-32331-2_20

individuals, companies, and the economy at large: in Germany the costs of treatment, rehabilitation, early retirement, and exclusion from the workforce are estimated at around 22 billion € (Allianz AG 2011) and thus create a burden on public finances. Companies must deal with the consequences of employees' inability to work and the reduced productivity of people with mental disorders (Wang 2005). The personal consequences for persons so affected are also immense: psychological stress can have a massive psychosocial impact, reducing quality of life (Spitzer et al. 1995), increasing mortality rates (Harris and Barraclough 1998), and often resulting in stigmatization (Corrigan 2004). Additionally, there are indications that mental disorders not only cause many problems in the work-place, but that problems in the work-place contribute to the development of these disorders (Melchior et al. 2007; Tennant 2001).

Thus, the prevention of risks to mental health and well-being in the work-place is becoming increasingly relevant. Many larger companies have already implemented occupational health management practices that far exceed the standards defined in legal rules and guidelines of western countries; however, many small to medium-sized enterprises (SMEs) lack the financial resources and personnel to offer such health management to their employees (see also chapter "Promoting Workers' Health in Small and Medium-Sized Enterprises: Designing and Evaluating a Concept for Preventing Occupational Skin Diseases"). Of note, the personnel of SMEs are at risk at least as much as their peers in larger organizations (Hodgins et al. 2010). Here, we present a Work Health Audit Instrument (WHAI), which provides 18 factors for measuring risks to mental health and well-being in the work-place. Selection of its items and scales was heavily influenced by our work with representatives of SMEs. We believe that this questionnaire is particularly suitable for SMEs.

To assist the design of the WHAI factors questionnaire, we researched the instruments most frequently used to measure mental stressors in German businesses. Our survey of current practices revealed that a new instrument would offer added value if it: (1) included both factors relating to the corporate environment (macro-factors) and individual factors at the employee level (micro-factors); (2) focused specifically on factors that the companies would be able to influence (intervention); (3) referred to current research on mental stressors in the work-place, such as the use of information technology or the psychosocial climate in the work-place; (4) offered a brief and easy-to-use instrument for all groups of staff; and (5) was suitable for assessing mental risk factors in the work-place in line with current regulatory requirements.

2 Theoretical Development of the WHAI Factors Questionnaire

The theoretical foundations for the WHAI factors questionnaire were drawn from the Job Demands Resources model (Bakker and Demerouti 2007; Demerouti et al. 2001), which assumes that people's well-being depends on the demands and resources they encounter in the work-place. "Demands", in this case, refers to stressors that require specific, sustained efforts, whereas "resources" are work-place-related factors that help achieve set goals, counter the negative effects of demands, or promote personal development or a combination of these (Bakker and Demerouti 2007). Using this concept, the questionnaire includes factors that measure demands (e.g. time pressures) as well as resources (e.g. autonomy). Noteworthy, each factor can be interpreted as representing a demand or a resource; depending on its level (e.g. a high workload may be a demand, whereas a lower workload may be seen as a resource).

The selection of the pool of items for the instrument was influenced by published models for explaining health risks at work, especially the gratification crises model proposed by Siegrist (1996), the demands/control model of Karasek et al. (1988), and the demands/resources model of Demerouti et al. (2001). In a systematic survey of current published reports, we used these models to facilitate identification of health-related stressors in the work-place and to source relevant factors in various databases. We paid particular attention to factors with high diagnostic validity. This search resulted in an initial pool of 20 factors (167 items) for measuring individual resources and risk factors on organizational or specific occupational levels. Additional factors were added to the questionnaire for validation purposes (59 items). These factors cover a wide range of indicators of employee mental health and well-being (e.g., general health, work engagement, job satisfaction). The initial item pool was completed with 37 general questions, including demographic markers and questions about the participants' area of work. The size of this initial item pool was subsequently reduced in factor analyses.

3 Collecting the Data

We recruited the sample for the construction and alignment of the instrument from April 2013 to April 2014, our specific intention being to reflect a corporate structure that was typical of Germany. To achieve this, we selected a random sample of companies from one region of Germany. We contacted 192 companies, 43 of which agreed to participate in the study. Another two companies self-selected into the sample after responding to media coverage on the project in local newspapers. Most of these companies were active in the services (seven) and healthcare/social sectors (seven). The make-up of the sample on the corporate level is presented in Tables 1 (sectors of industry) and 2 (company sizes).

We gave all 6824 members of the participating companies a code for accessing the online questionnaire. Of these, 2321 completed the questionnaire; thus, the participation rate was 34.01 %. The distribution of the study subjects' sexes and ages is shown in Table 3 and their occupational categories in Table 4.

Participation in the study by the employees of the participating companies was voluntary. The participating companies' managements sent all employees an invitation in writing accompanied by a brief description of the purpose of the study. An additional information leaflet reminded the participants that participation was

Table 1 Distribution of sectors of industry of participating companies

Sector of industry	Frequency	
	Absolute	Percentage
Service	7	15.56
Healthcare and social work	7	15.56
Retail	6	13.33
Construction, trades, manufacturing	3	6.67
Biotech, chemical, pharmaceutical, medical	3	6.67
Accounting, fiscal, legal	2	4.44
Electronics, electrical engineering, technology	2	4.44
Consumer goods, food, non-food	2	4.44
Business consulting	1	2.22
Catering, hotel, tourism	1	2.22
Mechanical and plant engineering	1	2.22
Aeronautical	1	2.22
Information technology, telecommunication	1	2.22
Media (print, film, radio, TV), publishing, printing	1	2.22
Other	6	13.33
Not specified	1	2.22

Note: Thirty-five sectors of industry and the category "Others" were available for selection. For clarity, all sectors not named by any company were excluded from the table

Table 2 Distribution of sizes of participating companies (number of regular insured employees in 2013)

Company size	Frequency	
	Absolute	Percentage
Smallest companies (up to 10 employees)	3	6.67
Small companies (up to 50 employees)	21	46.67
Medium-sized companies (up to 250 employees)	17	37.78
Large companies (more than 250 employees)	4	8.89

Note: No data concerning size were available for four companies. These companies were allocated to one of the above categories using the number of access codes for the online questionnaire

Table 3 Sex and age of participants

	Age category					
	17–29	30–39	40–49	50–59	60–69	70–79
Male	171	282	454	322	61	2
Female	181	247	314	232	35	1

Note: Sample size: $n = 2321$. Nineteen participants did not divulge their sex or age

voluntary, confidential, and strictly anonymous (specifically in all dealings with their employers). Before the survey started, the participants were informed about the criteria for inclusion and exclusion. The questionnaires were completed during working hours at web-connected workstations with no interruptions.

Table 4 Distribution of the respondents' occupational categories

Occupational category	Frequency	
	Absolute	Percentage
Service	647	45.63
Manufacturing/service delivery	328	23.13
Administration	220	15.51
Marketing and sales	175	12.34
Research and development	163	11.5
Accounting and controlling	97	6.84
Logistics and materials management	92	6.49
Human resources	58	4.09
Other	132	9.31

Note: Occupational categories were not available for 903 respondents because their companies requested company-specific terminology. Multiple responses were allowed; thus, the sum of categories exceeds 100 %

4 Data Analyses

To define the final WHAI factors and shorten the questionnaire, we performed a sequence of factor analyses. First we used exploratory factor analyses to gain an impression of the factor structure underlying the data. This was followed by confirmatory factor analyses to test the correlations and factors identified.

4.1 Exploratory Factor Analyses

All data for this study were sourced by means of questionnaires, a method that has a major impact on empirical correlations between data: because all data come from a single source, employee, halo effects or negative effects may obscure the true relationships between factors. For example, persons may evaluate all factors positively because they like their work-place overall, even though they sometimes experience difficulties which would, on careful inspection of the item, result in reduced item scores. In a dataset, such effects potentially obscure or amplify relationships between the theoretically assumed factors, a bias commonly known as "common method bias" (Podsakoff et al. 2012). It is therefore advisable to test for a potential common method factor problem before continuing with further analyses. There is a statistical way to test the size of the common method factor in exploratory factor analyses: if one factor ("common method factor") explains more than 30 % of the variances in the responses of the participants, then there is a strong common method bias in the data. In this study, the common method factor explained less than 30 % of this variance and could therefore be considered relatively insignificant.

As a second step, we used exploratory principal component analysis to assess the number of factors in the data. This process produced 16 factors with an eigenvalue >1 that we used in the subsequent confirmatory factor analysis.

4.2 Confirmatory Factor Analyses

We assumed that the instrument measures a range of factors that all relate to mental health at work. To facilitate the eventual interpretation of questionnaire results for employees or organizations, we aimed to specify these factors in this step. Because results of a questionnaire tool are usually interpreted factor by factor, final factors should differ from each other in a meaningful way to facilitate easy interpretation. However, there may be conceptual or statistical overlap of some factors, for example if they are basically measuring the same phenomena from slightly different angles. Such overlaps make interpretation of factor results difficult and may make some factors redundant. It was our goal to reduce such redundancy and arrive at a simple structure for the instrument, in which each item loads on only one factor.

Therefore, we performed confirmatory factor analyses (Muthén and Muthén 2010), based on the results of our exploratory factor analysis to establish the final factor structure of the questionnaire. To arrive at a simple structure for the instrument, we created new factors or removed items whenever single items correlated too directly with many factors. These factor analyses allowed a dramatic reduction in the number of items, producing a time-efficient but meaningful instrument. Confirmatory factor analysis produced a solution with 18 factors that fitted well with the data (better than the initial 16 factor solution obtained by exploratory factor analysis) and proved to be the most suitable for our purposes. We also developed and tested an alternative model with only four factors (organizational climate, occupational requirements and resources, personal resources, and consequences) and a single factor model for comparison, all of which showed reduced model fit, suggesting that the 18 factor model should be pursued because it best fit the data.

4.3 Naming the Factors of the WHAI

We defined the resulting WHAI factors as follows:

4.3.1 Participation in the Flow of Information

This refers to fairness in information management and opinion-forming processes at the organization in question. Items address the availability of information in decision-making processes, the presence of relevant standards, and the involvement of all parties affected by a decision.

4.3.2 Relationship with Superiors

This concerns the quality of mutual relationships between employees and their direct superiors. This quality depends on aspects of compassion, supervisor's respect, and the superior's loyalty in instances of conflict with third parties.

4.3.3 Psychological Safety Among Co-workers

This concerns whether the members of a team treat each other positively. Our items operationalize this by asking about supportiveness and tolerance for diversity or errors.

4.3.4 Solidarity

This concerns whether the climate at work is perceived as civil and engaging. It differs from *Psychological safety among co-workers* in that its items ask about friendly and empathetic attitudes in the general organizational environment and not about the specific psychosocial mechanisms at work in specific teams.

4.3.5 Workload

This defines whether time pressure at work is a stressor. Time pressure can be caused by disruption or interruptions, the sheer amount of work, and the need to complete different tasks simultaneously.

4.3.6 Rewards

This refers to the resources given to employees in return for their work and includes monetary compensation, career opportunities, and non-monetary rewards (including praise).

4.3.7 Job Security

This concerns the risk of redundancy as well as negative changes to the terms and conditions at work.

4.3.8 Feedback

This defines how much feedback employees receive for work done. The items for this factor include feedback from superiors, other team members, and colleagues.

4.3.9 Conflict Between Work and Family

This refers to the extent to which the requirements of work affect people's ability to fulfill the requirements of their private lives. The items concern issues of timing and the need to change planned family activities in response to job-related needs.

4.3.10 Clear Occupational Roles

This concerns the extent to which employees understand the purpose of their work and the ways of achieving that purpose. It also asks how well the employees know the scope of their area of responsibility.

4.3.11 Coherent Occupational Requirements

This factor differs from that of *clear occupational roles* in that it does not focus on unambiguous goals or work practices, but on possible conflicts in terms of those goals and practices. The items therefore ask whether employees are faced with requirements that cannot be reconciled with their outlooks or with conflicting requirements from different people.

4.3.12 Autonomy

This refers to how free employees are to decide how to perform the tasks required of them. It can include the choice of tasks or the individual steps and/or equipment used in the process, depending on the line of work and the qualifications of the employees in question.

4.3.13 Need for Cooperation

This concerns the extent to which work activities require contacts and cooperation with other people, including cooperation with colleagues and contact with clients or customers.

4.3.14 Environmental Forces

This means pressures from factors in the work environment that are not related to people. The items include disruptive factors such as noise, unpleasant temperature, or poor lighting.

4.3.15 Availability of Equipment

This concerns the extent to which the flow of work is disrupted or made more difficult by a lack of necessary equipment. Additionally, this factor concerns enforced waits.

4.3.16 Negative Emotionality

This addresses traits such as neuroticism. Its items cover experiences of sadness and doubt.

4.3.17 Resilience

This concerns the personalities of the respondents and addresses different aspects of resilience; resilient people are not discouraged by setbacks and consider themselves to have strong personalities.

4.3.18 Assumed Efficacy of Self-Control

This refers to particular aspects of resilience. It concerns respondents' belief in their ability to maintain concentration under pressure, belief in their ability to adjust to changes, and self-confidence gained from past achievements.

Table 5 Reliability of the factors

WHAI factor	Cronbach's alpha
Participation in the flow of information	++
Relationship with superiors	++
Psychological safety among co-workers	+
Solidarity	++
Workload	++
Rewards	–
Job security	–
Feedback	–
Conflict between work and family	++
Clear occupational role	+
Coherent occupational requirements	++
Autonomy	++
Need for cooperation	+
Environmental forces	+
Availability of equipment	+
Negative emotionality	++
Resilience	+
Assumed efficacy of self-control	+

Note: –, Alpha < 0.7; +, Alpha 0.7–0.8; ++, Alpha > 0.8

5 Analysis of Reliability

We evaluated the diagnostic accuracy of the WHAI factors by assessing their internal consistency (Demerouti 2012). A factor is deemed to have sufficient internal consistency to ensure reliability when Cronbach's Alpha is greater than or equal to 0.7. The results for the WHAI factors are presented in Table 5.

Fifteen of the 18 factors have Cronbach's Alphas higher than the threshold of 0.7, seven of them being greater than 0.8. The only factors falling below the threshold value are *Job Security*, *Rewards*, and *Feedback*. The comparatively low Alpha for *Job Security* can be explained by the small number of items (two) (Cortina 1993). However, the predictive validity of that factor (see Table 6) provided a strong reason for including it in the final instrument despite the poor internal consistency. The same is true for the factor *Feedback*, which again includes only two items.

6 Validating the Instrument by Regression Analysis

We designed the WHAI instrument to provide a detailed assessment of the risks to participants' well-being posed by stressors in the work-place. A diagnostic instrument is regarded as valid when it measures what it sets out to measure (Demerouti 2012). We therefore ascertained the validity of our instrument by exploring the

Table 6 Statistically significant correlations between WHAI factors and indicators of mental health and well-being

	Work satisfaction	Work engagement	Affective sense of belonging	Intention to leave	Cognitive irritation	Emotional irritation	Mental health
Participation in the flow of information	+	+	+	–		–	
Relationship with superiors	+	+	+				+
Psychological safety among co-workers		+					
Solidarity					–		
Workload					+	+	+
Rewards	+		+	–			
Job security	–	–		+	–		+
Feedback				+	–		–
Conflict between work and family	–			+	+	+	+
Clear occupational role	+	+	–	–		+	
Coherent occupational requirements	–			+		+	
Autonomy	+	+	+		+		
Need for cooperation	+	+	+		+		
Environmental forces					–		
Availability of equipment	+	+		–	+		
Negative emotionality	–	–	–	+	+	+	+
Resilience					–		
Assumed efficacy of self-control		+			+	–	
R^2	**0.55**	**0.44**	**0.39**	**0.41**	**0.48**	**0.62**	**0.58**

Note: +, significant positive correlation ($p < 0.05$); –, significant negative correlation ($p < 0.05$)

statistical link between the stressors measured by the instrument and various forms of health and well-being of the study participants. Note that we see well-being at work as a multidimensional concept. In the following, we therefore report relationships between the factors in our instrument and a number of concepts that measure work-related health and well-being (Diedrich et al. 2015; Quick et al. 2013).

6.1 Work Satisfaction

Work satisfaction is defined as the extent to which an employee likes or dislikes her work (Spector 1997). In our validation study, work satisfaction was measured by the question: "How satisfied are you with your job in general?", the responses being rated on a seven-point Likert scale from "not satisfied at all" to "very satisfied".

Table 6 shows statistically significant links between 11 WHAI factors and work satisfaction. The strongest links were with the subscales *Rewards*, *Job Security (negative)*, *Clear Occupational Roles*, and *Negative Emotionality* (negative). Taken together, the WHAI factors account for 55 % of the variance for the variable *Work Satisfaction* (cf. R^2 in Table 6).

6.2 Work Engagement

Work engagement can be defined as a positive and fulfilling acceptance of the given occupational circumstances and is characterized by vitality, involvement, and immersion in the occupational role(s) (Schaufeli et al. 2002). Vitality means a high level of energy, motivation for work, and the ability to remain undiscouraged by potential difficulties. Involvement refers to strong engagement with work roles and experiencing meaning, enthusiasm, inspiration, pride, and professional challenge in relation to them. Finally, immersion refers to a strong sense of concentration on work and experiencing satisfaction from being immersed in it. For employees who are in a state of immersion, time flies and it is difficult to stop working (Schaufeli et al. 2006). Work engagement can be promoted by autonomy, the experience of support, feedback, opportunities for learning and development, versatile jobs, and a fair organization (Schaufeli and Salanova 2007). We measured this with four items taken from the UWES-9 (Utrecht Work Engagement Scale; Schaufeli and Bakker 2004, unpublished manuscript: Department of Psychology, Utrecht University, Utrecht, the Netherlands); these items are rated on a seven-point Likert scale from "never" to "every day".

As shown in Table 6, there were statistically significant links between ten WHAI factors and work engagement. The strongest predictive effects were for the subscales *Need for cooperation*, *Negative emotionality (negative)*, and *Assumed efficacy of self-control*. Taken together, WHAI factors explain 44 % of the variance in work engagement.

6.3 Affective Sense of Belonging

An affective sense of belonging at work refers to the presence and strength of effective or emotional ties with the organization in which the respondent is working. People with a strong affective sense of belonging characteristically identify strongly with the organization, are strongly involved with its concerns, and enjoy their membership in it (Meyer and Allen 1991). An affective sense of belonging correlates positively with factors such as perceived psychological safety and perception of fairness. It is measured with three items developed by Meyer and Allen (1991).

As Table 6 shows, there are statistically significant links between seven WHAI factors and an affective sense of belonging. The strongest predictors for this variable were the WHAI factors *Participation in the flow of Information* and *Rewards*. One remarkable finding is the comparatively low (negative) regression coefficient for *Negative emotionality*. The extent of an affective sense of belonging seems less susceptible to criteria at the personal level. In total, WHAI factors account for 39 % of the variance of this variable.

6.4 Intention to Leave

The intention to leave for another job is another consequence of mental pressure that has considerable relevance for organizations (Quick et al. 2013). A desire to change jobs relates directly to a perceived sense of belonging (Meyer and Allen 1991): People who identify strongly with an organization and enjoy belonging to it are unlikely to move to another employer.

Table 6 shows statistically significant links between nine WHAI factors and participants' *Intention to leave*. The best predictors of this were the WHAI factors *Reward* and *Negative emotionality* (negative). In total, WHAI factors explain 41 % of the variance for this variable.

6.5 Irritation

Irritation refers to a "state of poor mental well-being caused by an experience of stress" (Müller et al. 2004, p 214). This state can be considered a transition phase from acute to chronic stress, making it an excellent indicator for risks to mental health and thus particularly important for the validation of our instrument. Irritation is caused by a perceived imbalance between the available personal resources and the stressors of everyday life. Work-place stressors are particularly relevant for the chosen construct. Here, irritation describes the subjective experience of *emotional* and *cognitive* stress (Mohr et al. 2005, 2006). Irritation distinguishes between two different consequences of stress: emotional irritation and cognitive irritation. Emotional irritation refers to a state of sensitive irritability, whereas cognitive irritation means an inability to detach from work stressors outside of actual working hours (Mohr et al. 2005, 2006). We measured the degree of irritation with the scale proposed by Mohr et al. (2005, 2006), using eight items rated with a seven-point Likert scale (from "no, does not apply at all" to "yes, applies completely").

As Table 6 shows, there are statistically significant links between 12 WHAI factors and cognitive irritation, with the WHAI factors *Workload*, *Conflict between work and family*, *Negative emotionality*, and *Resilience* (negative) being most effective at predicting cognitive irritation. In total, WHAI factors account for 48 % of the variance for this variable.

Table 6 also shows statistically significant links between the seven WHAI factors and emotional irritation. The strongest links here were for the WHAI factors *Workload* and *Negative emotionality*. WHAI factors in total account for 62 % of variance in the variable. The very high regression coefficient for the factor *Negative emotionality* also suggests that the high explained variation (R^2) is mainly attributable to this factor, whereas the predictive validity for more work-related resources and requirements is slightly lower. Nonetheless, the factors remaining when the scales for personal resources (*Negative emotionality*, *Assumed efficacy of self-control*, and *Resilience*) had been removed still present an R^2 of 0.42.

6.6 Mental Health

Because the purpose behind the design of the instrument was to assess work-place-related risks to participants' mental well-being, we expected particularly valid predictions in the area of mental health. There is some uncertainty concerning the terminology about mental health: it can refer to behaviors, attitudes, and feelings relating to personal satisfaction and effectiveness (Kornhauser 1965) or to the absence of clinically apparent mental disorders. To validate the WHAI factors, we referred to the General Health Questionnaire (GHQ) 12, a validated means for measuring clinical mental health in accordance with the diagnostic criteria laid out in the International Classification of Diseases-10 and Diagnostic and Statistical Manual-IV (Goldberg et al. 1997). Although the GHQ 12 was not developed specifically for use in the work-place, there is sufficient empirical evidence for its validity in the occupational context (Banks et al. 1980). Mental health expressed as a GHQ score was the sole criterion in our validation study that has no immediate link with occupational life. However, we can safely assume a statistically significant link between stressors at work and mental well-being. Numerous studies have shown that stress responses, which can be caused by an imbalance between occupational requirements and available resources, are linked with the development of psychological disorders (McEwen 2004). The most robust link identified so far concerns stress at work and the incidence of depression (Bonde 2008; Netterstrøm et al. 2008). According to a meta-analysis by Bonde (2008), the burden of occupational requirements, coupled with limited autonomy at work, increases the risk of depressive disorders. Kawakami et al. (1992) reported similar findings, having identified a link between limited freedom at work and lack of social support with depressive disorders. Kouvonen et al. (2008) also identified a link between reduced social capital in the work-place (e.g. durable personal relationships) and depression. Stansfeld and Candy (2006) performed a meta-analysis that revealed a link between the experience of professional stress and an imbalance between efforts and rewards with the incidence of depression or anxiety and compulsive disorders. Nieuwenhuijsen et al. (2010) were able to show that stress-related mental disorders relate to high workloads, limited autonomy, lack of social support, poor procedural fairness, and an imbalance between efforts and rewards.

Table 6 shows statistically significant correlations between seven WHAI factors and mental health. One particularly direct link is in the factor *Negative emotionality*. Taken together, the WHAI factors account for 58 % of the variance in mental health. The very high regression coefficient for the *Negative emotionality* factor suggests that the high R^2 is primarily attributable to this factor, with slightly lower predictive reliability for work-related resources and requirements. The factors as a whole, with the exception of those relating to personal resources (*Negative emotionality*, *Assumed efficacy of self-control*, *Resilience*), still have an R^2 value of 0.34, which is a remarkable finding because the items for the mental health scale have no immediate relationship with matters of employment. Mental health needs to be considered as an inherently complex issue that is subject to a wide range of influences and forces. The ability to predict 34 % of the variance for this variable with scales covering work-place resources and requirements is strong evidence for the validity of the proposed tool and underlines the urgent need for more scientific and practical attention to mental health risks in the work-place.

7 Diagnostic Significance

The diagnostic significance of a tool determines how effectively it distinguishes between the different origins or effects of occupational stressors (Demerouti 2012). In the case of our WHAI factors questionnaire, the diagnostic significance can be established by considering the results of multiple regression analyses: in some cases the regression coefficients of the individual subscales do vary considerably between the various criteria. The proposed instrument is therefore able to distinguish between the different effects of different stressors and allows a weighting of these forces.

8 Summary

There are other instruments for measuring the many risks to people's well-being posed by mental pressures in the work-place. To offer real added value, we began by defining five conceptual requirements for such an instrument: First, it had to cover stressors that would, in principle, be susceptible to outside interventions. In the final form of the WHAI factors questionnaire, a mere five of the 18 factors are only partially open to this type of influence, namely *Job Security*, *Need for cooperation*, *Negative emotionality*, *Resilience*, and *Assumed efficacy of self-control*. The other subscales are indeed perfectly suited to organizational interventions. We refer the reader here to the work on health in the work-place by Diedrich et al. (2015), which discusses which interventions could be taken and how they could be introduced on the basis of this instrument. Second, the instrument had to cover both micro and macro-level factors; this is achieved by the final choice of scales. Third, the instrument had to include currently relevant scales, which are provided by *Psychological safety among co-Workers* and *Solidarity*. Fourth, we intended to produce an instrument that could deliver meaningful results with minimal time investment. The final version of the questionnaire includes 18 scales with a total of 54 items, putting it in the average range for such instruments. We therefore succeeded in producing a practical, usable instrument that guarantees comprehensive coverage of the relevant influences and stressors in a time-efficient manner. Statistical assessment of the instrument's reliability, validity, and diagnostic significance showed that it is an ideal choice for monitoring mental health risks in the workplace.

Acknowledgement We would like to thank all members of the UBalance team at Leuphana University of Lueneburg.

References

Allianz Deutschland AG (ed) (2011) Depression: Wie die Krankheit unsere Seele belastet. Allianz Deutschland AG, Munich (in German)

Bakker AB, Demerouti E (2007) The job demands-resources model: state of the art. J Manage Psychol 22(3):309–328

Banks MH, Clegg CW, Jackson PR et al (1980) The use of the General Health Questionnaire as an indicator of mental health in occupational studies. J Occup Psychol 53(3):187–194

Bonde JPE (2008) Psychosocial factors at work and risk of depression: a systematic review of the epidemiological evidence. Occup Environ Med 65(7):438–445

Corrigan P (2004) How stigma interferes with mental health care. Am Psychol 59(7):614

Cortina JM (1993) What is coefficient alpha? An examination of theory and applications. J Appl Psychol 78(1):98

Demerouti E, Deutsches Institut für Normung (2012) Psychische Belastung und Beanspruchung am Arbeitsplatz: inklusive DIN EN ISO 10075-1 bis-3. Beuth, Berlin (in German)

Demerouti E, Bakker AB, Nachreiner F et al (2001) The job demands-resources model of burnout. J Appl Psychol 86(3):499–512

Diedrich L, Fischer S, Kleinlercher K-M et al (2015) Gesundheit im Unternehmen—Psychosoziale Ressourcen erhalten, Potenziale entwickeln. Kohlhammer, Stuttgart (in German)

Goldberg DP, Gater R et al (1997) The validity of two versions of the GHQ in the WHO study of mental illness in general health care. Psychol Med 27(1):191–197

Harris EC, Barraclough B (1998) Excess mortality of mental disorder. Br J Psychiatry 173(1): 11–53

Hodgins M, Battel-Kirk B, Asgeirsdottir AG (2010) Building capacity in workplace health promotion: the case of the Healthy Together e-learning project. Global Health Promot 17(1):60–68

Jacobi F (2009) Nehmen psychische Störungen zu. Rep Psychol 34(1):16–28 (in German)

Karasek RA, Theorell T, Schwartz JE et al (1988) Job characteristics in relation to the prevalence of myocardial infarction in the US Health Examination Survey (HES) and the Health and Nutrition Examination Survey (HANES). Am J Public Health 78(8):910–918

Kawakami N, Haratani T, Araki S (1992) Effects of perceived job stress on depressive symptoms in blue-collar workers of an electrical factory in Japan. Scand J Work Environ Health 18(3): 195–200

Kleinlercher K-M, Fischer S, Müller-Kanneberg B, Rössler W (2015) Psychische Belastungen am Arbeitsplatz erkennen. Kohlhammer, Stuttgar
Kornhauser A (1965) Mental health of the industrial worker: a Detroit study. Wiley, Oxford
Kouvonen A, Oksanen T, Vahtera J et al (2008) Low workplace social capital as a predictor of depression: the Finnish Public Sector Study. Am J Epidemiol 167(10):1143–1151
McEwen BS (2004) Protection and damage from acute and chronic stress: allostasis and allostatic overload and relevance to the pathophysiology of psychiatric disorders. Ann NY Acad Sci 1032(1):1–7
Melchior M, Caspi A, Milne BJ et al (2007) Work stress precipitates depression and anxiety in young, working women and men. Psychol Med 37(08):1119–1129
Menken M, Munsat TL, Toole JF (2000) The global burden of disease study—implications for neurology. Arch Neurol 57(3):418–420
Meyer JP, Allen NJ (1991) A three-component conceptualization of organizational commitment. Hum Res Manage Rev 1(1):61–89
Mohr G, Rigotti T et al (2005) Irritation—ein Instrument zur Erfassung psychischer Beanspruchung im Arbeitskontext. Skalen- und Itemparameter aus 15 Studien. Zeitschrift für Arbeits- und Organisationspsychologie 49(1):44–48
Mohr G, Müller A, Rigotti T et al (2006) The assessment of psychological strain in work contexts: concerning the structural equivalency of nine language adaptations of the irritation scale. Eur J Psychol Assess 22(3):198
Müller A, Mohr G, Rigotti T (2004) Differenzielle Aspekte psychischer Beanspruchung aus Sicht der Zielorientierung. Z Different Diagn Psychol 25(4):213–225 (in German)
Muthén LK, Muthén BO (2010) Mplus, 6th edn. Muthén & Muthén, Los Angeles
Netterstrøm B, Conrad N, Bech P et al (2008) The relation between work-related psychosocial factors and the development of depression. Epidemiol Rev 30(1):18–132
Nieuwenhuijsen K, Bruinvels D, Frings-Dresen M (2010) Psychosocial work environment and stress-related disorders, a systematic review. Occup Med 60(4):277–286
Podsakoff PM, MacKenzie SB, Podsakoff NP (2012) Sources of method bias in social science research and recommendations on how to control it. Annu Rev Psychol 63(1):539–569
Quick JC, Wright TA, Adkins JA et al (2013) Preventive stress management in organizations, 2nd edn. American Psychological Association, Washington
Schaufeli W, Salanova M (2007) Work engagement: an emerging psychological concept and its implications for organizations. In: Gilliland SW, Steiner DD, Skarlicki DP (eds) Managing social and ethical issues in organizations, vol 5, Research in social issues in management. Information Age, Greenwich, CT, pp 135–177
Schaufeli WB, Salanova M, González-Romá V et al (2002) The measurement of engagement and burnout: a two sample confirmatory factor analytic approach. J Happiness Stud 3(1):71–92
Schaufeli WB, Bakker AB, Salanova M (2006) The measurement of work engagement with a short questionnaire a cross-national study. Educ Psychol Meas 66(4):701–716
Siegrist J (1996) Soziale Krisen und Gesundheit. Hogrefe, Göttingen, pp 56–61
Spector PE (1997) Job satisfaction: application, assessment, causes, and consequences. Sage, Thousand Oaks
Spitzer RL, Kroenke K, Linzer M et al (1995) Health-related quality of life in primary care patients with mental disorders: results from the PRIME-MD 1000 Study. JAMA 274(19):1511–1517
Stansfeld S, Candy B (2006) Psychosocial work environment and mental health—a meta-analytic review. Scand J Work Environ Health 32(6):443–462
Tennant C (2001) Work-related stress and depressive disorders. J Psychosom Res 51(5):697–704
Wang JL (2005) Work stress as a risk factor for major depressive episode(s). Psychol Med 35(6): 865–871

Healthy Ageing in the Geriatric Nursing Profession: A Salutogenic Intervention Program

Anne-Katrin Haubold, Mary Kuntzsch, Wilhelm Beckmann, Ariunaa Ishig, Dany Kuntzsch, Harald Christa, and Rüdiger von der Weth

Abstract

Geriatric nursing is demanding in both physical and psychological senses. A 3-year intervention program aimed at preventing the health-related early retirement of geriatric nurses has been designed.

1 Introduction

The considerable stress and strain experienced by nurses has been well documented by a wide range of studies (e.g., Kato 2013). In a recently published Australian study of stress among nurses, high workload, unsupportive management, human resource issues, interpersonal issues, and shift work were among the most frequently mentioned stressors (Happell et al. 2013). In a similar vein, Gelsema et al. (2005) found in their cross-sectional study that working conditions have a significant impact on nurses' job satisfaction and distress. In addition, the big "nurses' early exit study" has shown that older caregivers are more heavily burdened than younger caregivers in the work environment; this cohort of caregivers, especially those in the 45 years and over age group, assess their own health as increasingly poor (Galatsch et al. 2011).

However, as Happell et al. (2013, p. 638) have noted, minimal research on reducing occupational stress among nurses is available. A Cochrane review of prevention programs for occupational stress in healthcare workers found some evidence for the effectiveness of such programs, but concluded that larger and better studies are needed (Marine et al. 2006). Other reviews have come to similar

A.-K. Haubold (✉) • M. Kuntzsch • W. Beckmann • R. von der Weth
Hochschule für Technik und Wirtschaft Dresden, Friedrich-List-Platz 1, 01069 Dresden, Germany
e-mail: haubold@htw-dresden.de

A. Ishig • D. Kuntzsch • H. Christa
Evangelische Hochschule Dresden, Dürerstraße 25, 01307 Dresden, Germany

M. Wiencke et al. (eds.), *Healthy at Work*, DOI 10.1007/978-3-319-32331-2_21

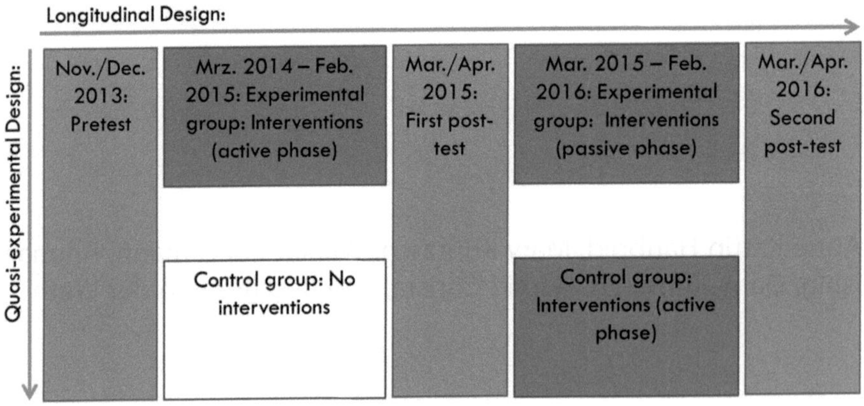

Fig. 1 Diagrammatic representation of the study design

conclusions (Mimura and Griffiths 2003; Ruotsalainen et al. 2008). In addition, in their systematic review of stress intervention studies among student nurses, Galbraith and Brown (2011, p. 709) pointed out that many of the available intervention studies are not theory based.

This contribution describes the first stage of a quasi-experimental longitudinal intervention study in the geriatric nursing sector (see Fig. 1).

The focus of this study is to develop and deploy an intervention program that will allow nurses in this sector to stay in their jobs until retirement rather than dropping out because of physical or mental exhaustion. The experimental group of this intervention program consists of the employees of three geriatric nursing homes in eastern Germany; the control group comprises the employees of a fourth geriatric nursing home. For ethical reasons, the interventions will also be made available to the control group in a time-lagged manner (after the first post-test). The presented intervention study is based on the salutogenesis approach of Antonovsky (1997). Antonovsky proposes moving away from the classical medical approach of looking at the root causes of a disease (pathogenesis) and instead focusing on the preconditions for and development of health (salutogenesis; see also Schliehe et al. 2000). In the salutogenesis concept, one of the most important predictors of health is a sense of coherence, which Antonovsky (1997, p. 13) defines as the general trusting of human beings that (1) the stimuli in their environment are structured, predictable and explainable; (2) they have enough resources to cope with these stimuli; and (3) the handling of these stimuli is perceived to be worthwhile.

The concept of salutogenesis has been frequently used in research projects in the health care sector, but generally with the focus on the health of the patient, not the health of the nurse (Brieskorn-Zinke 2010; Sullivan 1989). Another line of research applies the salutogenesis concept to the health of care-giving family members. Important pioneering research in the area of applying salutogenesis to the health of caregivers has been conducted by a research group around Ivars Udris (Rimann and Udris 1997, 1998; Udris 2006). Building on these and other studies, Höge and Büssing (Höge 2005; Büssing et al. 2004) conducted a cross-sectional study on the salutogenic status quo of employees of outpatient nursing services. They found correlations between a sense of coherence, organizational resources, perceived fairness, and the health of the nurses. However, because the study was cross-sectional, no conclusions on causes and effects can be drawn from its findings.

To achieve an integrated perspective on the health of employees, it is necessary to look at the topic from a managerial, or more precisely, a human resource managerial perspective. Basically, the primary, in some countries legally binding, obligation of human resource management within a company or organization is to support their employees to remain healthy and to look after them. We therefore based the present intervention study on a salutogenic human resource management approach in which we implemented the concept of salutogenesis in the following four intervention fields: (1) a training program for older employees (area human resource development); (2) a health circle (focusing on work design); (3) a salutogenic coaching program for managers (area leadership); and (4) reintegration of employees returning from long-term sickness (area personnel placement). These four interventions are presented in more detail below.

We postulated that the planned interventions would have a significant positive effect on subjective physical and mental well-being (assessed by the German version of the SF 12 of Bullinger and Kirchberger 1998). In addition, we used a sense of coherence scale (SoC 9; Singer and Brähler 2007) and a subjective salutogenic work-place analysis inventory (SALSA; Rimann and Udris 1997) in the pre- and post-tests.

2 Intervention 1: Healthy Ageing Training Program for Employees Aged 40+ Years

As mentioned in Sect. 1, there is evidence that older nurses, especially those aged 40 years and over, assess their own health as increasingly poor (Galatsch et al. 2011). Consistent with the basic premise of the salutogenic model, the goal of the healthy ageing training program is to increase participants' awareness of their own sense of coherence, thereby improving their health. A strong sense of coherence facilitates using personal resources to better handle the stresses of everyday life (Franke and Witte 2009).

The training module is based on the so-called HEDE training system developed by Franke and Witte (2009). Antonovsky uses the acronym HEDE to describe the two poles of the health–disease continuum: Health-Ease and Dis-Ease (cf. Antonovsky 1997, p. 36). HEDE training has been applied in various sectors, having previously been used in the automotive industry and now adapted to meet the specific needs of the geriatric care sector. The training module will be provided in eight 1.5-h-long group training sessions, each group consisting of eight to 12 employees.

The principal aspects of the training are to help each participant (Franke and Witte 2009): (1) examine their understanding of the concept of health; (2) gain an understanding of the salutogenic concept and its role in supporting personal health; (3) deal with burdens and stress without negatively impacting their overall health; (4) heighten their own sense of well-being; (5) improve handling of new and/or stressful situations; (6) become self-assured about their own capabilities; and (7) learn to better use their own resources.

During the introductory session, the participants will be introduced to the training module and its care-relevant topics, terminology will be explained, and they will reflect on their own state of health.

In the second training session, duties, burdens, and currently existing resources will be evaluated to prepare the ground for the development of suitable strategies for dealing with pre-existing burdens or stresses in the subsequent sessions.

The third session will address the greatest day-to-day burden: lack of time. Frequent interruptions, documentation/administrative responsibilities, and heavy workloads are often defined as problems. The participants will be introduced to better methods of time management, including setting priorities, a self-organized approach to work, and analysis of "time wasters"; these methods will be used both in the work-place and in their private lives.

The newly gained time could then be used for the recuperation and relaxation possibilities introduced in the fourth training session, in which the participants will learn how to regularly integrate relaxation into their everyday lives.

In the fifth training session, conflict situations with, for example, colleagues, superiors, family members, or residents will be addressed by means of targeted empathy and feedback exercises. Participants will learn to more effectively understand the behavior of other individuals and better and more objectively resolve conflicts.

Particularly in care professions, employees tend to place their own needs last, frequently meeting challenges or demands even when they seem overly burdensome. The sixth training session will therefore be supplemented by a further communication exercise in which the participants will be given the tools to refuse unwarranted or unjustified demands by purposefully and directly saying no.

Social support in the work-place is another means of reducing the burden of occupational stress (cf. Stadler and Spiess 2002). In the seventh training session participants will learn to recognize others' needs for support and to request assistance themselves.

The final training session will begin with revision of a comparison between burdens and resources (see second training session). Each participant will leave the training module with an individualized resource set. The participants will enter into a "contract with themselves" in which they will formulate their own personal health goals and establish concrete plans for meeting these goals. In addition, the training module will be evaluated.

The contents of the training module will be based on communication of theory, group exercises, role-play, and discussion. Hand-outs pertaining to theory, the contents of the training module and exercises will be provided to assist the participants. The sessions will be cumulative. Apart from the introductory and concluding training sessions, each session will be patterned on the following model (Franke and Witte 2009): (1) discussion of the homework from the previous training session; (2) theoretical introduction with deepening exercises; (3) assignment of new homework to test the learned material in everyday settings; (4) "flash feedback" (brief feedback from participants in an open setting); and (5) evaluation of the session regarding what has been learned and well-being.

Homework assignments are essential to achieving immediate success through the training. Simply attending training sessions does not improve participants' sense of coherence (Franke and Witte 2009): only testing the learned material in everyday situations enables the development of new means of dealing with stress.

Appropriate activities in both private and professional day-to-day life are necessary to maximize sustainability of the training and ensure that the achieved and/or enhanced resources are implemented in the most enduring way possible. The following three measures will be employed on a staggered schedule to achieve this:

(i) After 3 months, each of the participants will receive a copy of their "contract with themselves" (see eighth training session). This will both serve as a reminder and offer the opportunity to review and re-evaluate the self-determined health goals and revise them if necessary. At the same time, a short survey aimed at determining how much the material and skills learned in the HEDE training are being used in day-to-day life will be conducted.
(ii) After 6 months, the participants will receive a greeting by postcard as both a reminder of the content learned and also to motivate them to undertake something to promote their own health.
(iii) Finally, after approximately 1 year, a follow-up training session will take place, the purpose of which will be to exchange experiences relevant to each participant's health. The focus will be on the discussion of any problems encountered in implementing the skills learned during the course.

3 Intervention 2: Health Circle

The health circle provides an opportunity for the staff of geriatric nursing homes to cooperatively develop organizational and work-design means of preserving and improving their own health. By adding a participatory conditional preventive component to the salutogenic human resource management concept, such a circle maximizes and draws on the effects of Interventions 1 and 3 of the program. The increased sensitivity of the employees to health problems contributes to acceptance of and motivation for the circle work. Furthermore, the communication training supports open and constructive exchange in the circle.

The methodological basis of these circles is called "task-related information exchange" ("Aufgabenbezogener Informationsaustausch", also referred to as "AI", Neubert and Tomczyk 1986). It is a specific form of moderated group work for analyzing and intervening in given problems with a heterogeneous, cross-hierarchical group of interested employees. The focus of the method is problem solving using a common knowledge pool developed from the complementary individual knowledge of the group members. An appropriately structured process plan, a collection of methods for the realization of the process steps, and a number of working principles of the AI are described in detail in published reports (Pietzcker and Looks 2010). In the specific implementation of the AI as a health circle, the participants attend eight 90-min meetings on a 2- to 4-weekly cycle to: (1) formulate and prioritize work-related problems regarding the participants' health; (2) collect and evaluate information about problems encountered; and (3) develop and implement organizational and work-design solutions.

As a pre-analysis for capturing the status quo of work design in the geriatric nursing homes, observation and an employee survey were performed. Observation took place during both an early and late shift of each qualified nurse in each geriatric nursing home. Each period of observation was 8–9 h and incorporated comprehensive observation of the wide spectrum of nursing duties, cooperation, and disturbances.

In the circle, the participants acted as experts about their jobs and were encouraged to contribute and combine their experience and specialist knowledge. In addition, they formed a link between the circle and the other employees and departments. They were asked to update their colleagues about the circle work and complement their own input to the circle with combined feedback from their colleagues. In view of this double function, the representativeness of the circle for all relevant employee groups and departments is important. Ideally, every employee of the geriatric nursing home is represented by at least one participant from the same employee group and department to allow (1) an exchange of information to take place between the circle and all employees; (2) the circle to act in the interest of all employees; and (3) acceptance of the solutions that are developed.

The circles of six to nine participants therefore comprised representatives of the employee groups resident care, resident support, kitchen staff, and the nursing home director. Resident care is the biggest employee group. With regard to nursing competence, we divided this group into "basic" and "qualified" caregivers and concerning departmental affiliation into two personal and organizationally separate living areas. Ideally, all four sub-groups are represented in the circle.

Integration of non-participating employees in the geriatric nursing homes participating in the study was a challenge. In the three-shift system these nursing homes used, there was little overlapping of working hours within the groups of "basic" and "qualified" caregivers, which complicated personal exchanges between these groups and their representation in the circle. Therefore, to enable participants to collect more feedback on the content of the circle from their non-participating colleagues, the meeting cycle was changed from 2 to 4 weeks.

Though notices and official discussions at nursing home or living area level offer possibilities for exchange between overlapping groups, they do not support the feedback process optimally. It proved difficult to motivate health circle participants to inform their colleagues about the themes and findings of the health circle in official meetings.

The relatively high absenteeism rate in geriatric care also impacted the circle work in that the group composition varied across sessions. Thus, bigger working groups were more robust and remained more productive despite the sick-leave problem.

The initial session of the AI health circle was devoted to defining the relevant problems to be discussed in the subsequent sessions. A problem mentioned in several circles was the lack of support from colleagues and the low level of team spirit. In the subsequent problem solution process, the following two aspects were discussed in detail:

(i) How the organization usually handles conflicts: How is a conflict solved? What resources are available to assist problem solving? How does the organization deal with conflicts that target important organizational topics? Does the problem-solving process usually lead to improvements for the organization?
(ii) Organizational causes of conflicts: Do the root causes for conflicts lie in the processes or in the organization? Which process interfaces are especially prone to conflicts? Are there organizational rules and regulations that fuel conflicts? How can problems/deficits be fixed creatively?

4 Intervention 3: Salutogenic Coaching for Middle Managers

The main objectives of the salutogenic coaching program for the managerial staff (MS) were to improve the health and health resources of the involved managerial staff and to promote discussion of "health-oriented leadership behavior". The focus was on psychological health, because there is "no health without psychological health" (Prince et al. 2007, p. 859).

A substantial body of research documents the close relationship between leadership behavior on the one hand and the mental state, commitment, and performance of their employees on the other (e.g., Wegge et al. 2014). Building on this research, the coaching program for the MS focuses on the health-orientation of leadership behavior. Thus, the goals of the salutogenic coaching program were twofold: (1) to improve or maintain the health of the participants (leaders); and (2) to reflect on the health of the leaders' subordinates and how the leaders' behavior can impact their subordinates' health.

4.1 Structure

The coaching program consisted of several parts: an introductory plenary session followed by a series of five individual coaching sessions and a subsequent series of collective group coaching sessions.

To guarantee an optimal start to the health-coaching program, an introductory session for the MS was scheduled. Its main goals were to: (1) introduce the basic idea of a salutogenic coaching program; (2) make MS aware of the theoretical basis of the program; (3) develop the MS' first coaching aims; and (4) foster the MS' confidence and cooperativeness with the aim of optimizing their participation in the coaching program.

In the individual coaching, the MS worked with a professional health coach. The individual coaching relied on the following theoretical and methodological elements: (1) the salutogenesis approach was the theoretical basic of the work and consulting; (2) the coaching had a systemic framework (resources- and solution-oriented); and (3) five individual 60-min coaching sessions were conducted.

The collective coaching focused on the process in the group: the participants defined both their individual health goals and health-related leadership goals as a group and worked on them systematically. During this process, the participants learned from and with each other and developed the ability to deploy salutogenic aspects directly in their daily work and communication routines.

Two collegial groups were formed, each of nine persons from different geriatric nursing homes. Each of the five collective coaching sessions took 90 min except for the first meeting, which lasted 120 min because it included an introduction to the methods and the theoretical foundation of salutogenesis.

4.2 Evaluation

In addition to longitudinal assessment, the coaching program itself has been evaluated extensively. The content and results of the individual coaching process was confidential. It was anticipated that the two components of the coaching parts would have different effects on the MS; for example, on changes at the cognitive level regarding how health-related issues are evaluated and at the behavioral level in health-related leadership behavior. The following three dimensions of these effects were assessed: (1) goal evaluation—at the beginning of the coaching, coaching goals were formulated as concrete behavioral goals and were assessed based on target achievements; (2) process evaluation—the process and target achievements were assessed constantly and documented during the coaching; and (3) outcome evaluation—the MS were interviewed 6 months after completing the coaching to determine their personal views on the effectiveness of the coaching program.

5 Intervention 4: Salutogenically Oriented Integration Program for Employees Returning from Long-Term Sickness

Geriatric nurses who resume work after a long period of sickness represent another important group of workers on which salutogenic human resource management should focus. The aim of this intervention is for every such individual to develop a phased reintegration program, the progress of which is oriented towards developing a sense of coherence. This program is based on the legal requirements of German Social Law (§ 84 [2] SGB IX—"Occupational Integration Management"—as well as § 74 SGB V and § 28 SGB IX). These legal requirements to organize a reintegration program for employees returning after taking 6 or more weeks of sick leave are generally designated as the "Hamburg Model". It is anticipated that the salutogenically oriented program will go well beyond the scope of these minimum requirements. Therefore, we plan that one outcome of the research project will be the development of a "Dresden Model" for occupational integration management that will be analogous to the "Hamburg Model".

The first step was to hold semi-standardized guided interviews with company representatives (personnel management, facility management, and representatives of the staff association) and employees who have gone through reintegration. The aims were to gather information on the current status of reintegration and generate suggestions for ideas and implementation. Questions that could not be comprehensively answered or had newly emerged were clarified in the next step with the help of best practice interviews.

The following important findings have so far been generated from the interviews (sample): (1) At present, there is no occupational integration management or systematic reintegration for employees of the participating care institutions. (2) Contact with employees who are unable to work, conversations with them, and documentation during sickness, on return to the work-place, and during reintegration take place on a case-by-case basis. (3) Phased medical reintegration ("Hamburg model") is currently the only method used to reintegrate employees who have had a long period of sickness. Employees are familiar with this model. (4) During reintegration, activities that are considered inappropriate are transferred to other employees; however, the (still certified sick) employee may be placed on rota for full duties when there are personnel shortages. (5) Employees who are frequently sick and who have been unable to work for 6 weeks during a 12-month period because of repeated illness receive no special attention. (6) Reducing the number of or eliminating sick days, long-term preservation of the ability to work, and employee retention of jobs are regarded by the interviewees as the greatest priorities for the program. (7) Furthermore, the potential reintegration program is seen by the interviewees as an instrument for making employees more appreciative and thus improving staff loyalty and making the program attractive to employers. (8) The greatest barriers to reintegration are seen as the near absence of light-duty work-places/work-place alternatives in care facilities or in the company itself as well as insufficient time being allotted to carrying out everyday tasks. (9) The best practice experts we interviewed recommended ensuring commitment to the process by signing a company agreement that leaves enough room for individual arrangements in particular cases.

In summary, the parties involved identified many more opportunities than barriers and many more negative than positive evaluations of the current process. Consequently, there is a need for professionalization of the reintegration process in the facilities involved in this project.

In the next step, a new model for the reintegration process ("Dresden model"—see above in this section) will be developed. For this purpose, a reintegration team nominated by the parties who have been interviewed will be set up; it will include members of the human resource management team, a representative of the works council, and the heads of two geriatric nursing homes. The team can be added to as required by employees or other internal or external third parties. In this connection, the developed reintegration program will be piloted in the organizations taking part in the project and will incorporate the specific requirements of those particular nursing homes.

6 Challenges to the Current Intervention Program

Although the intervention program presented above has not yet been fully implemented, some challenges have already been identified.

A key challenge of the above program lies in its quasi-experimental set-up. To a greater degree than was expected, each geriatric nursing home in the study is perceived by its staff as a stand-alone organizational unit with a unique corporate culture. Thus, the effects of the interventions may differ substantially across these organizational units, mainly depending on how much attention is devoted to the "soft" topic of health at work. Furthermore, the studied nursing homes must contend with influences that are not program-related, such as changes in the system for planning shifts. Such external influencing factors cannot be fully captured by the longitudinal testing procedures and could override the effects of the intervention program.

A second challenge to the program concerns the participants' ages. The average age of participants in the coaching program, health circle and healthy ageing training program is over 50 years. As documented elsewhere, older employees characteristically participate significantly less in continuing education than the average employee (e.g., Bannwitz 2008 for Germany). What is more, Zwick (2011) analyzed some German data and concluded that trainings are less effective for older than for younger employees. As Callahan et al. (2003) concluded in their meta-analysis, older employees participate more fully in training when they see a clear link between the training measures and solutions to work-related problems. At the beginning of the intervention phase, many of the participants revealed considerable skepticism about the ability of the intervention program to solve their work-related health issues. A frequent comment was "I can tell from my work experience that none of this will make a difference". This skepticism may have been fueled by the age difference between the participants and the researchers who offered the interventions. Nonetheless, over the course of the intervention program, many participants have reported that they have gained something from the program; however, how much they will learn and implement into their daily work routines will only be known after all the data has been collected. Because the program is based on a theoretical approach that has already received a considerable amount of empirical support in various fields, the authors are confident that the salutogenesis approach will also prove to be beneficial for employees in the elder care sector.

References

Antonovsky A (1997) Salutogenese. Zur Entmystifizierung der Gesundheit. English edition: Antonovsky A (1997) Salutogenesis (trans. Franke A). dgvt, Tübingen (in German)

Bannwitz J (2008) Ältere Beschäftigte und betriebliche Weiterbildung. Results of CVTS 3, discussion paper 104, Bundesinstitut für Berufsbildung, Bonn (in German)

Brieskorn-Zinke M (2010) Salutogenese in der Pflege—zur Integration des Konzepts in pflegerische Handlungsfelder. In: Wydler H, Kolip P, Abel T (eds) Salutogenese und Kohärenzgefühl. Juventa, Weinheim, pp 173–184 (in German)

Bullinger M, Kirchberger I (1998) SF-36. Fragebogen zum Gesundheitszustand. Hogrefe, Göttingen

Büssing A, Glaser J, Höge T (2004) Psychische und physische Belastungen in der ambulanten Pflege: Ein Screening zum Arbeits- und Gesundheitsschutz. Zeitschrift für Arbeits- und Organisationspsychologie 48:165–180 (in German)

Callahan JD, Kiker DS, Cross T (2003) Does method matter? A meta-analysis of the effects of training method on older learner training performance. J Manage 29(5):663–680

Franke A, Witte M (2009) Das HEDE-Training. Manual zur Gesundheitsförderung auf Basis der Salutogenese. Huber, Bern (in German)

Galatsch M, Iskenius M, Hasselhorn HM (2011) Alter und Gesundheit—NEXT-Studie. http://www.next.uni-wuppertal.de. Accessed 5 June 2011 (in German)

Galbraith ND, Brown KE (2011) Assessing intervention effectiveness for reducing stress in student nurses: quantitative systematic review. J Adv Nurs 67(4):709–721

Gelsema TI, Van Der Doef M, Maes S et al (2005) Job stress in the nursing profession: the influence of organizational and environmental conditions and job characteristics. Int J Stress Manage 12(3):222–240

Happell B, Dwyer T, Reid-Searl K et al (2013) Nurses and stress: recognizing causes and seeking solutions. J Nurs Manage 21(4):638–647

Höge T (2005) Salutogenese in der ambulanten Pflege. Zum Zusammenhang zwischen organisationalen Ressourcen, erlebter Fairness, Kohärenzsinn und der psychophysischen Gesundheit von ambulanten Pflegekräften. Z Gesundheitspsychologie 13:3–11 (in German)

Kato T (2013) Coping with interpersonal stress and psychological distress at work: comparison of hospital nursing staff and salespeople. Psychol Res Behav Manage 7:31–36

Marine A, Ruotsalainen J, Serra C et al (2006) Preventing occupational stress in healthcare workers. Cochrane Database Syst Rev 4:CD002892

Mimura C, Griffiths P (2003) The effectiveness of current approaches to workplace stress management in the nursing profession: an evidence based literature review. Occup Environ Med 60(1):10–15 (in German)

Neubert J, Tomczyk R (1986) Gruppenverfahren der Arbeitsanalyse und Arbeitsgestaltung. Dt. Verl. d. Wiss, Berlin (in German)

Pietzcker F, Looks P (2010) Der Aufgabenbezogene Informationsaustausch. Zeitweilige partizipative Gruppenarbeit zur Problemlösung; mit besonderem Blick auf

Organisationsentwicklung, Wissensmanagement und betriebliche Gesundheitsvorsorge. vdf Hochschulverlag, Zürich (in German)

Prince M, Patel V, Saxena S et al (2007) No health without mental health. Lancet 370 (9590):859–877

Rimann M, Udris I (1997) Subjektive Arbeitsanalyse: Der Fragebogen SALSA. In: Strohm O, Ulich E (eds) Unternehmen arbeitspsychologisch bewerten. Ein Mehr-Ebenen-Ansatz unter besonderer Berücksichtigung von Mensch. Technik und Organisation, Zürich, pp 281–298 (in German)

Rimann M, Udris I (1998) "Kohärenzerleben" (Sense of Coherence). Zentraler Bestandteil von Gesundheit oder Gesundheitsressource. In: Schüffel W et al (eds) Handbuch der Salutogenese: Konzept und Praxis. Ullstein Medical, Wiesbaden, pp 351–364 (in German)

Ruotsalainen J, Serra C, Marine A et al (2008) Systematic review of interventions for reducing occupational stress in health care workers. Scand J Work Environ Health 34(3):169–178

Schliehe F, Schäfer H, Buschmann-Steinhage R et al (2000) Aktiv Gesundheit fördern. Verband Deutscher Rentenversicherungsträger, Köln (in German)

Singer S, Brähler E (2007) Die "Sense of Coherence Scale": Testhandbuch zur deutschen Version. Vandenhoeck & Ruprecht, Göttingen (in German)

Stadler P, Spiess E (2002) Mitarbeiterorientiertes Führen und soziale Unterstützung am Arbeitsplatz. Schriftreihe der Bundesanstalt für Arbeitsschutz und Arbeitsmedizin, Dortmund (in German)

Sullivan GC (1989) Evaluating Antonovsky's salutogenic model for its adaptability to nursing. J Adv Nurs 14:336–342

Udris I (2006) Salutogenese in der Arbeit—ein Paradigmenwechsel? Wirtschaftspsychologie 8 (2–3):4–14 (in German)

Wegge J, Shemla M, Haslam SA (2014) Leader behavior as a determinant of health at work: specification and evidence of five key pathways. Ger J Res Hum Res Manage 28:6–23

Zwick T (2011) Why training older employees is less effective. http://ftp.zew.de/pub/zew-docs/dp/dp11046.pdf. Accessed 30 Nov 2014

Mindfulness at Work: How Mindfulness Training May Change the Way We Work

Silke Rupprecht and Harald Walach

Abstract

Chronic stress, burnout, and major depression are currently major health challenges in Western cultures. Mindfulness is defined as the ability to voluntarily direct attention towards the present moment in an accepting and non-judgmental manner. This ability seems to help to train the mind to detach from the excessive or negative thinking (rumination) that plays a major role in the onset of burnout and depression. Having successfully been implemented in clinical settings, mindfulness trainings are becoming increasingly popular in organizational settings. In this article, we aim to provide a general map of mindfulness trainings in the work-place. Following some necessary background information on the roots of mindfulness, we continue with a short outline of some training formats that are being offered in work-place settings. We will then review recent, selected empirical findings on the impact of such training in work settings, and discuss the perils and promises of integrating mindfulness into the work-place.

1 Introduction to Mindfulness

Mindfulness-based trainings are becoming increasingly popular in both clinical and non-clinical settings. In clinical settings, they have been successfully used to treat affective disorders. For example, Mindfulness-Based Cognitive Therapy (MBCT) has been recommended by the British National Institute for Health and Clinical Excellence (2010) and granted "key priority status" for treating chronic depression.

S. Rupprecht (✉)
Leuphana Universität Luneburg, Luneburg, Germany
e-mail: silke.rupprecht@leuphana.de

H. Walach
Europa-Universität Viadrina, Frankfurt (Oder), Frankfurt (Oder), Germany
e-mail: walach@europa-uni.de

M. Wiencke et al. (eds.), *Healthy at Work*, DOI 10.1007/978-3-319-32331-2_22

Mindfulness training (MT) is also offered as a supportive treatment aimed at improving coping or symptom management in patients with somatic diseases such as cancer and fibromyalgia.

Recent neuroscientific research findings have sparked the interest of people who might not otherwise have been interested in contemplative practices. "The Mindful Revolution" was proclaimed by *Time Magazine* on its front page (Pickert 2014), the accompanying article stating that we need mindfulness to cope with the demands of the digital age. Increasing numbers of books market mindfulness as a self-enhancement tool and promise to show "How to avoid stress, achieve more, and enjoy life!" (McKenzie 2014). Not surprisingly, mindfulness has also attracted the interest of business leaders. Several years ago, the appearance of Buddhist monk Matthieu Ricard at the World Economic Forum in Davos would have been fairly unlikely. So-called Wisdom 2.0 Conferences that aim to bring contemplative practitioners and business leaders together attract leaders from major corporations such as Yahoo, IBM, and Google. To gain an understanding of the growing interest in mindfulness we will explore the roots and recent rise of this ancient concept.

1.1 Mindfulness Past and Present

1.1.1 Mindfulness in Buddhism

Mindfulness is derived from a 2500-year-old model that is rooted in Buddhist philosophy. In Buddhism, mindfulness is a translation of the Pali word "sati", which literally means "memory", denoting remembering to be aware in the present moment. In one of the most influential Buddhist texts, the Satipathana Sutta, mindfulness is considered to lead to "cessation of suffering" (Analayo 2010). Suffering ("dukkha"), in the Buddhist sense of the word, encompasses everything that causes us discomfort, such as stress, negative emotions, disease, and old age. According to Buddhist philosophy, the roots of suffering lie in our habitual craving or attachment to sensual objects, such as pleasure and success, and our aversion to negative objects, such as discomfort and failure. The former constitutes one of the "poisons of the soul", greed, and the latter the second poison, namely hate. The third poison is said to be ignorance. According to Buddhist teaching, all three of these poisons are remedied by the practice of mindfulness.

Mindful awareness is both a means of gaining insight into our mental habits and a tool for changing patterns of thinking or acting. The Satipathana Sutta also explains what we should be mindful of. Awareness may be focused on the body (e.g., the feet), emotions (e.g., happiness), mindset (e.g., boredom), and mind objects (e.g., negative thoughts). Meditation and mindful awareness in daily activities are tools for acquiring more mindfulness skills. The word "meditation" is a translation of the Pali word "bhavana", which literally means "getting to know", "familiarizing yourself with", or "cultivating". Meditation is therefore a practice for becoming more familiar with our mental, emotional, and physical habits and for cultivating our consciousness.

1.1.2 Contemporary Mindfulness

Mindfulness as we use it today is heavily influenced by the development of Mindfulness-Based Stress Reduction Training (MBSR). Twenty-five years ago, Jon Kabat-Zinn developed a secular training based on ancient Buddhist teachings. The contemporary and Buddhist models pursue similar goals, which include "diminishing suffering, enhancing positive emotions and improving the quality of life" (Vago and Silbersweig 2012, p. 1). Jon Kabat-Zinn defined mindfulness as "the awareness that emerges through paying attention on purpose, in the present moment, and nonjudgmentally to the unfolding of experience moment by moment" (Kabat-Zinn 2003, p. 145). He chose to focus on stress reduction as an umbrella term for all kinds of sufferings.

Trained as a molecular biologist at the Massachusetts Institute of Technology and a longtime practitioner of Buddhist meditation, he felt a call to share the "essence of meditation and yoga practices" (Kabat-Zinn 2011, p. 287) with people who might never be interested in attending a Buddhist meditation center. His intention to develop this training was "to recontextualize it (the Buddhist teachings) within the frameworks of science, medicine (. . .) and healthcare so that it would be maximally useful to people (. . .) whether they were doctors or medical patients, hospital administrators, or insurance companies" (Kabat-Zinn 2011, p. 288). Through the success of the training and subsequent research, mindfulness entered the academic arena. According to a search of the database Scopus with the keyword "mindfulness", there are now more than 700 scholarly mindfulness publications and their numbers have been increasing steadily since 2000 (Fig. 1).

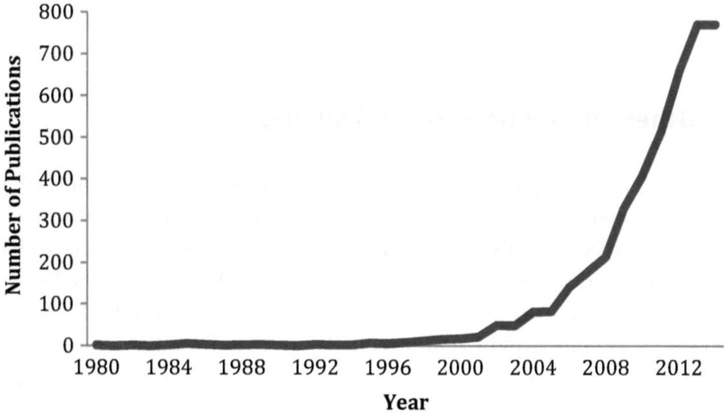

Fig. 1 Scopus search of the number of scholarly publications with the key word "mindfulness" by year

First behavioral medicine and psychology and now also corporate and educational settings are embracing this training.

1.2 Mindfulness-Based Stress Reduction

MBSR is a standardized group training delivered over 8 weeks. Its main goal is to aid participants in understanding and training mindfulness by letting them experience the benefit of regular meditation practice. Meetings are held once a week for 2.5–3 h and there is an additional retreat day, usually after the sixth course meeting. The total training time is 26–30 h. MBSR encompasses psychoeducational elements, for example teaching about the stress response, regulation of emotion, and maladaptive cognitive behavior such as rumination. However, the training's main objective is the practice of mindfulness. Formal meditation sessions in the group meetings last about 30–45 min, and participants are required to practice at home six times a week. Home practices include formal meditation exercises and informal mindful activities.

The main types of meditations offered are body scan, mindfulness of breath, thoughts and emotions, walking meditation, and compassion meditation (also called "metta" meditation), together with simple yoga exercises. These are all very simple and practical exercises with a focus on being aware of one's sensations in the present moment. Informal home practices include journaling about a pleasant or unpleasant experience each day and mindfully performing a daily activity, such as washing dishes or brushing teeth. Home practice is supported by audio material providing instructions for the daily meditations, and a course handout containing key learning points. Trainers encourage participants to share their experiences during group meetings, and inquire deeply about those experiences to enhance participants' understanding of the practice. Unlike other clinical trainings, MBSR-trainers are required to be experienced mindfulness practitioners themselves so that they can embody this simple but challenging concept that often takes years to acquire and understand.

1.3 Other Mindfulness-Based Trainings

An array of training adaptations based on the MBSR format have emerged, most of which address distinct clinical disorders or special populations. The most widespread adaptation is arguably MBCT for relapse prevention in subjects with chronic depression (Teasdale et al. 2000). Furthermore, Mindfulness-Based Relapse Prevention helps prevent substance abuse relapse in people with addictions (Bowen et al. 2009), Acceptance and Commitment Therapy is for a wide range of psychological problems (Hayes et al. 2006), and Dialectical Behavioral Therapy is for treating borderline personality disorder (Linehan 1993). There are educational trainings such as Mindfulness-Based Coping with University Life (Lynch et al. 2011) and Mindfulness in Schools Program that target elementary and secondary school students as well as teachers (Kuyken et al. 2013) and MindUp for elementary and secondary school students (Schonert-Reichl and Lawlor 2010). There are also increasing numbers of trainings aimed at enhancing work performance such as Google's Search Inside Yourself (SIY) Training for more creativity and success (Tan 2012) and Stanford University's Compassion Cultivation Training for improving resilience and well-being.

2 Examples of Mindfulness-Based Trainings at Work

There are different objectives for introducing MTs into the work-place. In accordance with findings of clinical research and experience, MBSR is effective in preventing stress-related diseases and promoting health (Fjorback et al. 2011). However, depending on the objectives or target group of the training, it is sometimes useful to tailor the training to specific needs. Some companies hope to enhance job performance and commitment through mindfulness practices. Another objective is to enhance care workers' empathy and compassion.

We will outline three different MTs that address specific demands of distinct settings: Google's Search Inside Yourself training for Google employees to improve emotional intelligence and creativity; .b (pronounced "dot be") Foundation's training for helping teachers to cope with stressors, and Mindfulness-Based Mind Fitness Training (MMFT) for preventing posttraumatic stress disorder in soldiers. There are various examples of trainings tailored to different subgroups, such as Mindfulness-Based Wellness Education for teacher trainees (Poulin 2014) and Cultivating Emotional Balance for health professionals. Many more trainings are currently being developed and piloted. However, only a few of them have been reported in detail and made available to the public. Generally, such adaptations build on the MBSR framework, either shortening or lengthening the duration of the training, home practice or both, and using different educational inputs or targeted meditation exercises depending on the need.

2.1 Search Inside Yourself

2.1.1 History

Based in California, Google was one of the first companies to embrace meditation and mindfulness in the work-place. Chade-Meng Tan, himself a Google pioneer, developed the mindfulness-based emotional intelligence curriculum by allocating 20 % of his work time outside his core job as an engineer. His opening question was "What if contemplative practices can be made beneficial both to people's careers and to business' bottom lines?" (Tan 2012, p. 3). The program he developed has been offered to Google employees since 2007 and has been gaining popularity outside of Google as well.

2.1.2 Goals

Drawing on the work of MBSR founder Jon Kabat-Zinn and in collaboration with emotional intelligence researcher Daniel Goleman, Tan devised a training the objectives of which include enhancing creativity and emotional intelligence to achieve more and be happier. Being an engineer himself, the program focuses on scientific evidence and employs language designed to be more appealing to the target group. "For example, where traditional contemplatives would talk about 'deeper awareness of emotion', I would say 'perceiving the process of emotion at a higher resolution'" (Tan 2012, p. 4–5). More generally, SIY promotes the concept of optimizing yourself by becoming more emotionally competent through the practice of mindfulness.

2.1.3 Content

The SIY curriculum comprises seven sessions amounting to 20 h of classroom time. The training may be attended during regular working hours and incorporates three steps: Trainees start with step one—focusing their attention—which is regarded as the basis of all higher cognitive abilities and is believed to make the mind calmer and more stable. With step two—self-knowledge and self-mastery—the new attention skills are used to gain a deeper understanding of thoughts and emotions. The third training step—creating useful mental habits—encompasses contemplative practices thought to cultivate compassion for others and happiness. All sessions begin with an overview of (neuro-)scientific evidence.

2.1.4 Evaluation

The author refers to the success of SIY, which he links to the company-wide popularity and personal success stories (Tan 2012). However, no formal evaluation based on empirical data is currently available.

2.2 .b (Pronounced "Dot Bee") Foundations Training

2.2.1 History

Richard Burnett and Chris Cullen founded the Mindfulness in Schools project in 2007. Their opening question was "When 25 teenagers tumble into your classroom, how are you going to interest them in mindfulness? They've never heard of mindfulness, it doesn't sound very exciting, and if you were to tell them that it involved periods of stillness and silence, you'd lose them before you began" (Burnett and Cullen 2014). They went on to develop mindfulness curricula for teenagers (.b training), and elementary school children (paws .b training), keeping in mind that it should be "fun, accessible and of genuine use". Therefore, trainings include the use of abbreviations (for example, .b is an abbreviation for stop, breathe and be). Students are encouraged to send daily text messages reading ".b" to friends who are then and there supposed to stop whatever they are doing and pause for a moment. With the increasing popularity of this UK-based program, there has been a growing need for experienced teachers to deliver the practice. In 2014, mindfulness trainer Sarah Silverton and teacher Dominic Morris therefore developed the .b foundation training for teachers.

2.2.2 Goals

The first objective is to introduce teachers to the practices of mindfulness and self-compassion. The aim of this is to support teachers to build resilience and cope with the demands of their profession through mindfulness practices. Second, participation in this course (or alternatively, in a MBSR training) is a prerequisite for teachers interested in teaching mindfulness to students. Through the .b foundation's training, teachers are familiarized with some of the specific components of the program that are also in the curriculum for students.

2.2.3 Content

The course comprises eight sessions of 1.5 hours amounting to 12 hours' teaching time plus additional home practice. It roughly follows the MBCT curriculum as laid out in the book *Mindfulness: A practical guide to finding peace in a frantic world* (Williams and Penman 2011) and includes course sessions entitled "Waking up to the autopilot" (session 1) and "Stepping back from thoughts and worries" (session 4). Audio-visual resources such as film clips and animations are provided during the sessions and home practice is supported by audio recordings of contemplative exercises (Morris and Silverton 2014).

2.2.4 Evaluation

In a non-randomized feasibility trial to assess the efficacy of ".b Foundations", 89 school teachers were either part of the experimental group (n = 49) or in a wait-list control group (n = 40) (Beshai S, McAlpine L, Weare K et al. 2014, unpublished manuscript). Self-report measures of stress, well-being, mindfulness, and self-compassion were obtained before and immediately after the training. Compared with the control group, experimental group teachers reported significantly less stress after the training and significantly more well-being. Large effects in all outcome measures were documented. Furthermore, changes in mindfulness and self-compassion were associated with changes in stress and well-being, indicating that mindfulness and self-compassion mediate the training effects.

2.3 Mindfulness-Based Mind Fitness Training (Pronounced M-Fit)

2.3.1 History

MMFT was created by Elizabeth Stanley, a former US army officer who has been practicing mindfulness for many years. Trained in MBSR herself, she saw a need to adapt the course for soldiers and other personnel working in similarly high-stress environments, such as fire-fighters or rescue workers. She found that soldiers receive physical training in preparation for deployment, but not mental training to cope with the profession's stressors. Stress is known to impair cognitive control and capacity to regulate emotions. The inability of soldiers to cope effectively with stress is linked with unethical behavior in combat, such as "unnecessarily damaging private property or insulting or physically harming noncombatants" (Stanley and Jha 2009, p. 145), and with the onset of affective disorders. Therefore, the question driving her work was "What can be done to enhance the military's capacities to operate in complex environments while simultaneously protecting against the stressors inherent in them?" (Stanley and Jha 2009, p. 145).

2.3.2 Goals

The main goal of the training is to prevent the development of posttraumatic stress disorder by improving soldiers' cognitive control and emotion regulation skills. Cognitive control encompasses the skill to willingly regulate attention and can be measured through working memory capacity (WMC), which is the ability to maintain a stable focus on a chosen task or object without getting distracted. People with better cognitive control simultaneously have greater emotion regulation skills (Jha et al. 2010).

2.3.3 Content

The course set-up is similar to that of MBSR: 24 h of training over 8 weeks with weekly 2-h meetings and a full-day silent retreat. Participants are asked to complete 30 min of home practice supported by CDs specifically created for this training. The course content is adapted to the needs of the participants. MMFT covers mindfulness skills and links them with concrete applications for the operational environment. In addition, psychoeducational information on stress, trauma and resilience is provided and discussed, for example the Trauma Resilience Model (Leitch 2007).

2.3.4 Evaluation

MMFT has been evaluated quantitatively and qualitatively. In a controlled study, improvement in cognitive control skills—measured through WMC—was linked to the amount of home practice. The frequent practice group showed modest improvement in WMC over time whereas the infrequent practice group showed a decrease in WMC, which is characteristic of the high-stress pre-deployment period (Jha et al. 2010). Furthermore, the greater the time spent on home practice, the greater the reduction in negative affect, presumably mediated through improved WMC. Participants reported increases in mindfulness that corresponded with decreases in perceived stress (Stanley et al. 2011). While some participants were resistant to the idea of practicing meditation in the army, others noticed improvements in the way they dealt with emotions, their relationships with team and family members, and their awareness of individual strengths and weaknesses (Stanley et al. 2011).

3 Effects of MTs in Work-Place Settings

Despite the notable interest in mindfulness practices, very little research has investigated their effects in work-place settings. In this chapter we will look at three types of work-places or goals: the impact of training in business settings, in educational or social settings and in the treatment of risk groups within work-places.

3.1 Effects of MT in Business Settings

In one of the earliest onsite work-place studies, Walach et al. (2007) investigated 29 call center agents in Germany. In a wait-list controlled trial, MBSR was found to enhance positive coping strategies while the demands of the job remained the same. Participants reported being more proactive in problem-solving and being able to disengage from work problems at home. Compliance with mindfulness practice was high during the course with 30 min practice about five times a week, but declined to once or twice a week by the 2-month follow-up. Participants reported high satisfaction rates with MBSR themselves but estimated it would be feasible for only about 56 % of staff. Qualitative interviews revealed an unintended effect of the training: Call center agents were becoming more aware of "problematic situations in the workplace conducive to stress" (Walach et al. 2007, p. 197), and as a consequence started procedures to change conditions within the company or considered leaving. Thus, correctly practiced mindfulness may lead to a greater awareness of unhealthy conditions.

Aikens et al. (2014) delivered a short online MT for anybody interested in a major chemical corporation. The 7-week training comprised weekly 1-h web-meetings and about 15 min of daily home practice. Ninety employees were randomly assigned to experimental and wait-list control conditions. The rate of satisfaction with the program was high and the intervention group dropout rate of 5.3 % was surprisingly low compared with other internet-based programs. The training reduced work-related stress by 33 % and improved mindfulness, resilience, and vigor. These benefits were maintained or improved at the 6-month follow-up. In addition, the study also identified effects on lifestyle choices that had not been specifically targeted during the course. Experimental group participants reported a 31 % reduction in the number of fast food meals eaten per week. However, no significant trends were reported for either amount of exercise or sleep quality. Interestingly, the study evaluated the return on investment based on the observed reduction of 1 day/week in self-reported burnout post-intervention. The authors calculated a potential 20 % increase in worker productivity. Based upon the average yearly wage in this company ($112,900) they estimated employer savings of up to $22,580 per employee year, provided the improvements in burnout rates persisted.

3.2 Effects of MT in Educational or Medical Settings

A randomized wait-list controlled MBSR study of 119 public school teachers in Canada and the USA found that MBSR reduced teacher occupational stress and burnout symptoms (Roeser et al. 2013). Post-test increases in mindfulness and self-compassion were believed to have mediated the reductions in stress and burnout found at the 3-month follow-up. Furthermore, attention tests administered at baseline and post-test showed improvements in WMC in the experimental but not in the control group. However, no changes were observed in biological measures linked to stress reduction, such as serum cortisol concentrations and heart rate.

Flook et al. (2013) developed a modified MBSR training (mMBSR) for training teachers to integrate mindfulness skills in the classroom. The total course time was comparable to that of MBSR. However, because of the participants' time constraints, the course content was delivered in shorter but more frequent meetings and home practice material offered a greater variety of possible practice durations (e.g., 15 min, 30 min, 45 min). Eighteen public elementary school teachers in the USA took part in this randomized controlled trial (RCT). The intervention group showed improvements in several self-report measures, such as a decrease in psychological symptoms and burnout and an increase in self-compassion and mindfulness, whereas no changes were observed in the wait-list control group. In addition, the experimental group improved in observer-rated classroom management, suggesting an effect of MT on teachers' core professional competence. Furthermore, the bias in attention towards negative stimuli decreased, indicating enhanced skills in processing and letting go of negative emotions. However, the study found no significant changes in the intervention group's morning cortisol concentrations. Cortisol was selected as an index of physiological stress. A medium-sized between-group effect was found for 12 of the 16 outcome variables, suggesting that mMBSR has the potential to support elementary school teachers' health and work performance.

Another MBSR adaptation for parents and educators of disabled children, called SMART-in-Education, yielded similar outcomes for stress reduction and positive psychological function (Benn et al. 2012). Moreover, this intervention helped improve relationship competence: both teachers and parents reported greater forgiveness and empathy skills. Together with the already mentioned .b foundation training, mindfulness seems to improve teachers' competence in dealing with stress, emotional and social skills, and job performance. On a similar note, a mindfulness-based training administered to health care professionals in Spain was effective in reducing burnout and increasing empathy (Asuero et al. 2014).

Krasner et al. (2009) invited 70 general practitioners to participate in a MT and measured their rates of burnout, empathy, mindfulness, and affective state 1 year after the training. They found large improvements in mindfulness, burnout, and affective state and moderate improvement in empathy. An interesting recent RCT involved 74 doctors working in various Mayo Clinics (West et al. 2014): The test participants attended biweekly meetings during their work time in which they discussed emotions and practiced mindfulness over 9 months, whereas doctors in the control group could take two paid hours off per week to do what they liked. Mindfulness home practice was not required, but was suggested. Outcomes of doctors in the study were compared with a control group and a database of non-participating doctors. Although job satisfaction, quality of life, depression, and stress did not change, empowerment and engagement at work increased significantly in the trial group compared with the control group, and depersonalization, and exhaustion decreased significantly. Additionally, the meaningfulness of work increased significantly for participants of the trial group, whereas it decreased for both control group and non-participating doctors. Although this study's potential impact may have been hampered by making regular practice optional, it nevertheless shows that spearheading institutions such as the Mayo Clinics have picked up on the possibility that investing company time and money into teaching employees mindfulness skills might benefit the system as a whole.

Several studies have focused on university faculty and staff (Klatt et al. 2009; Malarkey et al. 2013; Skarlicki et al. 2014; van Berkel et al. 2014). All of these trainings were abbreviated versions of MBSR and produced mixed results. Van Berkel et al. (2014) found no post-intervention effects on mental health, mindfulness, or work engagement in their study of teaching mindfulness to university staff. However, this lack of apparent effect may be attributable to the unusual measurement points of 6 and 12 months after the training: The training took 8 weeks and was followed by 6 months of personal e-counseling that only 6.3 % of participants complied with. Nevertheless, this study raises serious concerns about the sustainability of any of the aforementioned effects over a year.

A Mindfulness@work training for university employees comprising six weekly 2-h meetings and a 4-h silent mini retreat found short-term effects on work performance as indicated by improved results in a standardized test (Graduate Management Admission Test), creativity, and emotion regulation (Skarlicki et al. 2014). Improved regulation of emotion was also considered to have mediated changes in interpersonal conflict style.

3.3 Effects of MT in High-Risk Subjects

Malarkey et al. (2013) focused on a high-risk group of university staff ($n = 189$) with high markers of chronic inflammation. Inflammation is a risk factor for several diseases, including cardiovascular disease. To reduce selection bias, the program was advertised as a lifestyle intervention program; mindfulness was not mentioned until the start of training. The trial was actively controlled, the control group receiving an information-based training aimed at enabling participants to make healthy lifestyle choices based on scientific evidence. No effects of the training on either self-reported or biological measures such as indicators of inflammation and serum cortisol concentrations were found. Because obesity is associated with inflammation and may need specific treatment, the authors attributed the lack of changes in inflammatory markers to the high proportion of obese participants (body mass index > 30). Indeed, another similar training offered only to participants with body mass index ≤ 30 yielded more positive results (Klatt et al. 2009). Another possible explanation for the lack of effects is the high quality of the study: By excluding selection bias and using an active control, the researchers reduced effect sizes and thus likely had statistical power problems, because the active control is reportedly effective in itself (Chiesa and Serretti 2009; Fjorback et al. 2011).

Another at-risk group was broadly investigated in Denmark (Netterstrom et al. 2013). All general practitioners in the Copenhagen region were asked to refer patients who had been on stress-related sick leave and had no prior history of long-term psychiatric treatment. The 199 eligible participants, who had been absent from work for an average of 2 months, were randomized to an intervention group, treatment as usual (TAU) or wait-list-control group (WCG). The intervention consisted of eight 1-h individual stress treatment sessions and participation in a MBSR course. TAU comprised of 12 individual sessions with a psychologist. Compared with both control groups, the intervention group had significantly higher rates of returning to full-time work (67 % cf. 36 % TAU and 24 % WCG) and a greater number of participants increased their working hours (97 % cf. 71 % TAU and 64 % WCG). However, the intervention did not prove superior to TAU when reductions in stress-related symptoms or increases in general work ability were compared. Both intervention and TAU groups showed significant improvements compared with the wait-list control group.

A recent case report highlights the potential applicability of mindfulness therapy for workaholism (Shonin et al. 2014). Workaholism means an addictive over-commitment to work that results in neglect of family and social relationships, as well as individual needs. It affects about 10 % of the Western population (Sussman et al. 2011). Over 5 months, the subject's weekly working hours decreased from 65 to 50 and self-reported job performance improved.

4 Discussion

Although there are still only a few published studies investigating mindfulness at work, we have observed an increase in high-quality studies within the last 2 years, which is in line with an increased public interest in the field. Most of the studies reviewed employed research designs that increase the generalizability of outcomes, such as RCTs and active-control conditions. Furthermore, most studies recruited sample sizes based on prior power calculations. The published results are promising and mostly in line with previous research with clinical and non-clinical subjects. Evidence for the stress-reducing and life-enhancing quality of MTs in the work-place therefore seems promising, if not strong. Biological and other objective measurements have not always supported improvements achieved according to self-reporting (Malarkey et al. 2013; Roeser et al. 2013). However, evidence from neuroimaging studies indicating that MBSR is associated with biological changes is mounting. Hölzel et al. (2009, 2011) reported significant changes in the brain before and after a MBSR training of a non-clinical sample of adults, the changes comprising reduction in the size of the amygdala, the brain's stress center, and increased gray matter density in brain areas responsible for regulating emotions. The latter finding has also been supported by studies in the work-place: MT is linked to emotion regulation competences such as improved empathy and compassion. These traits are arguably useful in the social professions but also are generally desirable human capacities that are linked to positive health outcomes and more satisfying social relationships. More studies using biological measures, including neurological studies, could help to strengthen the evidence provided by subjective outcomes.

Some studies have reported improved job performance according to both self-report and task-based measures, including creativity and WMC. This is in line with studies showing that even short MTs increase WMC (Zeidan et al. 2010). WMC is crucial for sustaining attention: a high WMC increases the ability to block out unwanted distractors (Hasenkamp et al. 2012) and would therefore be advantageous in many modern work-place environments. Furthermore, a high WMC reduces the risk of mind wandering, which has been linked with negative mood and can be considered the opposite of attention (Smallwood and Schooler 2006; Smallwood et al. 2011). It is likely that the ability to pay attention will become even more important because electronic devices offer constant distractions, are often used in problematic ways, and require a new set of coping skills (Augner and Hacker 2011). Indeed, mindfulness is associated with less problematic mobile phone use such as texting while driving (Feldman et al. 2011). Mindfulness skills are likely to improve cognitive capacities (Skarlicki et al. 2014; Mrazek et al. 2013; Zenner et al. 2014).

MTs may also prove beneficial for treating high-risk groups in company settings and helping with reintegration after prolonged sick leave. However, it may be important for participants to have a general interest in the practice. A study in which the usual selection bias was avoided found MT ineffective (Malarkey et al. 2013), while another study that only invited people who were generally interested in mindfulness did find improvements in some outcome measures (Netterstrom et al. 2013).

Promising results have been reported for an attempt to measure the impact on group performance of students (Cleirigh and Greaney 2014). The effects of MT on team work or team cohesion in the work-place still need to be investigated. In addition, research regarding the impact on company culture, for example attitudes towards health promotion, on program results, and vice versa, would be an interesting future field.

Some studies have reported successful shortening of training hours and changes in the training content and style. An online study showed remarkably low dropout rates and participants were highly satisfied with shorter trainings (Aikens et al. 2014). However, a number of studies have relied on the traditional format; some have even extended the training hours. There is a belief that the number of necessary contact hours somehow correlates with the severity of the condition of the subjects.

Another trend reported by some studies was a reduction in practice time required at home; this does not necessarily decrease the effects. Some researchers have suggested that the depth of and motivation for practice, rather than its length, predict the outcome (Hölzel and Ott 2006), whereas others have found that practice compliance correlates with training effects (Carmody and Baer 2007; Hülsebusch and Michalak 2010; Jha et al. 2010). Nevertheless, a note of caution seems warranted: given that a major driver of stress in the work-place and other settings is the increasing urgency and need to be time-efficient, it may be worth asking whether an implicit adoption of this stance in trainings that are meant to combat exactly this habit is really productive. One very explicit finding of research into implementing mindfulness in university settings was students' consistent response that the most important experience for them was being allowed to have "me-time": "being told that it is okay to spend time with me and for myself, without having to meet a particular goal or target" (Lynch et al. 2011).

With all the "mindfulness hype" occurring in some circles, it may be worth raising a few critical issues: Mindfulness was originally one component of whole spiritual traditions. To take it out of such contexts and adapt it without the proper framework has been seen by some as abandoning the kernel of the concept (Grossman and van Dam 2011). While the successful adaptation of the mindfulness concept within clinical contexts shows that it definitely can be transferred into secular contexts and suggests that mindfulness is a psychological skill that does not need a spiritual backing, it is very likely that the fully fledged power and benefit of mindfulness is linked to personal motivation (Shapiro 1992) and can only be achieved by continuing dedicated practice within a broader framework of lifestyle choices (Grossman and van Dam 2011; Lutz et al. 2008).

These considerations lead to another critical point: Kabat-Zinn's considerable success with implementing MBSR was likely attributable to the fact that it was offered to severely ill patients who had tried everything else. These patients had made a conscious choice and committed themselves voluntarily to regular practice. The format of MBSR of 8 weeks is also designed to allow establishment of a habit of regular practice. Without such regular practice, benefits likely dwindle or even vanish. Therefore, any program that throws mindfulness at potential target participants, because someone has had a good experience with it and now wants to convert everybody else is bound to fail. Only if the ground is prepared, if participants are really willing to commit themselves to practice, will there be lasting effects. Therefore, participation in MT and associated programs should always be voluntary. Of course, managers and bosses who have undergone training and changed their behavior as a consequence are the best ambassadors. For this reason, and because company and corporation problems normally originate from problematic management, the leading strata of corporations should learn to practice mindfulness themselves before ordering the offering of corporation-wide trainings. More studies investigating the impact of MTs on management, leadership styles, and company culture are needed.

These considerations lead to the next critical point, which is methodological: because mindfulness depends on active cooperation and choice, randomized studies that do not emphasize this from the outset but offer mindfulness as a potential intervention like another medication for an ailment will only achieve minimal benefits. Meditation is not medication (Walach et al. 2014). Hence, such studies should only include participants who are committed to practicing and willing to make this choice. The only valid way to test this is by performing a randomized study with wait-list controls. Active-control trials, though more desirable from a scientific point of view, need to either depict the alternatives as somewhat equal or recruit participants with no significant predilection or bias. This approach precludes the potentially most important ingredient of mindfulness interventions: the choice to change habits and behaviors.

Finally, an ethical point should be kept in mind: During the Second World War, Zen training helped Japanese fighter pilots to become more effective weapons. Mindfulness is a skill that in itself is ethically neutral. It can improve effectiveness regardless of the values and goals of an action. Therefore, in its original context of Buddhism, it is embedded in a framework of ethics. When we secularize the concept and the skill, we need to be aware that the values and ethics of applying it must be provided by other routes. Using mindfulness to discipline disobedient pupils in school or to pacify dissatisfied workers in high-pressure environments without first investigating the root causes for the disobedience or dissatisfaction is a recipe for failure.

However, if these precautions are observed, mindfulness may be a very useful intervention that helps to increase well-being and health in the work-place and improve relationships and empathy, thus overall improving performance and reducing the cost of illness. MT may therefore be a good candidate for health promotion interventions in companies and complex corporations. However, more—and more solid—research is needed to determine its true value.

References

Aikens KA, Astin J, Pelletier KR et al (2014) Mindfulness goes to work: impact of an online workplace intervention. J Occup Environ Med 56:721–731

Analayo B (2010) Der direkte Weg. Beyerlein & Steinschulte, Stammbach

Asuero A, Moix Queralto J, Pujol-Ribera E et al (2014) Effectiveness of a mindfulness education program in primary health care professionals: a pragmatic controlled trial. J Contin Educ Health Prof 34:4–12

Augner C, Hacker GW (2011) Associations between problematic mobile phone use and psychological parameters in young adults. Int J Public Health 57:437–441

Benn R, Akiva T, Arel S et al (2012) Mindfulness training effects for parents and educators of children with special needs. Dev Psychol 48:1476–1487

Bowen S, Chawla N, Collins SE et al (2009) Mindfulness-based relapse prevention for substance use disorders: a pilot efficacy trial. Subst Abuse 30:295–305

Burnett R, Cullen C (2014) Mindfulness in schools project. Available from http://bit.ly/1oyzHk3. Accessed 21 Mar 2015

Carmody J, Baer RA (2007) Relationships between mindfulness practice and levels of mindfulness, medical and psychological symptoms and well-being in a mindfulness-based stress reduction program. J Behav Med 31:23–33

Chiesa A, Serretti A (2009) Mindfulness-based stress reduction for stress management in healthy people: a review and meta-analysis. J Altern Complement Med 15:593–600

Cleirigh DO, Greaney J (2014) Mindfulness and group performance: an exploratory investigation into the effects of brief mindfulness intervention on group task performance. Mindfulness. doi:10.1007/s12671-014-0295-1

Feldman G, Greeson J, Renna M et al (2011) Mindfulness predicts less texting while driving among young adults: examining attention- and emotion-regulation motives as potential mediators. Pers Individ Dif 51:856–861

Fjorback LO, Arendt M, Ørnbøl E et al (2011) Mindfulness-based stress reduction and mindfulness-based cognitive therapy—a systematic review of randomized controlled trials. Acta Psychiatr Scand 124:102–119

Flook L, Goldberg SB, Pinger L et al (2013) Mindfulness for teachers: a pilot study to assess effects on stress, burnout, and teaching efficacy. Mind Brain Educ 7:182–195

Grossman P, van Dam NT (2011) Mindfulness by any other name: trials and tribulations of Sati in Western psychology and science. Contemp Budd 12:219–229

Hasenkamp W, Wilson-Mendenhall CD, Duncan E et al (2012) Mind wandering and attention during focused meditation: a fine-grained temporal analysis of fluctuating cognitive states. Neuroimage 59:750–760

Hayes SC, Luoma JB, Bond FW et al (2006) Acceptance and commitment therapy: model, processes and outcomes. Behav Res Ther 44:1–25

Hölzel BK, Ott U (2006) Relationships between meditation depth, absorption, meditation practice, and mindfulness: a latent variable approach. J Transpers Psychol 38:179–199

Hölzel BK, Carmody J, Evans KC et al (2009) Stress reduction correlates with structural changes in the amygdala. Soc Cogn Affect Neurosci 5:11–17

Hölzel BK, Carmody J, Vangel M et al (2011) Mindfulness practice leads to increases in regional brain gray matter density. Psychiatry Res 191:36–43

Hülsebusch J, Michalak J (2010) Die Rolle der Übungshäufigkeit in der Achtsamkeitsbasierten Kognitiven Therapie. Z. Für Klin Psychol Psychother 39:261–266 (in German)

Jha AP, Stanley EA, Kiyonaga A et al (2010) Examining the protective effects of mindfulness training on working memory capacity and affective experience. Emotion 10:54–64

Kabat-Zinn J (2003) Mindfulness-based interventions in context: past, present, and future. Clin Psychol Sci Pract 10:144–156

Kabat-Zinn J (2011) Some reflections on the origins of MBSR, skillful means, and the trouble with maps. Contemp Budd 12:281–306

Klatt MD, Buckworth J, Malarkey WB (2009) Effects of low-dose mindfulness-based stress reduction (MBSR-ld) on working adults. Health Educ Behav 36:601–614

Krasner MS, Epstein RM, Beckman H et al (2009) Association of an educational program in mindfulness communication with burnout, empathy, and attitudes among primary care physicians. JAMA 302:1284–1293

Kuyken W, Weare K, Ukoumunne OC et al (2013) Effectiveness of the Mindfulness in Schools Programme: non-randomised controlled feasibility study. Br J Psychiatry 203(2):126–131

Leitch ML (2007) Somatic experiencing treatment with tsunami survivors in Thailand: broadening the scope of early intervention. Traumatology 13(3):11–20, http://doi.org/10.1177/1534765607305439

Linehan MM (1993) Skills training manual for treating borderline personality disorder. Guilford, New York

Lutz A, Brefczynski-Lewis J, Johnstone T, Davidson RJ (2008) Regulation of the neural circuitry of emotion by compassion meditation: effects of meditative expertise. PLoS One 3(3):e1897

Lynch S, Gander ML, Kohls N et al (2011) Mindfulness-based coping with university life: a non-randomized wait-list controlled pilot evaluation. Stress Health 27:365–375

Malarkey WB, Jarjoura D, Klatt M (2013) Workplace based mindfulness practice and inflammation: a randomized trial. Brain Behav Immun 27:145–154

McKenzie S (2014) Mindfulness at work: how to avoid stress, achieve more, and enjoy life! Gildan Media, New York

Morris D, Silverton S (2014) .b Foundations Teacher Pack. Mindfulness in Schools Project, London

Mrazek MD, Franklin MS, Phillips DT et al (2013) Mindfulness training improves working memory capacity and GRE performance while reducing mind wandering. Psychol Sci 24: 776–781

National Institute for Health & Clinical Excellence (2010) Depression. The NICE guideline on the treatment and management of depression in adults, National Clinical Practice Guideline. The British Psychological Society and Royal College of Psychiatrists, Leicester

Netterstrom B, Friebel L, Ladegaard Y (2013) Effects of a multidisciplinary stress treatment programme on patient return to work rate and symptom reduction: results from a randomised, wait-list controlled trial. Psychother Psychosom 82:177–186

Pickert K (2014) The mindful revolution. TIME Mag. Retrieved from http://time.com/1556/the-mindful-revolution/. Accessed 16 Apr 2015

Poulin PA (2014) Mindfulness-based wellness education: a longitudinal evaluation with students in initial teacher education. Dissertation, ProQuest Information & Learning, USA

Roeser RW, Schonert-Reichl KA, Jha A et al (2013) Mindfulness training and reductions in teacher stress and burnout: results from two randomized, waitlist-control field trials. J Educ Psychol 105:787–804

Schonert-Reichl KA, Lawlor MS (2010) The effects of a mindfulness-based education program on pre- and early adolescents' well-being and social and emotional competence. Mindfulness 1(3):137–151, http://doi.org/10.1007/s12671-010-0011-8

Shapiro DH (1992) A preliminary study of long-term meditators: goals, effects, religious orientation, cognitions. J Transpers Psychol 24(1):23–39

Shonin E, Gordon WV, Griffiths MD (2014) The treatment of workaholism with meditation awareness training: a case study. Explore (NY) 10:193–195

Skarlicki DP, Kay A, Diamond A, Soloway G (2014) The effect of mindfulness training on employee performance: Does it work and why? Working paper, Sauder School of Business, University of British Columbia

Smallwood J, Schooler JW (2006) The restless mind. Psychol Bull 132:946–958

Smallwood J, Mrazek MD, Schooler JW (2011) Medicine for the wandering mind: mind wandering in medical practice. Med Educ 45:1072–1080

Stanley EA, Jha AP (2009) Mind fitness. Improving operational effectiveness and building warrior resilience. Joint Force Q 55:144–151

Stanley EA, Schaldach JM, Kiyonaga A et al (2011) Mindfulness-based mind fitness training: a case study of a high-stress predeployment military cohort. Cogn Behav Pract 18:566–576

Sussman S, Lisha N, Griffiths MD (2011) Prevalence of the addictions: a problem of the majority or the minority? Eval Health Prof 34:3–56

Tan C-M (2012) Search inside yourself: the unexpected path to achieving success, happiness (and world peace). HarperOne, New York

Teasdale JD, Segal ZV, Williams JMG et al (2000) Prevention of relapse/recurrence in major depression by mindfulness-based cognitive therapy. J Consult Clin Psychol 68:615–623

Vago DR, Silbersweig DA (2012) Self-awareness, self-regulation, and self-transcendence (S-ART): a framework for understanding the neurobiological mechanisms of mindfulness. Front Hum Neurosci 6:1–30

Van Berkel J, Boot CRL, Proper KI et al (2014) Effectiveness of a worksite mindfulness-related multi-component health promotion intervention on work engagement and mental health: results of a randomized controlled trial. PLoS One 9(1):e84118

Walach H, Nord E, Zier C et al (2007) Mindfulness-based stress reduction as a method for personnel development: a pilot evaluation. Int J Stress Manage 14:188–198

Walach H, Schmidt S, Esch T (2014) Meditation intervention reviews: comments and responses. JAMA Intern Med 174(7):1193–1194

West CP, Dyrbye LN, Rabatin JT et al (2014) Intervention to promote physician well-being, job satisfaction and professionalism: a randomized clinical trial. JAMA Intern Med 174(4): 527–533

Williams M, Penman D (2011) Mindfulness: a practical guide to finding peace in a frantic world. Piatkus, London

Zeidan F, Johnson SK, Diamond BJ et al (2010) Mindfulness meditation improves cognition: evidence of brief mental training. Conscious Cogn 19:597–605

Zenner C, Herrnleben-Kurz S, Walach H (2014) Mindfulness-based interventions in schools—a systematic review and meta-analysis. Front Psychol 5:603

Promoting Workers' Health in Small- and Medium-Sized Enterprises: Designing and Evaluating a Concept for Preventing Occupational Skin Diseases

Mirella Cacace, Bettina Riegel, and Valentina Leier

Abstract

This chapter describes the development and evaluation of a concept for trainings in skin-care tailored to the specific needs of small and medium-sized enterprises (SMEs) in rural regions. It covers SME in industries in which employees work in wet or damp environments or are exposed to strong chemical or allergenic substances or both; these risk factors frequently lead to work-related skin problems. Florists, outpatient and inpatient care providers, hairdressers, bakers, and metalworkers were recruited to our study. Overall, 72 SMEs participated. The first research step was to explore company owners' expectations of the training and the requirements particular to their organizations. Trainings on skin-care were developed and tailored to the specific needs of the SMEs and surveyed industries. For evaluation purposes a randomized controlled trial was conducted in which the effectiveness of the trainings was quantitatively assessed by employees of the participating SMEs completing three waves of written surveys. A significant reduction in skin-related symptoms, a positive effect on skin-care and cleaning, and a significant positive effect on participants' knowledge in the area of occupational skin protection were found. However, the dermatological training did not have a positive effect on the participants' satisfaction with their work. The average cost of the intervention per participant was approximately 63.50 €; foregone working time not included.

M. Cacace (✉)
SOCIUM Research Center on Inequality & Social Policy, University of Bremen, Bremen, Germany
e-mail: cacace@uni-bremen.de

B. Riegel
Leuphana University Luneburg, Luneburg, Germany
e-mail: bettina.riegel@inkubator.leuphana.de

V. Leier
University Medical Center Hamburg-Eppendorf (UKE), Hamburg, Germany
e-mail: v.leier@uke.de

M. Wiencke et al. (eds.), *Healthy at Work*, DOI 10.1007/978-3-319-32331-2_23

1 Introduction

Work-related skin diseases are one of the most common ailments currently affecting employees (Rustemeyer 2012). According to the Federal Office for Occupational Safety and Health (BAuA), there were 24,805 cases of dermatological illnesses with suspected occupational causes in 2012 alone in Germany (BAuA 2014). Accounting for approximately 34% of all reported cases, this would make dermatological problems the most frequently reported occupational illness in Germany. Even if only a minute share of around 2% of the reported cases were indeed truly work-related conditions,[1] dermatological problems caused by occupational factors would still cause lengthy periods of incapacity in significant numbers of people and force some to change their occupations (Schliemann and Elsner 2007; Wulfhorst and Schwanitz 2001). The damage to the economy as a whole in terms of decreased productivity and lost working hours is estimated at more than 1.5 billion € per annum (Batzdorfer and Schwanitz 2004; John 2008).

Skin diseases tend to particularly affect younger workers (median age for women, 22 years; men, 31 years, see Diepgen 2012), especially those who work in wet or damp environments or are in contact with strong chemical or allergenic substances (Breuer and John 2011). High-risk occupations include hairdressing, metalworking, caring and medical professionals, food workers, construction workers, cleaning professionals, and painters, almost 80% of occupational dermatological problems being reported in these occupations (Diepgen 2012; Elsner and Wigger-Alberti 1997). Many professionals in these occupations tend to work in small or medium-sized enterprises (SMEs[2]) with fewer than 250 employees. Thus, preventative measures are particularly important for SMEs that want to protect the health of their workers. However, only a minor share of SMEs have implemented adequate preventative measures (Zelfel et al. 2011). SMEs in rural regions are particularly under-represented in this respect (Gieseke 2005).

Several intervention studies on preventing occupational skin diseases have confirmed the positive impact of training that covers both the basic theory and pragmatic means of achieving skin-care and protection (Bauer et al. 2002; Diepgen et al. 2004; Held et al. 2002; Löffler et al. 2006; Pohrt 2007; Schliemann and Elsner 2007; Schwanitz et al. 2003). Research has not yet ascertained how such concepts would have to be developed to match the specific circumstances and needs of SMEs, which characteristically lack financial and human resources for prevention.

It is here that the project "Healthy at Work" ("Gesund im Beruf" [GiB])[3] comes into the equation with its mission to help fill the remaining blank spots in coverage

[1]See Diepgen (2012) for the criteria for diagnosing skin diseases as work-related conditions.

[2]SMEs are a mainstay of Germany's economy. Around 99% of all companies in Germany are SMEs; 69% of all employees in regular employment work for companies in this class (Zelfel et al. 2011).

[3]The GiB project is part of the innovation incubator, an EU flagship project for regional economic development. In this project, the Leuphana University of Luneburg and the State of Lower Saxony are contributing to the development of the former district of Luneburg.

of rurally based SMEs. This project has designed measures to prevent occupational skin diseases in SMEs. Whereas most previous studies focused on single industries, the GiB study covered companies from a range of sectors. This chapter showcases the intervention provided by the GiB project for the primary prevention of occupational skin diseases. In Sect. 2, an overview of the objectives is followed by a look at how the concept was adapted to high-risk occupations and the specific needs of SMEs, using the findings from the accompanying surveys. Section 3 sets out how the effectiveness and cost of the training are being quantitatively assessed in a randomized controlled trial (RCT). Section 4 introduces the results of the effectiveness evaluation and cost measurement. The chapter ends with a discussion of the findings from the point of view of the problem outlined in this section.

2 Developing the Concept for the GiB Skin Protection Training

2.1 Objectives and Approach

The mission of the GiB training is to empower the participants to practice primary prevention: to maintain their skin health or, at the first signs of dermatological problems, to stabilize the disorder or restore their dermatological health. With the training, the participants are encouraged to become aware of and take action to protect and care for their skin.

Training courses as a means of improving knowledge about skin protection and skin-care are known to reduce the incidence of occupational skin problems and diseases (Williams and Snow 2012; Wulfhorst and Schwanitz 2001). Changing skin-care practices can reduce clinical symptoms and prevent the onset of work-related dermatological diseases (Schwanitz et al. 2003). The GiB training operates on the level of imparting knowledge, changing attitudes, and affecting people's behavior. It should be remembered, however, that the simple acquisition of knowledge does not automatically lead to changes in attitudes or behavior (Pohrt 2007). For this reason, the actual effects on behavior and practices were a major part of the evaluation of the GiB training (see Sect. 3).

2.2 The GiB Skin Protection Training

The design of the GiB skin protection training was based on the tried and tested concepts of the institution for statutory accident insurance and prevention in health and welfare services (Berufsgenossenschaft für Gesundheitsdienst und Wohlfahrtspflege [BGW]) and adjusted to accommodate specific industries (see Sect. 2.3) and the specific needs of SMEs (see Sect. 2.4). The practice-oriented training was developed to introduce the participants to the fundamental know-how in four distinct modules. These include the structure and function of human skin, occupational skin disorders (causes and types of dermatitis), skin-care and

Table 1 Concept and contents of module 1

Module 1	Contents
The structure and function of human skin	• The layers of the skin • The skin's job

protection, responses to work-related skin problems, and applicable legal requirements. In addition to presenting the theory, practical exercises and tests were included to allow visualization of the content and involve the participants on a practical level. Such practical elements and exercises have a proven positive impact on prevention practices (Wulfhorst 1996a).

The first module covers the structure and function of human skin, as outlined in Table 1.

This information represents the first step towards learning more about dermatology, which can provide a basis for changes in actual habits and behaviors (Wulfhorst 1996a). It also enables participants to better understand the subsequent modules.

The second module refers to the different types of work-related skin disorders and their possible consequences for employees. Its contents are covered in Table 2.

Module 2 reviews the various stressors that could affect participants in their work and private lives and might encourage the development of occupational skin disorders. One special focus here is damp work,[4] which is one of the most common risk factors for work-related skin problems. Carers are often exposed to the considerable stressor of moisture-proof gloves and frequent contact with water, whereas the skin of metalworkers is affected primarily by the often very intensive cleaning required to remove the grease and dirt they pick up in the work-place. The participants are also taught about different types of dermatitis, which is very relevant to occupational dermatology. Finally, the participants are given the opportunity to use the corneometry method to test the humidity levels of the uppermost layer of their skin (the horny layer). This simple and rapid test could be repeated in various skin regions exposed to different stressors (e.g., palms, back of the hand, upper arm) to show the typical differences.

The third module constitutes the core of the GiB training and revolves around the topic of skin protection. This module introduces the participants to ways of preventing the development of work-related skin disorders. To that end, it covers the essential pillars of systematic skin protection: protection, care, gentle hygiene (cleaning and disinfection), and the use of protective gloves. Table 3 outlines the contents of this module.

[4]Damp work relates to all activities that fulfil one or more of the following criteria: work that is regularly performed to a substantial degree in damp environments; work that requires the lengthy or frequent use of moisture-proof gloves; and work that requires frequent or intensive hand washing (BAuA 2008).

Table 2 Concept and contents of module 2

Module 2	Contents
Occupational skin diseases (causes and types of dermatitis)	• Stressors at work and at home and how they affect the skin (focus: damp work) • Possible consequences of work-related illnesses • Different forms of dermatitis (focus: chronic skin irritation on the hand and irritant contact dermatitis) • Highly allergenic substances • Practice test: Corneometry

Table 3 Concept and contents of module 3

Module 3	Contents
Skin-care and skin protection	*Skin-care and skin protection:* • Using skin-care and protection; the effects of specific care and protection balms • Exercise: Sugar cube test to demonstrate the effects of skin protection; Dermalux (Whiteley, North Sydney, Australia) test to observe the effects of incomplete application; systematic application *Gentle hand hygiene:* • The impact of hand-washing • When is hand-washing necessary? • Introducing less damaging cleaning products • Methods for gentler hygiene • Washing vs. disinfecting (differs by sector of industry) • Exercise: Mirror test to test degreasing (hand washes vs. disinfectants) *Gloves:* • The function of protective gloves • The right glove for the right job • Minimizing dermatological damage when wearing gloves • Exercise: Selecting and testing the right gloves for the job

Practicing personal protection, such as by using skin-care products or wearing protective globes, is considered the most immediately effective decision in favor of dermatological health (Wulfhorst and Schwanitz 2001), which is why these practices are an important part of the GiB training. While skin protection refers mostly to activities before or during work that protect the skin against the stresses and dangers at work, skin-care happens primarily after work or in leisure hours and helps the skin regenerate itself. Module 3 also introduces the participants to the optimal timing and application of different skin-care and protection products (see Pohrt 2007). The topic of hygiene was chosen primarily to introduce the participants to gentler cleaning and disinfection options. Finally, this central module addresses the question of protective gloves. The participants tried on different gloves to ascertain the right size and fit for them. This is expected to encourage their use of gloves in the work-place: ill-fitting gloves can affect their ability to work and

Table 4 Concept and contents of module 4

Module 4	Contents
Responding to potential work-related skin disorders and legal requirements	• Responding to potential work-related skin disorders • GP treatment • Legal rights and responsibilities of employers and employees according to occupational health and safety laws (ArbSchG)

eventually result in workers not wearing gloves despite the danger posed by the hazardous materials they might be using (Wulfhorst 1996b).

The fourth module introduces the participants to the ways in which they can respond to any occupational skin problems and to the relevant legal regulations. Table 4 presents the content of this module.

This module covers the dermatological procedure[5] and protocol to be followed, which was designed collaboratively by doctors and statutory accident insurers to encourage the immediate and effective use of suitable means of preventing and treating occupational skin disorders, thus allowing those exposed and/or affected to continue working in their occupations (John et al. 2006).

On completion of the training sessions, all participants are given a so-called starter package with skin-care, hand-washing, and disinfectant products that are appropriate for their line of work as well as the right-sized gloves, which were selected during training. Equipping participants with these products ensures that the intervention continues beyond the training and thus achieves a more lasting effect. It facilitates the participants using their new know-how immediately in the work-place and encourages them not to use inadequate or unsuitable products that were previously available to them.

2.3 Adaptation According to Industry

The success of the GiB training depends substantially on whether the recommended skin-care and protection practices are appropriate for the participants' work-places (cf. Pohrt 2007). The participants in the training were introduced to products relevant to the issues arising in their specific occupational activities (skin-care, protection, gentle hygiene products, and protective gloves). As an example of this, both hairdressers and carers frequently use single-use gloves for cleaning and disinfection purposes; such gloves tend to offer insufficient protection from the hazardous substances used by these workers. Participants from these fields were therefore introduced to the preferable alternative of thicker household gloves. In the baking business, it is essential that the gloves used when handling foodstuffs are

[5]The dermatological procedure (=Hautarztverfahren) aims at preventing occupational skin disorders and is only used in Germany.

rated food-safe and do not bleed any potentially hazardous substances into the handled ingredients or baked goods when exposed to fat or grease. Again, these participants were introduced to suitable types of gloves. In the metalworking and floristry industries, employees often perform dirty tasks, making frequent and very intensive hand-washing necessary. These participants were shown protective gloves that can protect against dirt and small-scale injury. They were also introduced to suitably heavy-duty hand washes that used alternative, gentler means than the common abrasives, like walnut or wood shavings, to prevent skin lesions on the microscopic level.[6]

2.4 Adaptation to the Specific Needs of SMEs

One particular feature of the GiB training is its adjustment to the specific needs of SMEs. The executive managers of the participating companies were surveyed to facilitate understanding of these needs. Figure 1 shows the SMEs' expectations concerning the benefits of preventative efforts.

Less absenteeism and fewer accidents were seen as very important advantages of prevention work by 68% and 67%, respectively, of the surveyed entrepreneurs. Another almost 60% anticipated a substantial improvement in the company's climate, 58% of the sample anticipated a marked improvement in the employees' knowledge of the subject, and 57% anticipated a very positive impact on actual health-care and prevention practices among their employees.

As Fig. 2 shows, a full half of the SMEs surveyed as part of the GiB study stated that a lack of time was a strong or even very strong force stopping them from offering their personnel the means of preventing skin problems. Insufficient human resources were a strong or very strong inhibitor for 40% of the surveyed companies. Thus, alongside limited financial support and a lack of qualified contacts to turn to,[7] time and money are the two most common factors at work here.

A first change to the BGW concept related to the length of the GiB courses. By contrast to the original 8.5–13.5 h of the BGW seminars, the model for SMEs covered a condensed selection of topics to allow for shorter sessions of 120 min. The minimum time for a meaningful training was suggested by the experience of Bauer et al. (2002), whose intervention comprised two 60-min seminars 4 weeks apart with instruction videos, a lecture, and practical training units covering skin-care and protection. This intervention, designed for apprentices in the baking

[6]The information regarding the sector-specific stressors is derived from the factors mentioned by the SMEs themselves and from a selection of publications of the relevant institutions of the statutory accident insurers (see Berufsgenossenschaft Holz und Metall [BGHM] 2011; Berufsgenossenschaft Handel und Warendistribution [BGHW] 2009, 2012; Berufsgenossenschaft für Gesundheitsdienst und Wohlfahrtspflege [BGW] 2012, 2014; Vereinigung der Metall-Berufsgenossenschaften [VMBG] 2008).

[7]See Cacace et al. (2016), chapter 16; in this volume.

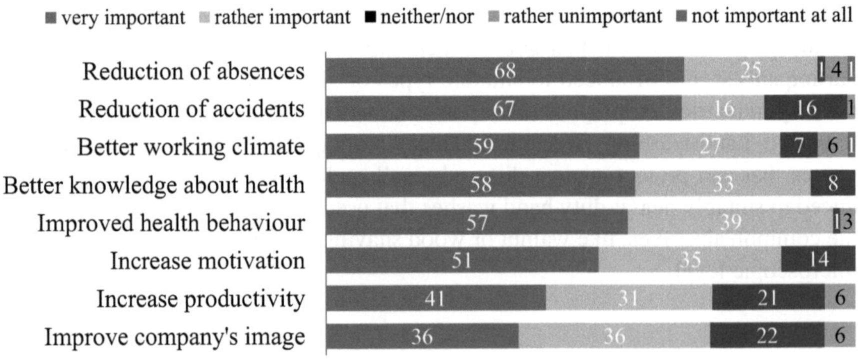

Fig. 1 The benefits of prevention from the point of view of SMEs (all figures in %, N = 72)

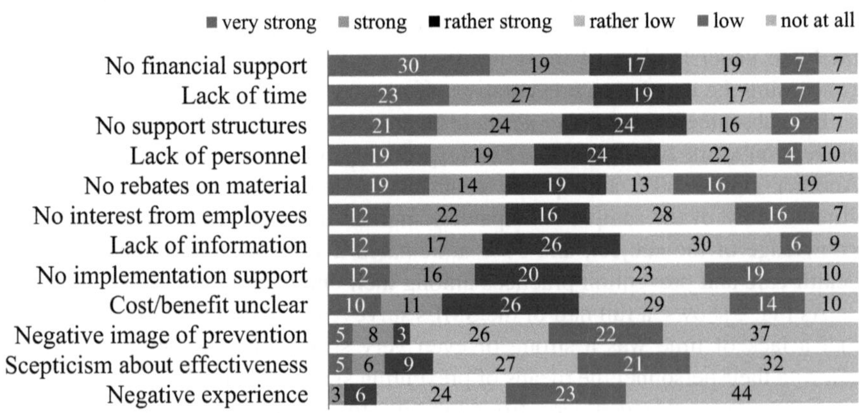

Fig. 2 Factors inhibiting the provision of preventative measures (all figures in %, N = 72)

industry, has been shown to have a positive impact in a dedicated RCT (Bauer et al. 2002).

In contrast to the BGW training, which entails substantial secondary individual interventions for participants already affected by medical conditions, the GiB training solely covers primary prevention. Individual elements of the BGW concept, such as medical check-ups or theoretical input on suitable treatments, were therefore excluded because they were considered irrelevant for the new design of the GiB training.

The concept was revised further to match the specific needs of the industries and companies. For this purpose, the trainers consulted the companies' executive managers before conducting the actual training courses to inform development of individual adaptations to their structure and contents. The GiB training differed from the BGW concept in that the aim of the former was to provide access to training in a rural region and to offer in-house training in response to the unique needs of the participating SMEs. The strong preference of the participating SMEs for in-house training was confirmed in another study that formed part of the wider GiB project (see Cacace et al. 2014). The training courses were therefore offered on the companies' premises, obviating the need for participants to travel long distances to and from the training venues. This was particularly advantageous for their employers, because releasing employees from their duties for the duration of the training would usually have meant a direct loss of revenue. Providing training on site kept this "loss of labor" to a minimum.

Conducting the GiB training in-house had several distinct advantages and drawbacks. One negative element was that there were typically no seminar rooms with the necessary technical equipment. In special cases, usually in the smallest companies (with fewer than ten members of staff), the training had to be offered during regular operations and was therefore subject to frequent interruptions. At the same time, the in-house approach allowed the trainers to become more strongly acquainted with the participants' actual knowledge. The participants' key occupational responsibilities and any previous dermatological know-how were reviewed in a prior meeting with their managers, at which time it was ascertained whether they had already received training on this issue and what type it had been. The resulting information allowed the trainers to focus the eventual training's contents and choose an emphasis that matched the conditions in the company in question.

Another crucial advantage of the chosen in-house method was the opportunity to offer advice directly in the participants' work environment (Schwanitz et al. 2003). The actual conditions on site could be referred to in the training, and possible flaws and poor practices addressed directly with the participants. Additionally, because of the specific circumstances under which the SMEs operate, on-site delivery of the training improved acceptance and, by implication, the success of the training. At the same time, this approach required considerable improvisation skills and flexibility on the part of the trainers.

3 Evaluation Method

For the RCT evaluating the costs and effectiveness of the GiB skin protection training, approximately 1400 SMEs from a total of 7960 registered in the target region of north-eastern Lower Saxony were contacted by mail or phone. Seventy-two of these agreed to take part in the study, including florists (32%), outpatient and inpatient care providers (24%), hairdressers (22%), bakers (17%), and companies in the metalworking industry (6%). A random selection of 23 SMEs (around 32% of the sample) were defined as the intervention group; the remaining 49 SMEs served as the control group. The size of the intervention group was determined by the availability of time and personnel to conduct the trainings according to the research plan. The companies chosen as the intervention group were offered 120-min dermatology training as outlined in Sect. 2.

The effect of the GiB training was evaluated by means of three waves of written surveys that were completed by the employees of the 72 participating SMEs. Previous studies on the impact of dermatological training have focused primarily on the dermatological health of the participants and on changes in their knowledge, attitudes, and care and protection habits. Several controlled intervention studies by the University of Osnabruck have, for instance, tracked skin changes and changes in the attitudes and primary prevention practices of their participants.[8] The randomized controlled study by Held et al. (2002) similarly concentrated on the participants' knowledge and changes in their dermatological symptoms. Building on these models, the effect of the GiB training was measured in terms of the following criteria: (1) the knowledge of the participants in the area of skin protection at work; (2) skin-care and protection habits; (3) dermatological health or medical symptoms; and (4) satisfaction with the working climate.

The questionnaires were structured in line with the recommendations of the practitioners' manual "Evaluation 2/2008" of the Working Group of Germany's Joint Health Insurers (Arbeitsgemeinschaft der Spitzenverbände der Krankenkassen 2008). The proposed items concerning satisfaction with the "working climate/company climate" were modified to offer a more direct bridge between employee satisfaction and the topic of (dermatological) health. The items relating specifically to dermatological health in the work-place were derived from the Nordic Occupational Skin Questionnaire[9] (Flyvholm 2002). Finally, the sociodemographic questions complied with the standards of Germany's Federal Statistical Office (Hanefeld and Hoffmeyer-Zlotnik 2010).

After the baseline had been established by completion of the first questionnaire (t0), the intervention groups were formed and the training courses conducted immediately upon receiving the participants' responses. A second wave of written surveys (t1) followed 2 months after the GiB skin protection training had been completed, a third wave following 6 months thereafter (t2). Pseudonyms were used in the survey, the participants being asked to encrypt their responses themselves with a given code. To keep the impact of possible environmental factors within a manageable range, the surveys and training courses were grouped by sectors of industry.

The costs of the measure were calculated with all economically relevant costs, including all resources used for the purposes of the GiB training being incorporated. In particular, this meant that all consumed resources such as skin-care and protection products were valued at current market prices, even if they had been sourced from sponsors. There is a conceptual difference between fixed and variable costs, and a further distinction needs to be made between human resources (HR) and material expenses. Some costs are fixed irrespective of the actual number of GiB courses. Such fixed material costs refer to the longer-term liabilities of office leases, insurance premiums, vehicles, or the procurement of consumable materials, such as office equipment. Fixed HR costs relate primarily to the trainers conducting the courses. Variable costs however, differ depending on the number of training

---

[8]For an overview, see Schwanitz et al. (2003).

[9]Translated and adapted for use in German language by the Institute and Polyclinic for Occupational and Social Medicine of the Technical University of Dresden.

courses actually completed. As for fuel costs, all other consumables were also included in the variable material costs. Fixed and variable costs were finally calculated to arrive at the mean cost per participant and training course.

4 Results

4.1 Results of the Survey on Dermatological Risks

Five hundred and thirty-eight valid responses were received from the employees of 72 participating SMEs. This included 279 responses at the first survey point (t0), 166 at the second point (t1), and 93 at the (final) third point (t2). The SMEs contacted in the follow-up attributed this considerable drop in the rate of returns to a lack of time. The participants also mentioned that the issue of health-care and prevention had not (yet) been prioritized by their executives and employees. Because of the numerous dropouts between points t1 and t2, the data from these two points were combined wherever feasible. The data were processed by means of MS Excel and SPSS. The results of the employee surveys are discussed here before analyzing the evaluation in Sects. 4.2 and 4.3.

The first survey was conducted at point t0 (prior to training) and was intended primarily as a means of sourcing general information about the survey subjects and their work-related dermatological conditions. At t0, outpatient and inpatient care workers were the best-represented profession (at 48%), followed by florists and bakers (both at 21%) and then hairdressers (at 9%) and metalworking companies (at 1%[10]). Almost half of the respondents (44%) were in full-time employment at the time of the survey, with a substantial gap between sexes: whereas 81% of the male participants were in full-time employment, this was true for only 37% of the female respondents. Ten percent of the participants were in part-time employment with more than 30 h/week, 28% worked between 20 and 30 h/week, and 18% worked fewer than 20 h/week. The majority (88%) had permanent contracts. Thirty-six percent of the respondents had spent more than 20 years in their occupations, which implies that they had been exposed to similar or related dermatological risk factors, such as damp environments, on a long-term ongoing basis. Forty-seven percent named no injuries or disabilities, whereas 29% stated that they were completely incapacitated by dermatological problems. Another 7% were restricted to part-time work as a result of such conditions. Thus, more than a third of the sampled subjects had severe dermatological problems. At the time of the survey, 27% of them were receiving medical treatment because of damage to their hands or forearms. Twenty-eight percent of the respondents considered their

[10]Approximately the same number of SMEs were contacted in each of the five industries while recruiting for the study. The different response rates are in part attributable to the different company sizes and seasonal differences in typical workloads.

general health to be excellent, 60% reported good health, and 13% reported less good or poor health.

Twenty-one percent of the respondents who had previously received dermatological training did not follow the practices that had been recommended or only followed them infrequently. When asked about the reasons for this (with multiple answers allowed), 37% said that the quality of their work was affected by the recommended measures, 31% blamed a lack of time for regular skin-care, 20% reported that their employers did not provide the necessary products, 17% could not see any way of combining the practices with their workflows, and 11% were concerned about negative reactions from clients/patients. A further 6% considered the recommended practices prohibitively expensive.

Fifty-four percent of respondents believed that their employers provided them with sufficient skin-care and protection products, whereas 20% stated that such materials were usually available, 15% mentioned infrequent availability, and 11% were provided with no materials at all. The private spending on skin-care for hands and forearms was estimated at an average of approximately 9 € per month.

At point t0 there were no significant differences between intervention and control group in sex, age, family status, education, household income, or type of contract (permanent/non-permanent). A significant difference was found in the number of hours worked (part-time/full-time): 22% of the control group and 36% of the intervention group worked 20–30 h/week. Thus, overall, the groups were quite homogenous with regards to socio-economic variables and work status.

4.2 Effectiveness of the GiB Training

A positive effect on skin-care, cleaning, and protection practices was identified when the mean scores for the trial and control groups were tracked over time. As Fig. 3 shows, the participants in the training group made more use of skin-care and protection products than their peers in the control group almost across the board. In particular, there were significant differences between these groups for wearing gloves in damaging environments and for using special products for gentle hand cleaning.

Analysis of developments in the trial group revealed that the participants' knowledge in the area of occupational skin-care/protection had improved significantly, from an average of 5.5 to 6.3 points on an 11-point scale (from 0 = no knowledge to 11 = very good knowledge) in the control group, compared with a greater increase from an average of 6.9 to 8.3 points in the trial group.

The symptoms that can indicate stressed or already infected skin were assessed at different points in the process. Figure 4 shows dermatological symptoms in the trial and the control groups before and after the training. The most definite and significant positive effects of the training were in for redness, rapidly appearing

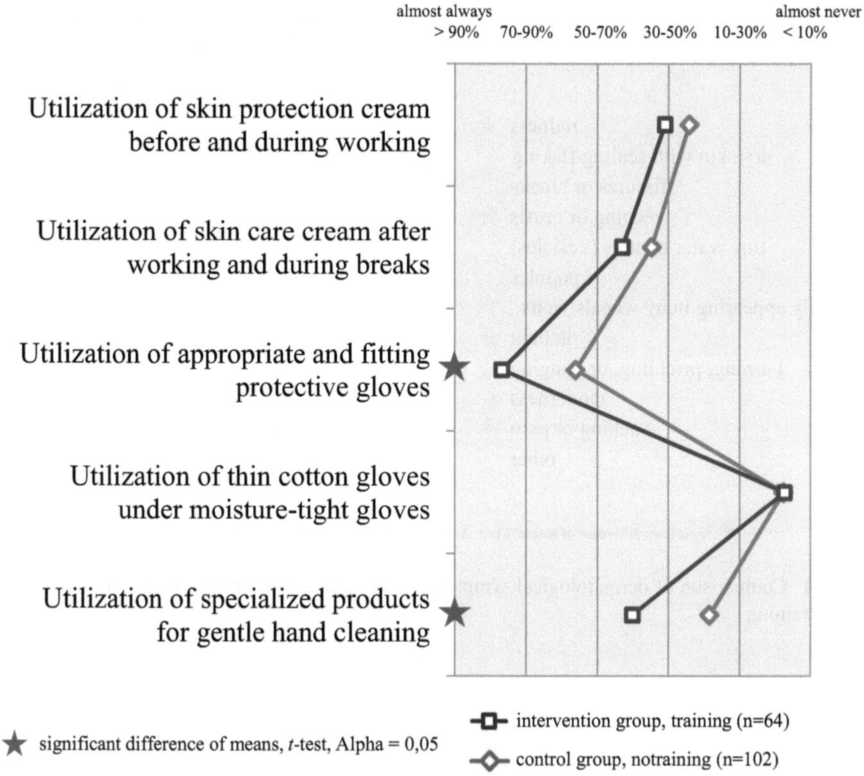

Fig. 3 Different usage practices in the trial and control groups (t1)

itchy wheals/welts (urticaria), burning, prickling, stinging, and skin tenderness. There was also a significant reduction in symptoms of weeping or crusts, itching, and general aching or pain. The control group, which received no training, showed no significant change in the severity or frequency of these symptoms.

The GiB skin protection training did not have a positive effect on the participants' satisfaction with their work, the atmosphere in their companies, or their superiors in general. The opposite is true: there were negative trends in degree of satisfaction, albeit not at a significant level. As Fig. 5 shows, after training, the trial group showed less satisfaction for the items "the option to address superiors about health problems", "support by the employer for the prevention of illnesses", and "support by the employer for the prevention of skin diseases". The general climate at work was also rated slightly worse after attending training.

In almost all surveyed items, the control group showed a lesser negative or, in some cases, a slightly more positive trend than the trial group. One explanation for this is that the reduced satisfaction of the trial group was attributable to their perception of certain failings on the part of their employers. As described in Sect. 2.2, the GiB training covers the legal requirements and employers' obligations

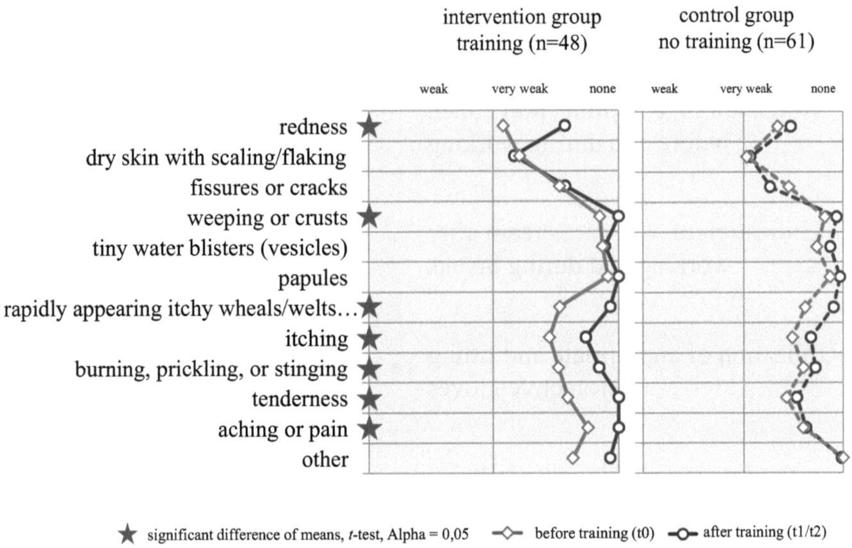

Fig. 4 Comparison of dermatological symptoms in the trial and control groups before and after GiB training

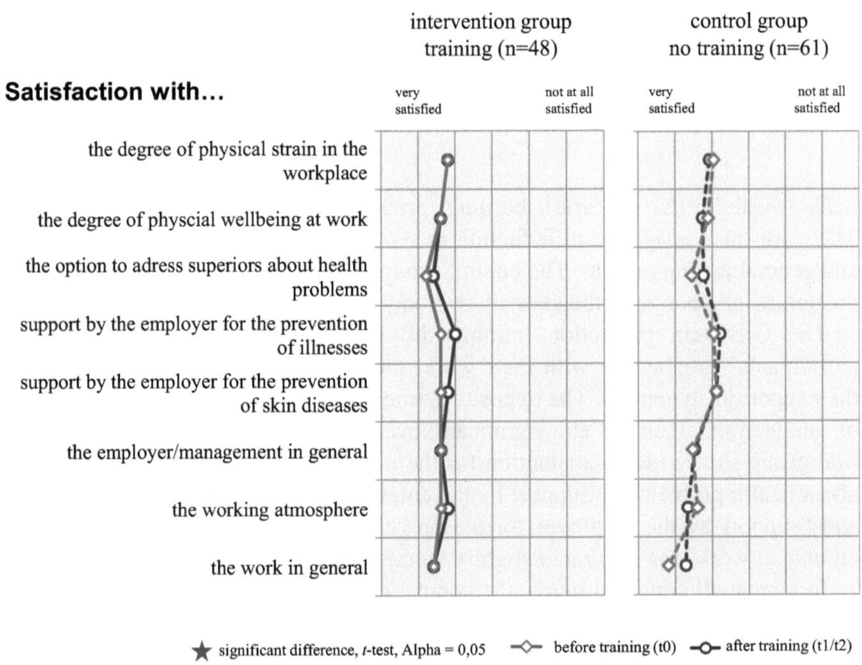

Fig. 5 Comparison of satisfaction with the working atmosphere in trial and control groups before and after GiB training

regarding protecting the health and wellbeing of employees. The GiB training may therefore have drawn attention to previously unnoticed shortcomings.

These findings imply that an improvement in participants' knowledge about skin-care, cleaning, and protection does translate into an improvement in individual skin-care practices. The improved dermatological condition of the trial group seemed to be directly related to the positive impact of this change in behavior.

4.3 Costs of GiB Training

Average costs per participants amounted to approximately 63.50 € per training, with this figure including the average fixed and variable material and HR costs. The calculation is based on 6.5 attendees per training group, which was the average during the trial. Training was provided at 23 SMEs.

When calculating the HR cost it was assumed that one single person, fully employed or self-employed, could plan and deliver the trainings. A fictional annual salary of 50,000 € was used for estimating these cost. At an average of 220 8-h working days, this meant an hourly rate of approximately 28.41 €. The GiB training in north-eastern Lower Saxony required travel of an average of 123 km per training event. Thus, the rural nature of the study region made the cost of traveling to the participating SMEs particularly relevant. The variable material costs for travelling accounted for 2.35 € per participant. They were mainly attributed to fuel cost. The variable costs of consumables (at 15 € per participant) represent the costs for the starter packs, including information material, sample gloves, and skin-care, cleaning, and protection products. The fixed overheads, including lease of an office space, a company car, office materials, and general overheads, amounted to 73.00 € per training or 11.18 € per participant and training. Table 5 shows the average costs for the GiB skin protection training, broken down by cost types.

Because the cost of labor varied considerably from company to company, the calculated amount does not include expenses in terms of the time spent by the participants.

Table 5 Costs of the GiB skin protection training per participant (in Euro)

Cost type		Costs per participant
Fixed costs	HR costs: preparation & administration, delivery of the training, travel	34.97
	Attributable overheads: leases, vehicles etc.	11.18
Variable costs	Travel expenses/fuel	2.35
	Consumables: e.g. sample gloves, skin-care products	15.00
Total		**63.50**

5 Discussion

The GiB project aimed to contribute to filling the current gaps in primary prevention provided by SMEs, in particular in rural regions, by developing and evaluating a prevention measure. The focus was on preventing work-related skin diseases because of their importance and potential costs to the economy at large. The training concept paid heed to the specific needs of different professions and the unique nature of SMEs.

A parallel survey (see chapter 16 in this volume) showed that lack of time and limited finances and HR are currently stopping SMEs from rolling out preventative initiatives on a larger scale. The unique quality of the GiB training lies in its specific adaptations to the circumstances under which SMEs operate; these adaptations include provision of in-house training and a focused and condensed course that covers only the key elements. The in-house training provided several benefits to the SMEs hosting them, including saving the time away from work due to employees traveling to training centers. Provision of on-site advice is a plus for SMEs, but requires considerably greater flexibility on the part of the trainers.

Because there are no reliable data about implementation in SMEs in rural regions, the GiB project also evaluated the measure with the aim of making it a dependable basis for decision-making by these companies. The evaluation revealed positive and, generally, significant outcomes in terms of awareness of skin protection, relevant practices, and improvements in current symptoms or illnesses in the trial group, indicating that, in general, the risk of developing occupational skin diseases and the subsequent costs are reduced considerably by the GiB skin protection training. The intervention had a negligible, even negative, effect on the satisfaction of the surveyed persons with the support to prevent such illnesses or medical problems they received from their employers and direct superiors. The increased dissatisfaction with their employers' presumed failings may have been attributable to the employees' growing awareness of the issue.

As to the expectations of the employers (see Fig.1), the companies did reap the expected benefits in terms of improvements in their employees' health-care knowledge and subsequent practices in this area. What was not confirmed, however, was the expected improvement in employee satisfaction and the general atmosphere in the organizations. This discrepancy should not be seen as a criticism of the companies' failings, but as an opportunity for them to address health-care and prevention in their organizations and thus improve the climate at work. Also, there is an opportunity for more research exploring how prevention initiatives could be designed to lead to greater staff satisfaction.

As to the costs, these were about 63.50 € per participant for a 120-min training. One limitation of this study is that the effects of the intervention in monetary terms were not ascertained, preventing a meaningful comparison with the required investments. This would have required more surveys than could be completed within the limits of the project. An investment in preventing skin diseases would be more appealing for employers if they anticipated benefits in return for the costs. Of note, in this study additional costs, such as the temporary loss of labor, were not considered because they differ considerably from company to company. However, these costs are minimized by providing the training in-house. It should also be noted that the evaluated preventative measures also offer health and occupational accident insurers opportunities to reduce the costs of dermatological illnesses and their complications. It is therefore beneficial to them to co-sponsor such activities.

Another limitation of the study is the considerable decline in the number of participants responding to the follow-up surveys, which the participating SMEs attributed to a lack of time and human resources. Nevertheless, the responses we did receive provided some empirically significant findings.

While occupational skin diseases are a major threat to employment, they are far from being the only such danger. More studies are needed to understand and evaluate preventative concepts in other areas in terms of their suitability for the unique circumstances in SMEs, potential costs, and outcomes.

References

Arbeitsgemeinschaft der Spitzenverbände der Krankenkassen (2008) Anwenderhandbuch Evaluation Teil 2: Evaluation von betrieblicher Gesundheitsförderung [Manual evaluation part 2: evaluation of occupational health promotion]. Available from URL www.gkv-spitzenverband.de/media/dokumente/krankenversicherung_1/praevention__selbsthilfe__beratung/praevention/praevention_evaluation/betriebl_gesundheitsfoerderung/Praev_Eva_Handbuch_2_BGF_2008-06.pdf. Accessed 11 Mar 2015 (in German)

Batzdorfer H, Schwanitz J (2004) Direkte und indirekte Kosten berufsbedingter Hauterkrankungen [Direct and indirect costs of work-related skin diseases]. Arbeitsmed SozialmedUmweltmed 39:578–582 (in German)

Bauer A, Kelterer D, Bartsch R et al (2002) Skin protection in bakers' apprentices. Contact Dermatitis 46:81–85

Berufsgenossenschaft für Gesundheitsdienst und Wohlfahrtspflege (BGW) (2012) Gesunde Haut mit Schutz und Pflege. Tipps und Informationen für Pflegeberufe [Healthy skin due to protection and care. Tips and information for care professions]. Available from URL https://www.bgw-online.de/DE/Medien-Service/Medien-Center/Medientypen/bgw-themen/TP-HAP-11_Gesunde_Haut_durch_Schutz_und_Pflege.html. Accessed 11 Mar 2015 (in German)

Berufsgenossenschaft für Gesundheitsdienst und Wohlfahrtspflege (BGW) (2014) Schöne Hände—gesunde Haut. Pflegetipps und Informationen für das Friseurhandwerk [Beautiful hands—healthy skin. Care tips and information for hairdressers]. Available from URL https://www.bgw-online.de/DE/Medien-Service/Medien-Center/Medientypen/bgw-themen/TP-HAP-9_Schoene_Haende_Gesunde_Haut_Friseure_Arbeitnehmer.html. Accessed 11 Mar 2015 (in German)

Berufsgenossenschaft Handel und Warendistribution (2009) Merkblatt "Blumen und Pflanzen" (M57) [Leaflet "Flowers and plants" (M57)]. Available from URL medien-e.bghw.de/bge/pdf/m57.pdf. Accessed 9 Mar 2015 (in German)

Berufsgenossenschaft Handel und Warendistribution (2012) M 101 Hautschutz beim Umgang mit Lebensmitteln [Skin protection when handling food (M101)]. Available from URL medien-e.bghw.de/bge/pdf/m101.pdf. Accessed 11 Mar 2015 (in German)

Berufsgenossenschaft Holz und Metall (BGHM) (2011) BG 80.19. Tipps für eine gesunde Haut. Hautschutz für Beschäftigte in Metallbetrieben und in der Holzbranche [Tips for a healthy skin. Skin protection for metal- and woodworking industries]. Available from URL http://bghm.de/fileadmin/user_upload/Webshop/Webshopmedien/Flyer/BG_80.19.pdf. Accessed 11 Mar 2015 (in German)

Breuer K, John SM (2011) Occupational aspects. In: Werfel T, Spergel JM, Kiess W (eds) Atopic dermatitis in childhood and adolescence. Pediatr Adolesc Med 15:133–148

Bundesanstalt für Arbeitsschutz und Arbeitsmedizin (2008) Technische Regel für Gefahrstoffe 401: Gefährdung durch Hautkontakt—Ermittlung, Beurteilung, Maßnahmen [Technical rule for hazardous substances 401: Risk due to skin contact—Assessment, evaluation, measures]. Available from URL http://www.baua.de/cae/servlet/contentblob/666104/publicationFile/56314/TRGS-401.pdf. Accessed 11 Mar 2015 (in German)

Bundesanstalt für Arbeitsschutz und Arbeitsmedizin (2014) Sicherheit und Gesundheit bei der Arbeit 2012. Unfallverhütungsbericht Arbeit [Safety and health at work 2012. Accident prevention report]. Bundesministerium für Arbeit und Soziales, Dortmund (in German)

Cacace M, Franz I, Ratz D (2014) Using conjoint analysis to elicit preferences for occupational health services in small and micro-enterprises. Athens J Health 1(4):237–254

Cacace M et al (2016) Small- and medium-sized enterprises' preferences for occupational health services and willingness to pay. In: Wiencke M et al (eds) Healthy at work: interdisciplinary perspectives. Springer, Heidelberg

Diepgen TL (2012) Berufsbedingte Hauterkrankungen [Work-related skin diseases]. JDDG 10:297–316 (in German)

Diepgen TL, Schmidt A, Kresken A (2004) Prävention berufsbedingter Handekzeme durch Hautschutzmaßnahmen—Ergebnisse einer betrieblichen Interventionsstudie [Prevention of work-related hand eczema through skin protection measures—Results of an occupational intervention study]. Arbeitsmed Sozialmed Umweltmed 39:306–314 (in German)

Elsner P, Wigger-Alberti W (1997) Der Hautschutz in der Prävention der Berufsdermatosen [Skin protection in the prevention of occupational dermatitis]. Deut Ärzteblatt 22(37): 1489–1492 (in German)

Flyvholm M-A (2002) Nordic Occupational Skin Questionnaire—NOSQ-2002. Nordic questionnaire for surveying work-related skin diseases on hands and forearms and relevant exposures. Available from URL www.arbejdsmiljoforskning.dk/~/media/Spoergeskemaer/nosq/nosq-uk-long-2002-03-01.pdf. Accessed 7 Mar 2015

Gieseke O (2005) Betriebliche Gesundheitsförderung in kleinen und mittleren Unternehmen [Occupational health promotion in small and medium-sized enterprises]. Verlag Hans Huber, Bern (in German)

Hanefeld U, Hoffmeyer-Zlotnik JHP (2010) Demographische Standards [Demographic standards], 5th edn. Statistisches Bundesamt (Statistik und Wissenschaft), Wiesbaden (in German)

Held E, Mygind K, Wolff C et al (2002) Prevention of work related skin problems: an intervention. Occup Environ Med 59:556–561

John SM (2008) Optionen für eine vernetzte interdisziplinäre Prävention am Beispiel berufsbedingter Hautkrankheiten [Options for networked and interdisciplinary prevention using the example of work-related skin diseases]. GMS Ger Med Sci 6:Doc06 (in German)

John SM, Skudlik C, Römer W et al (2006) Hautarztverfahren [Dermatological procedure]. Available from URL http://www.dguv.de/medien/inhalt/praevention/themen_a_z/haut/documents/hautarztverfahren.pdf. Accessed 9 Mar 2015 (in German)

Löffler H, Bruckner T, Diepgen T et al (2006) Primary prevention in health care employees: a prospective intervention study with a 3-year training period. Contact Dermatitis 54(4):202–209

Pohrt U (2007) Skin protection training. The route of practical applications. In: Schliemann S, Elsner P (eds) Skin protection. Curr Probl Dermatol 34:161–170

Rustemeyer T (2012) Kanerva's occupational dermatology, 2nd edn. Springer, Berlin

Schliemann S, Elsner P (2007) Skin protection. Practical applications in the occupational setting. Curr Probl Dermatol 34:120–132

Schwanitz HJ, Riehl U, Schlesinger T et al (2003) Skin care management: educational aspects. Int Arch Occup Environ Health 76:374–381

Vereinigung der Metall-Berufsgenossenschaften (VMBG) (2008) BG-Information: Hautschutz in Metallbetrieben (BGI 658). Information of the institutions of the statutory accident insurers: skin protection in the metalworking industry (BGI 658). Available from URL http://publikationen.dguv.de/dguv/pdf/10002/bgi658.pdf. Accessed 15 Nov 2014 (in German)

Williams JS, Snow MD (2012) Promoting health in small and medium-sized enterprises. J Small Bus Enterprise Dev 19(4):729–744

Wulfhorst B (1996a) Hautschutzseminare [Skin protection trainings]. In: Schwanitz HJ, Uter W, Wulfhorst B (eds) Neue Wege zur Prävention—Paradigma Friseurekzem. Universitätsverlag Rasch, Osnabrück (in German)

Wulfhorst B (1996b) Schutzhandschuhe [Protective gloves]. In: Schwanitz HJ, Uter W, Wulfhorst B (eds) Neue Wege zur Prävention—Paradigma Friseurekzem. Osnabrück Universitätsverlag, Rasch (in German)

Wulfhorst B, Schwanitz HJ (2001) Hautkrankheiten und Hautschutz. Für Unternehmer, Beschäftigte, Betriebsärzte, Sicherheitsfachkräfte, Betriebs- und Personalräte [Skin disease and skin protection. For entrepreneurs, workers, company doctors, safety experts, works and staff councils]. GUV-Informationen: Theorie und Praxis der Prävention. GUV-I 8559 (bisher GUV 50.0.11), Bundesverb der Unfallkassen, München (in German)

Zelfel RC, Alles T, Weber A (2011) Gesundheitsmanagement in kleinen und mittleren Unternehmen—Ergebnisse einer repräsentativen Unternehmensbefragung [Health management in small and medium-sized enterprises—Results of a representative company survey]. Das Gesundheitswesen 73:515–519 (in German)

Is Psoriasis Associated with Specific Occupations?

Anca Chiriac and Caius Solovan

Abstract

Psoriasis, a systemic, immune-mediated disease associated with significant comorbidities, has a strong negative impact on quality of life. The aim of this study was to evaluate the occupational category of individuals with psoriasis and establish a correlation between this disease and patients' occupations. Selected clinical variables and characteristics of work-places of 1236 individuals diagnosed with psoriasis in an outpatient clinic over 8 years (2004–2011) were assessed and documented. Of the 1236 study patients, 669 (54.13 %) were men and 567 (45.87 %) women, the male to female ratio being 1.18/1. The median age at diagnosis was 29.34 ± 15.24 years (mean ± standard deviation [SD]). Retired persons were the most strongly represented category (149, 12.05 %), followed by primary and secondary school students (112, 9.06 %), unskilled workers/laborers (108, 8.74 %), engineers (87, 7.04 %), students (70, 5.66 %), managers (56, 4.53 %), self-employed persons working from home (54, 4.37 %), professors (53, 4.28 %), drivers (46, 3.72 %), salespersons (39, 3.16 %), economists (38, 3.07 %), health-care professionals (32, 2.59 %), unemployed persons (71, 8.77 %) and all others types of work (321, 25.97 %). The present study yielded no conclusive evidence that psoriasis is an occupational or occupation-induced disease. Thus, we believe that psoriasis is a constitutional disease that may be triggered by occupational factors.

A. Chiriac (✉)
Apollonia University, Iaşi, Romania

Nicolina Medical Center, Iaşi, Romania

P. Poni Research Institute of Macromolecular Chemistry, Romanian Academy, Iaşi, Romania
e-mail: ancachiriac@yahoo.com

C. Solovan
Department of Dermatology, Victor Babes University of Medicine and Pharmacy, Timisoara, Romania

M. Wiencke et al. (eds.), *Healthy at Work*, DOI 10.1007/978-3-319-32331-2_24

1 Introduction

Psoriasis is a systemic, autoimmune, inflammatory disease that significantly impairs work productivity and quality of life (Armstrong et al. 2012). The development of psoriasis involve interactions between genetic background and various environmental factors including infections, stress, drugs (non-steroidal anti-inflammatory agents, beta-blockers, lithium salts), and physical triggers (Kalayciyan et al. 2007).

The great majority of individuals with psoriasis report psychological issues such as depression, social isolation, stigmatization (Schmid-Ott et al. 1996; Gupta et al. 1998; Solovan et al. 2014) and significant work-related impediments (Pearce et al. 2006; Horn et al. 2007).

An occupational disease can simply be defined as one that is caused, or worsened, by exposure to some factor at work (Cherry 1999). The terminology of occupationally induced psoriasis was introduced by Moroni et al. (1998) who reported that 1.2 % of all occupational dermatoses (3000 patients enrolled in the study) were work-related psoriasis. From 1995 to 2010, the German statutory accident insurance body recognized 130 cases of psoriasis as occupationally related; this was only 0.1 % of all occupationally related dermatoses (Mahler et al. 2014).

A study of 1901 patients with occupational dermatoses performed in 2008 by Skudlik et al. showed that 73 patients (3.8 %) had psoriasis of the hands; 42 of these 73 cases (2.2 %) reported work-place factors (repeated mechanical microtrauma, wet conditions of work, contact with irritants). It is now recognized that pressure, trauma and friction may cause psoriasis of the hands and fingers (Kanerva and Estlander 2001). *Palmar psoriasis*, which is characterized by hyperkeratosis and fissures, is functionally incapacitating and markedly stigmatizes the affected individual. Individuals with palmoplantar psoriasis report higher Dermatology Life Quality Index scores than those with moderate to severe plaque psoriasis and use larger quantities of topical treatments (Chung et al. 2014).

A number of papers concerning occupational psoriasis (mostly palmar type) have been published during the last decades (Fisher 1979; Hill and Ostlere 1998; Kanerva et al. 1998). Lists of professions/jobs with a high incidence of work-related psoriasis have been published (Fisher 1979; Hill and Ostlere 1998; Kanerva et al. 1998) and include the following risk activities: baker, barkeeper, bookbinder, bus driver, cashier, cleaner, construction worker, dentist, draftsman, electrician, joiner, manufacturing/packaging worker, masseur, mechanic, newspaper-vendor, pharmacist, seamstress and tailor (Menne and Hjorth 1984).

Smoking and alcohol use have been studied as triggers for psoriasis; however, varying and controversial findings have been reported. A positive correlation between *smoking* and severity of psoriasis has been proved in some studies (Fortes et al. 2005; Gerdes et al. 2010; Asokan et al. 2014); some studies emphasize an association between smoking and onset of psoriasis, but not with the severity of the disease (Wolk et al. 2009). An association between *alcohol* intake and severity of psoriasis is also debatable. In 1986 Monk and Neill reported that the index severity of psoriasis was strongly connected with alcohol use. However, a few years later a population-based study from Sweden failed to connect alcohol use and severity of psoriasis, although an association with onset of psoriasis was demonstrated among male subjects (Wolk et al. 2009). In contrast, a German study has confirmed a relationship between alcohol consumed and severity of psoriasis, but only in female subjects (Gerdes et al. 2010). More controversial data have continued to be published.

A slight increase in the prevalence of psoriasis in farmers exposed to organophosphoric *pesticides* was found in India in 2011 (Jawahar 2011); however, further studies are required concerning this factor.

Geographic variations and *climatic factors* such as sun exposure and humidity could influence the prevalence of psoriasis and explain (among other factors) the variability of incidence of the disease throughout the world (Balato et al. 2013). Prolonged periods of cold and low humidity have a negative impact on psoriasis, in contrast to sunbathing which relieves symptoms.

Work-related physical activity is reportedly correlated with the onset of psoriatic arthritis (Eder et al. 2011).

A study conducted from 2003 to 2011 by the National Psoriasis Foundation was recently published in the USA (Armstrong et al. 2012). It reported that psoriasis is prevalent in working people (64 %) (full-time, 48 %; part-time, 11 %; homemaker, 5 %). Retired people accounted for 22 % of their cases, jobless people 12 % and children attending school 2 %. No details about the specific activities of individuals with psoriasis were reported.

Psychosocial stress may play a role in exacerbations of psoriasis (Hunter et al. 2013). *Stress* and psoriasis are linked: psychological (alexithymia, personality, affect) and biological (cortisol, epinephrine, neurogenic inflammation) factors reportedly have a great impact on the disease (Kálmán et al. 2014). Daily stress increases pruritus and correlates with index severity of psoriasis (Verhoeven et al. 2009), especially in younger patients. A correlation between psychological factors and psoriasis is now considered indisputable. Anxiety can persist despite improvement in the psoriasis (Sampogna et al. 2007). Patients with psoriasis of long duration reportedly have a lower level of distress that is considered attributable to their acceptance of and adaptation to the disease and to avoidance of trigger factors (Devrimci-Ozguven et al. 2000). This is the case for older patients.

Patients' *family incomes* correlate negatively with psoriasis severity (Hawro et al. 2014). An inverse relationship between psoriasis severity, employment and income has been recently confirmed: 601 patients were studied by Horn et al., who showed that 31.2 % of individuals with severe psoriasis had low incomes, compared with 18.1 % of those with mild psoriasis and 20 % of unemployed persons (Horn et al. 2007).

Rotating night-shift work has been noted as a risk factor for psoriasis (Li et al. 2013): this could at least partially explain the incidence of severe psoriasis among medical staff.

In a survey of 117 musicians from a professional orchestra, psoriasis was reported as an occupational and stress-related skin problem (Onder et al. 2000).

Because the impact of psoriasis on quality of personal and professional life is well known, the aim of this study was to investigate possible correlations between the disease and the occupation of the patients. This is the first study on psoriasis and its relation to work, in a time of many socio-economic changes when new skills and professions appeared. There is a reconfiguration of jobs on the market and that could impact the psoriasis patient.

2 Patients and Methods

We here report a single-center retrospective study of 1236 patients with psoriasis seen between 2004 and 2011 in north-eastern Romania, which contains approximately one third of Romania's population. All patients were examined as outpatients by trained dermatologists and the following data noted: age, sex, age at diagnosis, family history of psoriasis, residence, level of education and type of work. The patients had also been asked to fill in questionnaires to acquire relevant personal data (Table 1).

The diagnosis of psoriasis was based on clinical grounds and, when necessary, verified by skin biopsy. Severity of the disease was assessed according to the Psoriasis Area Severity Index score (PASI) and Dermatology Life Quality Index (DLQI) scores (Feldman and Krueger 2005). The World Health Organization (WHO) defines quality of life as "individuals' perception of their position in life in the context of the culture and value system in which they live and in relation to their goals, expectations, standards and concerns". PASI scores were used to categorize patients as having mild (0–4.0), moderate (4.1–9.9) or severe (10 or higher) psoriasis. Linear regression was used to perform multivariate analysis of factors affecting psoriasis.

3 Results

Recent official data regarding the distribution of the population in Romania according to sex and occupation were published by Romanian National Institute of Statistics (Table 1). Selected clinical variables and characteristics of work-places of 1236 individuals diagnosed with psoriasis were assessed and documented and the results statistically analyzed.

3.1 Age at Time of Clinical Examination

Age must clearly be considered when assessing associations between occupation and psoriasis; children (under 6 years) with psoriasis were excluded from present study. At the time of clinical examination, the largest age categories were 30–40 and 40–50 years (43.12 %) and the smallest categories over 70 years (5.83 %) and

Table 1 Romanian demographic data

		% of total population
Total population Romania 2011	20,121,641	*100*
Women	10,333,064	*51.4*
Retired persons	5,487,000	*27.2*
Not working/unemployed persons	744,000	*3.69*

Table 2 Age distribution at time of clinical examination

Age at time of diagnosis (years)	N	%
≤10	3	0.24
10–20	65	5.26
20–30	178	14.40
30–40	265	21.44
40–50	268	21.68
50–60	224	18.12
60–70	161	13.03
70–80	58	4.69
80–90	13	1.05
90–100	1	0.08
	1236	

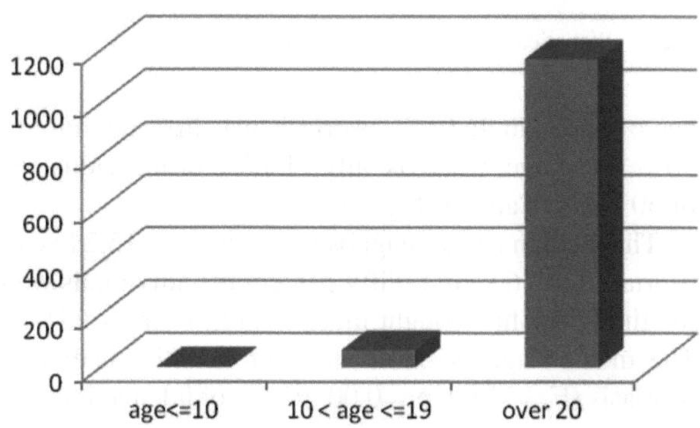

Fig. 1 Childhood versus adult psoriasis. Age distribution at time of clinical examination

under 20 years (5.5 %) (Table 2; Figs. 1 and 2). Additionally, individuals with more severe disease tended to be older than those with milder involvement (Table 4).

Type 2 psoriasis ("late-type" psoriasis), which is defined as psoriasis diagnosed after the age of 40 years (the incidence peaking in the sixth decade) and tends to be mild and have a negative family history and no genetic markers, was well represented (Table 2).

3.2 Age at Onset of Psoriasis

The onset of psoriasis was under the age of 10 years in 104 individuals (8.41 %) and between the ages of 10 and 19 years in 263 (21.28 %). The majority of individuals had onset in adulthood (869, 70.31 %). A few individuals (7.77 %) did not recall the

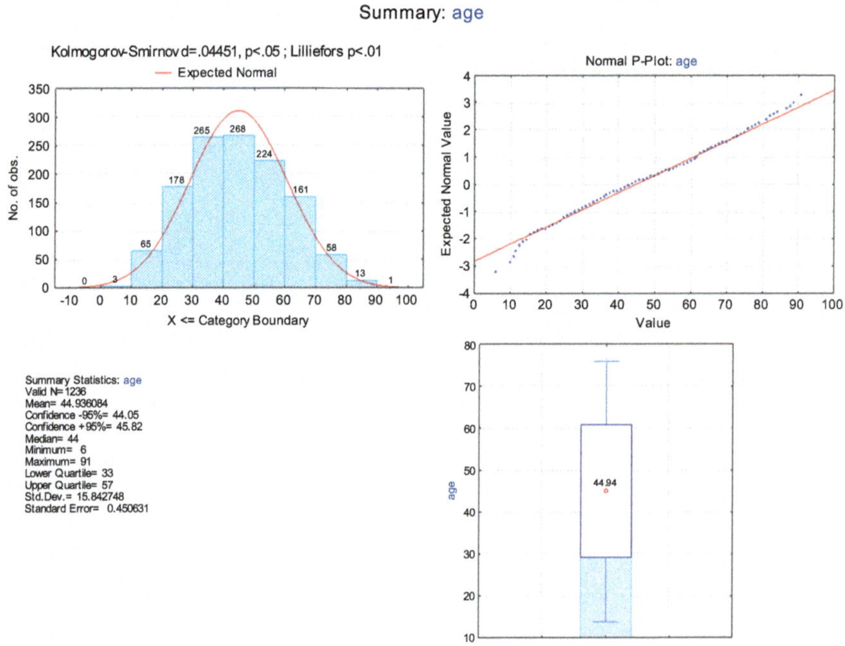

Fig. 2 Statistical analysis of age at time of diagnosis

age of onset. In 46.04 %, psoriasis was first diagnosed somewhere between 10 and 30 years of age, whereas only 11.17 % of individuals were diagnosed over the age of 50 years (Table 3; Fig. 3).

The median age at diagnosis was 29.34 ± 15.24 years (range 6 months [neonatal psoriasis] to 76 years). Fifty percent of individuals were aged less than 27 years at the time they first sought medical aid, whereas 25 % were over 39 years (Fig. 3). The median age of onset was significantly older for patients with more severe psoriasis ($F = 11.69$, $p = 0.00$, 95 % confidence interval [CI]) (Table 4).

3.3 Occupational Category at Time of Admission to Study

Working individuals (65.74 %) were the largest occupational category, followed by *retired persons* (12.06 %), *students* (9.06 %), *self-employed persons working from home* (4.37 %) and *unemployed persons* (8.77 %) (Fig. 4). Further analysis of occupations at the time of admission to the study revealed the following (Table 5): unskilled workers (laborers) (8.74 %), engineers (7.04 %), students (5.66 %), managers (4.53 %), professors (4.28 %), drivers (3.72 %), salespersons (3.16 %), economists (3.07 %), and professional healthcare (2.59 %). A high proportion of students had mild psoriasis (66.48 %), severe disease being reported in only 2.75 %

Table 3 Age at onset of psoriasis

Onset age (years)	N	%
≤10	104	8.41
10–20	263	21.28
20–30	306	24.76
30–40	198	16.02
40–50	131	10.60
50–60	103	8.33
60–70	30	2.43
70–80	5	0.40
Unknown	96	7.77
	1236	

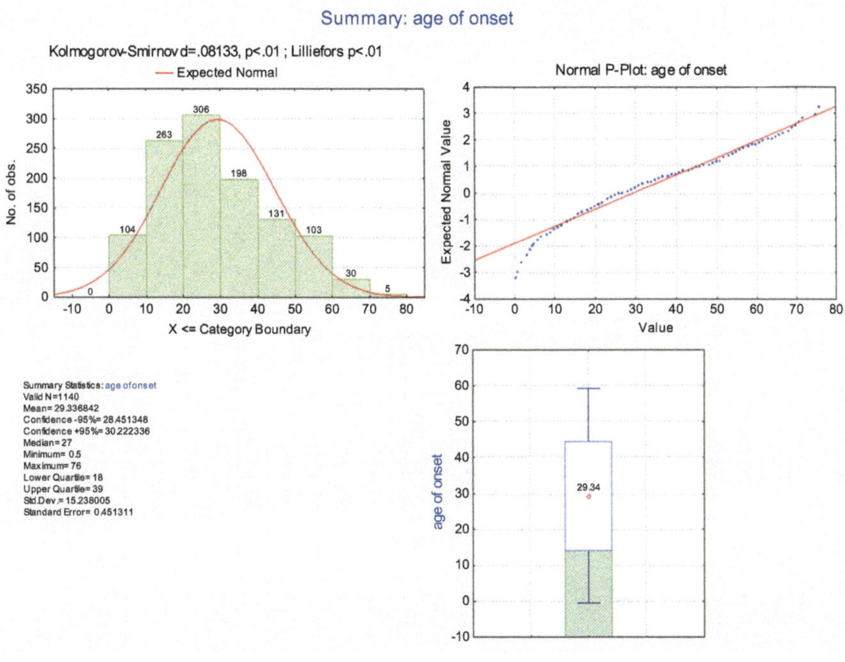

Fig. 3 Statistical analysis of age at onset

of students, whereas 24.16 % of retired and 24 % of unemployed persons had severe psoriasis. Moderate psoriasis was seen in 43.62 % of retired and 42.4 % of unemployed persons. The highest proportion of severe disease was diagnosed in unemployed persons ($\chi^2 = 66.67$, $p < 0.01$, 95 % CI) (Fig. 5).

Table 4 Severity of psoriasis according to various patient variables

	Mild		Moderate		Severe		Total	
Male	266	39.76 %	270	40.36 %	133	19.88 %	669	54.12 %
Female	270	47.62 %	230	40.56 %	67	11.82 %	567	46.09 %
Student	121	66.48 %	56	30.77 %	5	2.75 %	182	0.97 %
Employee	313	41.62 %	316	42.02 %	123	16.36 %	752	60.84 %
Retired	48	32.21 %	65	43.62 %	36	24.16 %	149	12.05 %
Receiving social assistance	12	42.86 %	10	35.71 %	6	21.43 %	28	2.26 %
Unemployed	42	33.60 %	53	42.40 %	30	24.00 %	200	16.20 %
Middle school	19	35.19 %	23	42.59 %	12	22.22 %	54	4.37 %
College	46	31.08 %	72	48.65 %	30	20.27 %	148	11.97 %
Vocational school	13	27.08 %	23	47.92 %	12	25.00 %	48	3.88 %
High school	124	35.94 %	158	45.80 %	63	18.26 %	345	27.91 %
Tertiary school	20	42.55 %	15	31.91 %	12	25.53 %	47	3.80 %
University graduate	193	46.84 %	153	37.14 %	66	16.02 %	412	33.33 %
Family history unavailable	361	41.45 %	368	42.25 %	142	16.30 %	871	70.47 %
First-degree relative affected	95	47.50 %	79	39.50 %	26	13.00 %	200	16.18 %
Second-degree relative affected	60	52.17 %	29	25.22 %	26	22.61 %	115	9.30 %
Third-degree relative affected	14	38.89 %	19	52.78 %	3	8.33 %	36	2.91 %
Fourth-degree relative affected	6	42.86 %	5	35.71 %	3	21.43 %	14	1.13 %
Urban	469	45.27 %	406	39.19 %	161	15.54 %	1036	83.82 %
Rural	67	33.50 %	94	47.00 %	39	19.50 %	200	16.18 %
Smoker	122	35.99 %	157	46.31 %	60	17.70 %	339	24.43 %
Non-smoker	414	46.15 %	343	38.24 %	140	15.61 %	897	75.57 %
Alcohol consumed	140	34.15 %	183	44.63 %	87	21.22 %	410	33.17 %
Alcohol not consumed	396	47.94 %	317	38.38 %	113	13.68 %	826	66.83 %
Single location at onset	492	43.66 %	456	40.46 %	179	15.88 %	1127	91.2 %

Multiple locations at onset	44	40.37 %	44	40.37 %	21	19,27 %	109	9.8 %
Comorbidities absent	353	48.22 %	278	37.98 %	101	13.80 %	732	59.23 %
Comorbidities present	183	36.31 %	222	44.05 %	99	19.64 %	504	40.77 %

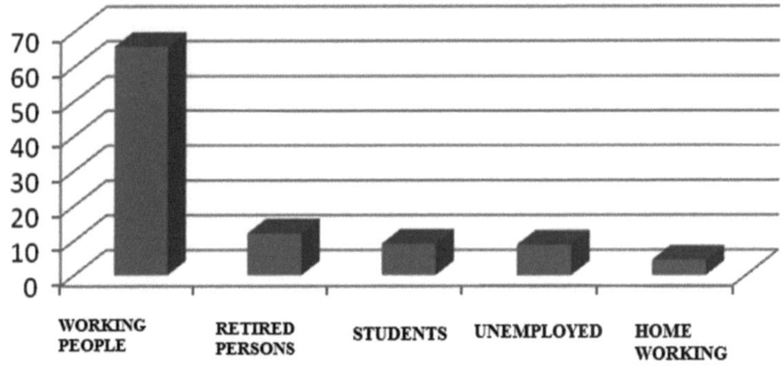

Fig. 4 Work status at time of admission to study

3.4 Correlations Between Severity of Psoriasis and Family History

Genetic factors (genetic predisposition) are reportedly responsible for 54 % (30–73 %) of susceptibility to psoriasis, common environmental triggers for 12 % (0–29 %) and individual environmental factors for 34 % (27–43 %) (Mahler et al. 2014; Lønnberg et al. 2013). A Romanian study reported a positive family history of psoriasis is 29.5 % of subjects, 16.2 % of first-degree relatives, 9.3 % of second degree, 2.9 % of third degree and 1.1 % of fourth degree. In that study, a family history of psoriasis was significantly associated with greater disease severity ($r = -0.448$, $\chi^2 = 18.32$, $p = 0.018$, 95 % CI) (Chiriac et al. 2014; Fig. 6).

3.5 Correlations Between Psoriasis and Occupation

The present data confirm a high prevalence of psoriasis in working people, especially those with stressful, anxiety-provoking occupations such as engineers, students, professors, managers, drivers, salespersons and medical staff. Retired persons were frequently diagnosed with psoriasis after a long evolution of the disease (psoriasis march), possibly because in their retirement they could afford the time and money to seek medical help; alternatively, chronic infections, comorbidities or drug administration may have triggered psoriasis flares.

3.6 Correlations Between Alcohol Consumption, Smoking and Psoriasis

Unlike alcohol, smoking was associated with psoriasis severity for all individuals with psoriasis that was unrelated to work activity (Table 5).

Table 5 Occupations of persons with psoriasis

Job	Nr	%	Job	Nr	%	Job	Nr	%
Retired	149	12.06	Police officer	5	0.40	Camera person	1	0.08
Primary and secondary students	112	9.06	Nurse	4	0.32	Railway worker	1	0.08
Laborer	108	8.74	Social worker	3	0.24	Legal adviser	1	0.08
Engineer	87	7.04	Attorney	3	0.24	Construction worker	1	0.08
Student	70	5.66	Librarian	3	0.24	Builder	1	0.08
Manager	56	4.53	Chef/cook	3	0.24	Cosmetician	1	0.08
Homemaker	54	4.37	Registrar	3	0.24	Typist	1	0.08
Professor	53	4.28	Inspector	3	0.24	Butcher	1	0.08
Driver	46	3.72	Schoolteacher	3	0.24	Designer	1	0.08
Unemployed	71	8.77	Business person	3	0.24	Dispatcher	1	0.08
Salesman	39	3.16	Post person	3	0.24	Physiotherapist	1	0.08
Economist	38	3.07	Foreman	3	0.24	Sketcher	1	0.08
Health care	32	2.59	Welder	3	0.24	Plumber	1	0.08
Accountant	22	1.78	Adviser/ consultant	3	0.24	Planner	1	0.08
Clerk	22	1.78	Veterinary assistant	2	0.16	Warrant officer	1	0.08
Technician	21	1.70	Bartender	2	0.16	Storekeeper	1	0.08
Physician	17	1.38	Trades person	2	0.16	Nun	1	0.08
Mechanic	17	1.38	Security agent	2	0.16	Meteorological	1	0.08
Electrician	13	1.05	Pharmacist	2	0.16	Miner	1	0.08
Farmer	10	0.81	Physicist	2	0.16	Curator	1	0.08
Secretary	9	0.73	Barber	2	0.16	Forester	1	0.08
Administrator	8	0.65	Judge	2	0.16	Guardian	1	0.08
Cleric	8	0.65	Machinist	2	0.16	Fireman	1	0.08
Computer	8	0.65	Supervisor	2	0.16	Driving instructor	1	0.08
Lawyer	7	0.57	Waiter	2	0.16	Psychologist	1	0.08
Cashier	7	0.57	Carpenter	2	0.16	Inspector	1	0.08
Sales agent	7	0.57	Forester	2	0.08	Reporter	1	0.08
Chemist	6	0.49	Insurance agent	1	0.08	Sociologist	1	0.08
Officer	6	0.49	Analyst	1	0.08	Trainer	1	0.08
Clerk	6	0.49	Town clerk	1	0.08			
Child care worker	5	0.40	Pharmacy assistant	1	0.08			
Analyst	5	0.40	Biologist	1	0.08			
Soldier	5	0.40	Furrier	1	0.08			
Operator	5	0.40	Baker	1	0.08			

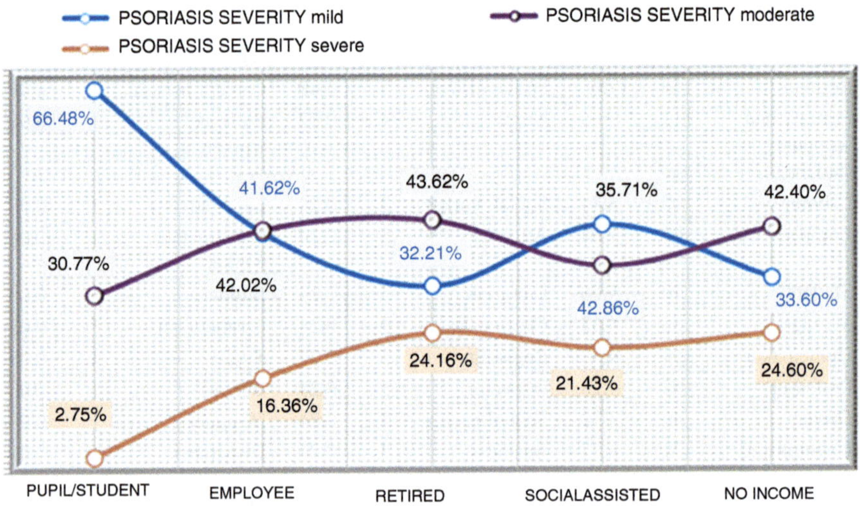

Fig. 5 Correlations between severity and occupational category

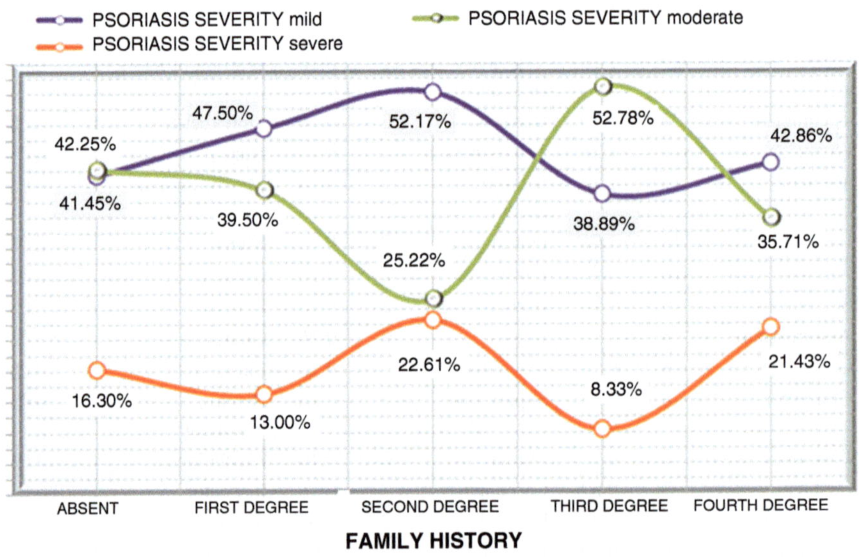

Fig. 6 Correlations between severity of psoriasis and family history

3.7 Types of Psoriasis

Wearing a special helmet in the construction sector or industrial activity may be a mechanical trigger factor for scalp psoriasis. Scalp psoriasis was diagnosed in more than half the subjects of the present study (681/1236, 55.10 %); however, we did not assess a possible association with helmet-wearing. The Koebner phenomenon (isomorphic response) was observed in 173 patients (14 %) (Fig. 8; Tables 6 and 8).

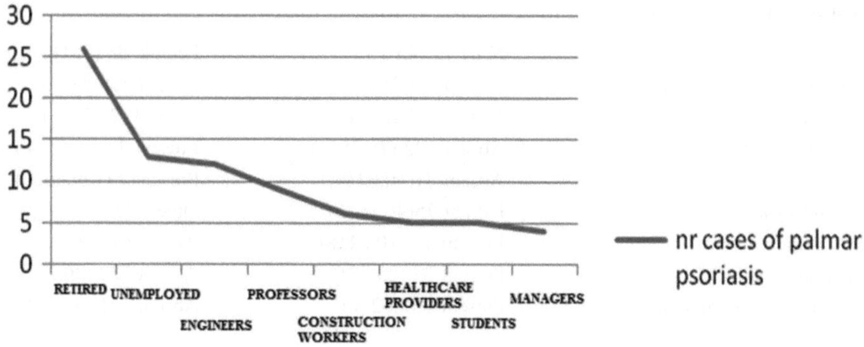

Fig. 7 Occupational categories of subjects with palmar psoriasis

Table 6 Type of psoriasis

	N	%
Nail psoriasis	165	13.35
Psoriatic arthritis	309	25.00
Koebner phenomenon	173	14.00
Scalp psoriasis	681	55.10
Guttate psoriasis	146	11.81
Upper limbs	788	63.75
Lower limbs	736	59.55
Trunk	462	37.38
Face	55	4.45
Palmoplantar/palmar	205	16.59
Other	265	21.44
Total	**1236**	

In the present study, palmoplantar psoriasis was recognized in 205 patients (16.59 %) (Table 7). It was frequently seen in women aged over 20 years and had no proven association with exposure to mechanical and/or irritant factors. Palmar psoriasis was linked to occupational activities in 159/1236 individuals (12.86 %), with a female predominance of 1.2/1. Alcohol use and smoking were denied by 79.75 % of subjects with palmar psoriasis. There was no family history of psoriasis in 76.25 % of them. Comorbidities were present at the time of clinical examination in almost half the cases. Psoriatic arthritis, nail involvement and the Koebner phenomenon were not diagnosed in the majority of cases (Table 7).

Work status (at the time of examination) of patients with palmar psoriasis was as follows: 26 retired persons (16.25 %), 13 unemployed persons (8.12 %), 109 low skilled workers (68.12 %), 12 engineers (7.5 %), nine professors (5.62 %), six construction workers (3.75 %), five health-care providers (3.12 %), five primary or secondary school students (3.12 %) and five tertiary students (4.37 %) (Fig. 7).

Table 7 Palmar psoriasis

Sex	Male 72 (45 %)	Female 88 (55 %)
Alcohol consumption	Yes 34 (21.25 %)	No 126 (79.75 %)
Smoker	Yes 37 (23.12 %)	No 123 (76.78 %)
Family history of psoriasis	Absent 122 (76.25 %)	Present 38 (24.75 %)
Comorbidities	Absent 81(50.62 %)	Present 79 (49.38 %)
Onset location	Palmar 75(46.86 %)	Other (53.14 %)
Psoriatic arthritis	Present 17 (10.62 %)	Absent (88.38 %)
Psoriatic nails	Present 38 (23.75 %)	Absent 122 (76.25 %)
Koebner phenomenon	Present 9 (5.62 %)	Absent 151 (94.38 %)

4 Discussion

> An occupational disease may be defined simply as one that is caused, or made worse, by exposure at work (Cherry 1999).

Epidemiological data concerning the prevalence and incidence of work-related skin diseases vary considerably; mostly, they represent only the tip of the iceberg. This variability is partly because every country has its own legislation and registration system for work-related skin diseases, list of recognized occupational skin disorders, and legal procedures for obtaining compensation for such disorders Fig. 8; Table 8. Additionally, physicians are key determinants of what data is registered; occupational physicians are legally obliged to report the work-related skin diseases that they diagnose.

The following factors can interfere with monitoring individuals in risk occupations: (1) variations in financial compensation; (2) the risk of becoming unemployed if a work-related skin disease is reported; (3) cumbersome, lengthy reporting procedures and administrative processes; (4) lack of anonymity of employer and employee if work-related skin disorders are reported; and (5) failure to legally recognize certain skin diseases as work-related.

Because of differences in distribution of occupations in different countries, statistical data cannot accurately be compared. Most countries maintain registers of notified work-related skin disease; however, under-diagnosis and under-reporting of these conditions for reasons listed above can result in a falsely low incidence.

Although psoriasis is a constitutional disease that may be triggered by occupational factors, it is not considered a work-related skin disease. A causal relationship between daily activities and the development of psoriasis is contentious. Onset of the disease is attributable to a combination of genetic and environmental factors.

Occupational activities that involve intrinsically irritating and mechanical factors could explain, at least partially, palmar type of psoriasis in professors, engineers, construction workers, health-care providers, and managers. The Koebner phenomenon can be regarded as triggered by occupational factors (Cherry 1999).

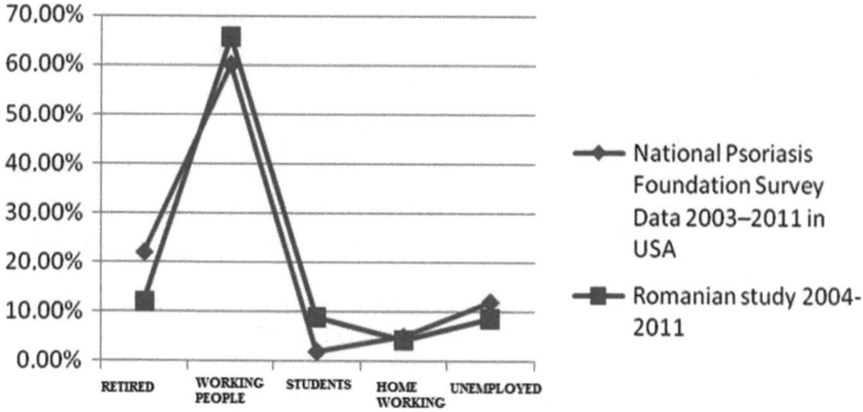

Fig. 8 Comparison between findings of USA National Psoriasis Foundation Survey and Romanian study regarding occupational status

Table 8 Comparison between findings of USA National Psoriasis Foundation Survey and Romanian study regarding occupational status

	USA National Psoriasis Foundation Survey Data 2003–2011	Romanian study 2004–2011
Retired	22 %	12.06 %
Working	48 % full-time + 11 % part-time = 59 %	65.74 %
Student	2 %	9.06 %
Homemaker	5 %	4.37 %
Unemployed	12 %	8.77 %

Contact with chalk, cement, cleaning chemicals, paper or pens are likely the most important trigger factors. Occupational trigger factors can be classified as follows: (1) chemical agents, both direct irritants (which induce chemical and irritative skin reactions) and sensitizers to allergic reactions; (2) physical agents, including extreme temperatures (hot and cold) and radiation; and (3) mechanical agents, including trauma (friction, pressure, abrasions, lacerations, and contusions).

An occupational history, medical examination and documentation are of paramount importance. The following aspects are crucial: medical history of the disease, age of onset and first location of skin lesions, the presence or absence of psoriasis at the time of beginning an occupational activity, family history, environmental factors, and correlation between timing and location of skin lesions and work. Adequate skin protection measures are necessary; namely, wearing gloves and avoiding irritant and mechanical factors in the work-place, especially in subjects with palmar psoriasis.

Physical activity, alcohol consumption, smoking, work-place pollution, repeated trauma, and irritants are all linked to psoriasis in workers (Horn et al. 2007).

The following are highlights of the findings of our study:

- The highest incidence of psoriasis was in the 30–50 year age group (43.12 %) and the lowest in patients aged over 70 years (5.83 %) and under 20 years (5.5 %)
- The median age at the time of clinical examination was 29.34 ± 15.24 (SD) years; the median age of onset was higher for patients with more severe psoriasis ($F = 11.69$, $p = 0.000009$, 95 % CI)
- Working persons (65.74 %) were the largest occupational category, whereas retired persons comprised 12.06 % and unemployed persons 8.77 %
- Moderate forms were seen in retired persons (43.62 %) and severe forms in unemployed persons (42.4 %) ($\chi 2 = 66.67$, $p << 0.01$, 95 % CI);
- Palmar psoriasis was linked to occupational activities in 159/1236 individuals (12.86 %), with a female predominance of 1.2/1.

There are no officially registered data regarding either the prevalence of psoriasis or its relationship with work in Romania, limiting the ability to assess risk, recognize the condition early and prevent it. Individuals with risk factors at work, such as those with palmar psoriasis, would benefit the most from early detection, which would improve their prognoses.

Dermatologists, occupational health providers, and workers know less about skin damaging factors and occupational risks than they do about other work-related disorders.

There are no means of quantifying work-related skin damage or risk in vulnerable subjects, such as those with psoriasis. Screening of individuals in risky occupations can be achieved by clinical examination, questionnaires, and regular medical monitoring of employees.

Approved registers that record data about occupational skin diseases are necessary for epidemiological studies and financial compensation.

References

Armstrong AW, Schupp C, Wu J et al (2012) Quality of life and work productivity impairment among psoriasis patients: findings from the National Psoriasis Foundation Survey Data 2003–2011. PLoS One 7(12):e52935. doi:10.1371/journal.pone.0052935

Asokan N, Prathap P, Rejani P (2014) Severity of psoriasis among adult males is associated with smoking, not with alcohol use. Indian J Dermatol 59(3):237–240

Balato N, Di Costanzo L, Patruno C et al (2013) Effect of weather and environmental factors on the clinical course of psoriasis. Occup Environ Med 70:600. doi:10.1136/oemed-2013-101505

Cherry N (1999) Occupational disease. BMJ 318(7195):1397–1399

Chiriac A, Solovan C, Chiriac AE et al (2014) A case-control study and analyze the epidemiological importance risk of family history of psoriasis. Our Dermatol Online 5(1):90–91

Chung J, Callis Duffin K, Takeshita J et al (2014) Palmoplantar psoriasis is associated with greater impairment of health-related quality of life compared with moderate to severe plaque psoriasis. J Am Acad Dermatol 71(4):623–632

Devrimci-Ozguven H, Kundakci TN, Kumbasar H et al (2000) The depression, anxiety, life satisfaction and affective expression levels in psoriasis patients. J Eur Acad Dermatol Venereol 14(4):267–271
Eder L, Law T, Chandran V et al (2011) Association between environmental factors and onset of psoriatic arthritis in patients with psoriasis. Arthritis Care Res 63:1091–1097
Feldman SR, Krueger GG (2005) Psoriasis assessment tools in clinical trials. Ann Rheum Dis 64: ii65–ii68
Fisher AA (1979) Occupational palmar psoriasis due to safety prescription container. Contact Dermatitis 5(1):56
Fortes C, Mastroeni S, Leffondré K et al (2005) Relationship between smoking and the clinical severity of psoriasis. Arch Dermatol 141(12):1580–1584
Gerdes S, Zahl VA, Weichenthal M et al (2010) Smoking and alcohol intake in severely affected patients with psoriasis in Germany. Dermatology 220(1):38–43
Gupta MA, Gupta AK, Watteel GN (1998) Perceived deprivation of social touch in psoriasis is associated with greater psychologic morbidity: an index of the stigma experience in dermatologic disorders. Cutis 61(6):339–342
Hawro T, Zalewska A, Hawro M et al (2014) Impact of psoriasis severity on family income and quality of life. J Eur Acad Dermatol Venereol. doi:10.1111/jdv.12572
Hill VA, Ostlere LS (1998) Psoriasis of the hands köbnerizing in contact dermatitis. Contact Dermatitis 39:194
Horn EJ, Fox KM, Patel V (2007) Association of patient-reported psoriasis severity with income and employment. J Am Acad Dermatol 57:963–971
Hunter HJ, Griffiths CE, Kleyn CE (2013) Does psychosocial stress play a role in the exacerbation of psoriasis? Br J Dermatol 169(5):965–974
Jawahar S (2011) Prevalence of psoriasis linked to pesticide in farmers. Int J Pharm World Res 2(3):1
Kalayciyan A, Aydemir EH, Kotogyan A (2007) Experimental Koebner phenomenon in patients with psoriasis. Dermatology 215:114–117
Kálmán LJ, Gonda X, Kemény L et al (2014) Psychological and biological background of the interaction between psoriasis and stress. Orv Hetil 155(24):939–948
Kanerva L, Estlander T (2001) Occupational post-traumatic psoriasis. Contact Dermatitis 44: 317–318
Kanerva L, Talvi A, Estlander T (1998) Occupational contact psoriasis. Eur J Dermatol 8:217–218
Li WQ, Qureshi AA, Schernhammer ES et al (2013) Rotating night-shift work and risk of psoriasis in US women. J Invest Dermatol 133(2):565–567
Lønnberg AS, Skov L, Skytthe A et al (2013) Heritability of psoriasis in a large twin sample. Br J Dermatol 169:412–416
Mahler V, Diepgen T, Skudlik C et al (2014) Psoriasis predisposition and occupational triggering factors in the appraisal of occupational medical expertises. J Dtsch Dermatol Ges 12(6): 519–529
Menne T, Hjorth N (1984) Mechanically provoked contact dermatitis and psoriasis. Z Hautkr 59: 647–653
Monk BE, Neill SM (1986) Alcohol consumption and psoriasis. Dermatologica 173(2):57–60
Moroni P, Cazzaniga R, Pierini F et al (1998) Occupational contact psoriasis. Dermatosen 36:163–164
Onder M, Cosar B, Oztas MO et al (2000) Stress and skin diseases in musicians: evaluation of the Beck depression scale, general psychologic profile (the brief symptom inventory [BSI]), Beck anxiety scale and stressful life events in musicians. Biomed Pharmacother 54(5):258–262
Pearce DJ, Singh S, Balkrishnan R et al (2006) The negative impact of psoriasis on the workplace. J Dermatolog Treat 17:24–28

Sampogna F, Tabolli S, Abeni D, IDI Multipurpose Psoriasis Research on Vital Experiences (IMPROVE) investigators (2007) The impact of changes in clinical severity on psychiatric morbidity in patients with psoriasis: a follow-up study. Br J Dermatol 157(3):508–513
Schmid-Ott G, Jaeger B, Kuensebeck HW et al (1996) Dimensions of stigmatization in patients with psoriasis in a "Questionnaire on Experience with Skin Complaints". Dermatology 193: 304–310
Skudlik C, Wulfhorst B, Gediga G et al (2008) Tertiary individual prevention of occupational skin diseases-a decade's experience with recalcitrant occupational dermatitis. Int Arch Occup Environ Health 81(8):1045–1058
Solovan C, Marcu M, Chiticariu E (2014) Life satisfaction and beliefs about self and the world in patients with psoriasis: a brief assessment. Eur J Dermatol 24(2):242–247
Verhoeven EW, Kraaimaat FW, Jong EM et al (2009) Effect of daily stressors on psoriasis: a prospective study. J Invest Dermatol 129(8):2075–2077
Wolk K, Mallbris L, Larsson P et al (2009) Excessive body weight and smoking associates with a high risk of onset of plaque psoriasis. Acta Derm Venereol 89:492–497

Assessment of Age-Correlated Occupational Strain as a Prerequisite for Age-Appropriate Work Organization

Friedrich Müller and Nadine Kakarot

Abstract

Because of demographic changes in Germany, the average age of the workforce is progressively increasing. In an attempt to reduce pressure on the labor market, Germany recently changed the retirement age to 67 years. At the same time, economic constraints are forcing companies to reduce the number of employees. Thus, older workers are required to do physically demanding work involving heavy weights. In response to these factors, the Hamburg Sewage Company recently set a goal of creating working conditions that would enable their existing aging personnel to fulfill their duties efficiently and to a high quality while ensuring that their employees' health would not be at a risk. As a prerequisite to achieving this goal, quantitative data on work load, physical and mental effort required to meet job demands, and age-related changes in physical and mental strain were required. To this end, a combination of complementary measures aimed at achieving an in-depth analysis that would enable appropriate adaptations of working conditions for the aging workforce have been implemented.

1 The Problem

The Hamburg Sewage Company (Hamburger Stadtentwässerung; HSE) is one of the leading sewage corporations in Europe. Although employing the most modern working techniques, the maintenance of the 5400-km sewage system still requires extensive heavy labor, which is carried out by about 330 workers.

F. Müller (✉) • N. Kakarot
Institute of Experimental Industrial Psychology, LüneLab, Leuphana University Lüneburg, Lüneburg, Germany
e-mail: f.mueller@leuphana.de

M. Wiencke et al. (eds.), *Healthy at Work*, DOI 10.1007/978-3-319-32331-2_25

Among the physically demanding tasks are the yearly inspection of more than 100,000 manholes, clearing of gullies, and repair and rebuilding of manholes. Workers in charge of manhole inspections open and close more than 100 manholes a day, which requires repeated lifting of 93-kg manhole covers. Heavy, oversized vacuum cleaners are manipulated by hand to remove greenery, mud and other impurities from gullies. Rebuilding manholes involves extremely heavy work in a confined space and is accompanied by dust, loud noises and often nauseating odors. These tasks are usually performed by teams of two laborers, one of whom safeguards the workplace and does the administrative work while their partner performs the heavy work. In earlier times, teams consisted of a younger and an older worker, with the younger colleague doing most of the heavier work. Most workers were allocated less demanding tasks at the age of about 58–60 years.

Because of dramatic demographic changes in Germany, where the birth rate is <1.4 per family, and constraints forcing reductions in the number of employees, the average age of the workforce of the HSE has increased to 52 years. The share of employees aged between 50 and 65 years increased during the last decade from 32 to 51 %. In the future, the HSE will be compelled to further reduce the number of employees. Additionally, the German pension scheme now mandates that present workers must work up to the age of 67 years; new, younger employees cannot be taken on. As a consequence, the average age of the employees will increase even more and these progressively older workers will be forced to continue performing physically heavy and demanding jobs.

With these factors in mind, the HSE set a goal of creating working conditions that would enable existing employees to fulfill their duties in this leading sewage company economically and efficiently to a high standard, while simultaneously ensuring that the employees' health is not at risk and they will be in good health when they do retire.

Prerequisites for achieving this goal included acquiring quantitative data on age-related changes in the work load associated with tasks, the physical and mental effort required to meet job demands, and the physical and mental strain and stress caused by the tasks together with reliable data on the employees' constitutions, mental and physical fitness, and state of health. We used a combination of complementary measures to obtain hard data on the work-related burden to serve as a basis for an in-depth analysis and adaptation of working conditions to the requirements of an aging workforce.

The conceptual basis for our studies is the stress-strain model developed by Rohmert (1984), which went into EN ISO 10075, defining ergonomic principles related to mental work-load. This defines mental stress as "the total of all assessable influences impinging upon a human being from external sources and affecting it mentally." In contrast, mental strain is defined as "the immediate effect of mental stress within the individual, (not the long term effect) depending on his/her individual habitual and actual preconditions, including individual coping styles"; the term "mental" refers to cognitive, informational and emotional processes in human beings. To what extent stress, which is usually defined in physical terms and measures, is perceived as strain depends on individual characteristics such as abilities, skills, knowledge, level of aspiration, attitudes, coping styles and, regarding older people in particular, age, experience, general condition, physical constitution, confidence in own capabilities and health.

2 Measures Describing Different Aspects of Strain

Because strain is related to the individual experiences of each person, primarily valid and reliable phenometric measurement techniques are required, to which might be added psychophysiological measures such as heart rate and oxygen consumption. We selected the phenometric measure known as the category partitioning procedure (Heller 1982; Müller 1996), which has proven highly successful at quantifying perceived physical exertion (Müller et al. 1995; Kakarot et al. 2012; Kakarot and Müller 2014) and psychological tension (Heller 1981). According to a proposal by Bartenwerfer (1969), psychological tension can be used as a phenomenological measure of mental strain.

Additionally, physiological indicators that reflect work load such as heart rate and electromyographic measures were recorded, posture was assessed and a standardized psychological job analysis was performed. The measures on load and strain were complemented by a detailed assessment of each worker's health status.

2.1 Assessment of Physical Exertion and Mental Strain Associated with the Tasks Required to Maintain a Sewage System

Sixty-three sewer workmen and one sewer workwoman were assigned to eight discussion groups of eight workers each, such that either eight who were very experienced or eight who were less experienced at performing one of the four assessed working activities comprised a group. Within each of the approximately 6-h discussions, all steps required to perform the assessed activities were isolated and time ordered, after which category partitioning scores for physical exertion (PE) and mental workload, experienced as psychological tension (PT) were assigned for each of the individual work steps. Provided the related measurement regulations are followed, the category partitioning technique provides absolute quantitative scales for perceived strain that are minimally affected by the scaling context.

The scaling of exertion and mental strain provides a precise picture of the perceived burden that accompanies each of the various work steps. As an example, Fig. 1 shows how the various work steps involved in manhole inspection were assessed by the workers.

Of note, compared with younger workers, older workers reported experiencing more physical exertion for work steps like lifting covers off manholes, clearing gullies, and lifting and carrying grit collectors; however, they reported more

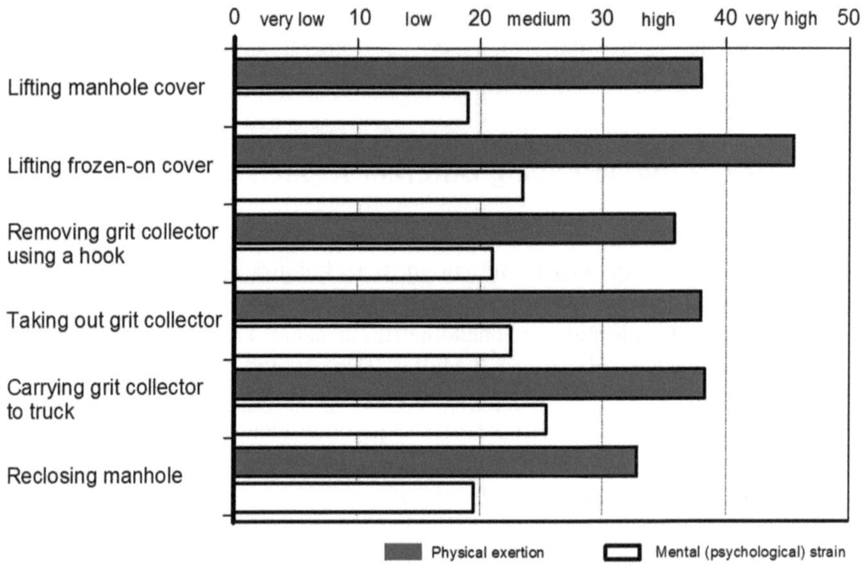

Fig. 1 Exertion and strain caused by various work steps involved in inspecting manholes. Note the remarkably high physical exertion when manhole covers are frozen. *Grey*, physical exertion; *white*, mental strain

composure (less mental effort) when performing administrative and intellectual activities like planning and preparing reports. Interindividual variation in perceived exertion, effort and strain increased with increasing age of workers. Additionally, older workers had wider variations in physical fitness and resilience. Conversely, younger workers occasionally reported greater physical exertion than their older colleagues for tasks that required expert knowledge and experience (e.g., reconstruction work).

2.2 Alternative Means of Assessment of Work-Related Strain

The association between age and perceived exertion for manhole inspection and gully clearing was assessed by another method: the workers were asked to compare their work with the duties of a caretaker at a small elementary school, a job familiar to all the participants. The instruction was: “Imagine the duties of a caretaker in an elementary school. How much strain do you experience in sewage duct inspection and gully clearing compared with a caretaker’s job? Please mark the point on the given line that corresponds to the relationship between your job and that of a caretaker.” (Fig. 2).

It was found that both tasks are highly demanding and clearly age dependent. The older the worker, the more demanding the job (Fig. 3).

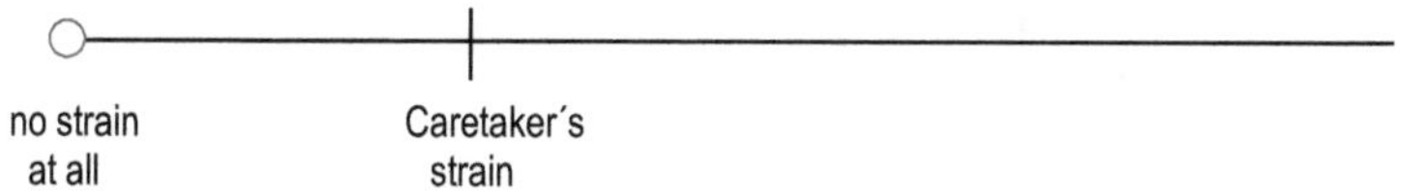

Fig. 2 Comparison of experienced strain with that of an imaginary school caretaker. Subjects were asked to indicate their experienced strain by marking the *line*

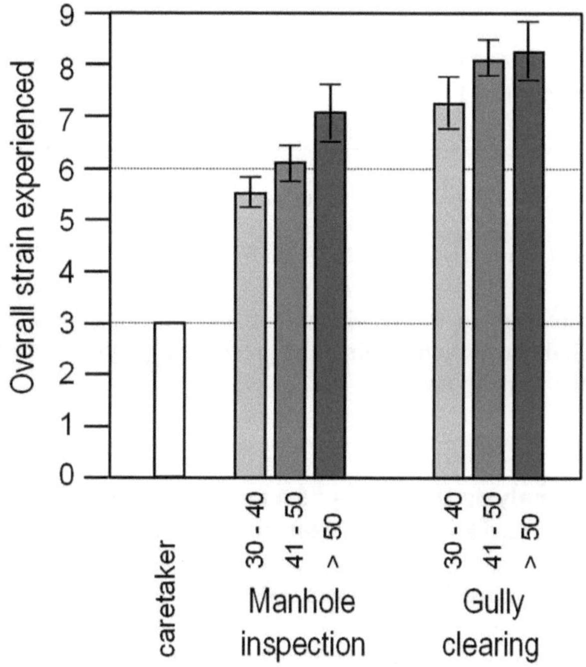

Fig. 3 Average line length as marked by the workers, indicating the overall strain caused by manhole inspection and gully clearing

2.3 Analysis of Posture

In a further step, the workers were videoed from different angles and with varying detail while performing their routine tasks. A subset of participants was equipped with electrodes and a storage unit for continuous monitoring of heart rate and tension in the erector spinae using a VarioPort system (Becker Meditec, Karlsruhe)

The videotaped work steps were used to analyze posture according to the Ovako Working Posture Analyzing System (Louhevaara and Suurnäkki 1992). The total time required for each of 14 postures (e.g., back straight, bending, turning, turning and bending simultaneously, one or both arms below or above shoulder; legs bending, standing, walking and kneeling) and four categories of weights carried (<10 kg, 10–20 kg, 20–40 kg and >40 kg) was measured to determine the relative

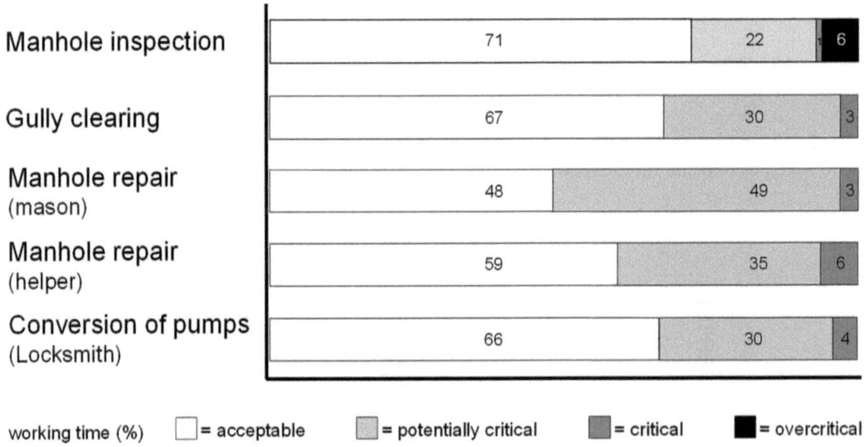

Fig. 4 Results of posture analysis using the Ovako Working Posture Analyzing System

amount of time spent in acceptable, potentially critical, critical or overcritical postures. Manhole inspection required the greatest proportion of critical postures; about 6 % of the working time (Fig. 4).

2.4 Task Analysis Using the "Screening Instrument for Assessment and Design of Humane Working Conditions"

The screening instrument for assessment and design of humane working conditions (SIGMA) task analyzing system (Windel 1998; Salewski-Renner 1998) considers demands of the task, work organization, work environment and specific burdens such as requirement for vigilance, shift work or time pressure. The findings are summarized in graph form in Fig. 5.

2.5 Assessment of Workers' Health Status

The health status of every worker and hundreds of control subjects were ascertained by careful history-taking and recorded. Each individual's reported symptoms were then related to that worker's age, predominantly performed job and job tenure. Significant interrelations were found between job, job tenure, age and incidence of musculoskeletal complaints. A significant increase in the incidence of, and treatment for, symptomatic musculoskeletal disorders was observed with increasing age and years of employment, particularly in those whose tasks had been gully clearing and inspection. This association was not found for those who had predominantly performed gully repairs; presumably, this task does not put constant strain on the musculoskeletal system.

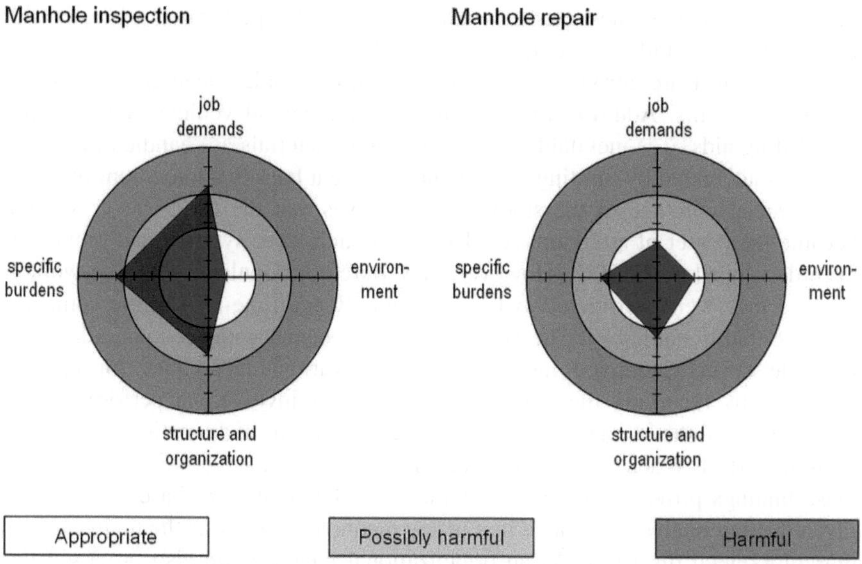

Fig. 5 Overall findings for manhole inspection according to SIGMA. The greater amount of shadowed *gray* in the outer area, the more critical the analyzed task

3 Recommendations

Once precise data had been acquired, measures to promote the safety and health of the employees were self-evident. The resultant extensive catalog of measures designed to reduce strain and improve the workers' health included the following: (1) technical and organizational measures for avoiding or reducing undue and one-sided strain; (2) measures for preserving and improving the performance and well-being of the employees; (3) practical tips and aids for fostering beneficial practices; and (4) closer analysis of particular activities and organizational procedures. Some of the proposed measures were already in place or were well known but had not yet been implemented. For the latter, the data provided indisputable arguments for their immediate implementation.

However, there were some unexpected findings that astonished management, researchers and employees.

3.1 Examples of Recommendations and Their Implementation

It was immediately apparent that various strains associated with the required procedures were too great. This was true of the following tasks: (1) opening manhole covers and gully grids, particularly when frozen; and (2) clearing out dried-up gullies or certain types of gullies with small openings. A decision was therefore made that these tasks would no longer be performed when outside temperatures had fallen close to freezing point.

Another measure introduced was the retrofitting of older cleaning devices with watering systems. Additionally, although all commercial vehicles are equipped with lifting aids, it is inevitable that some heavy materials are handled manually. This was addressed by limiting the weight of cement bags to a maximum of 25 kg.

A crucial outcome of the study's findings was that the use of lever systems became the center of attention. A selection of such lever systems had previously been developed at the HSE. However, they were not popular among the workers because they were considered awkward, unmanly and time-consuming compared with the traditional hooks. The critical strain associated with the manipulation of manhole covers prompted the assessment of various techniques for opening manholes by recording the tension of the muscles involved in performing this task. Figure 6 shows that a gentle approach results in a dramatic reduction of muscular effort. When combined with statistical data on the workers' health status, these findings provided convincing evidence in favor of lever-based techniques, given that our analysis indicated that lifting manhole covers was the main cause of musculoskeletal disorders. As an organizational measure, we also proposed that performance bonuses be paid only to workers who used lever systems (Fig. 6).

Another efficient measure implemented was job rotation to prevent skewed loads. This measure had already been the subject of a works agreement; however, because shift schedules are the responsibility of the working teams, some employees tended to bypass the regulations already in place. Because it was

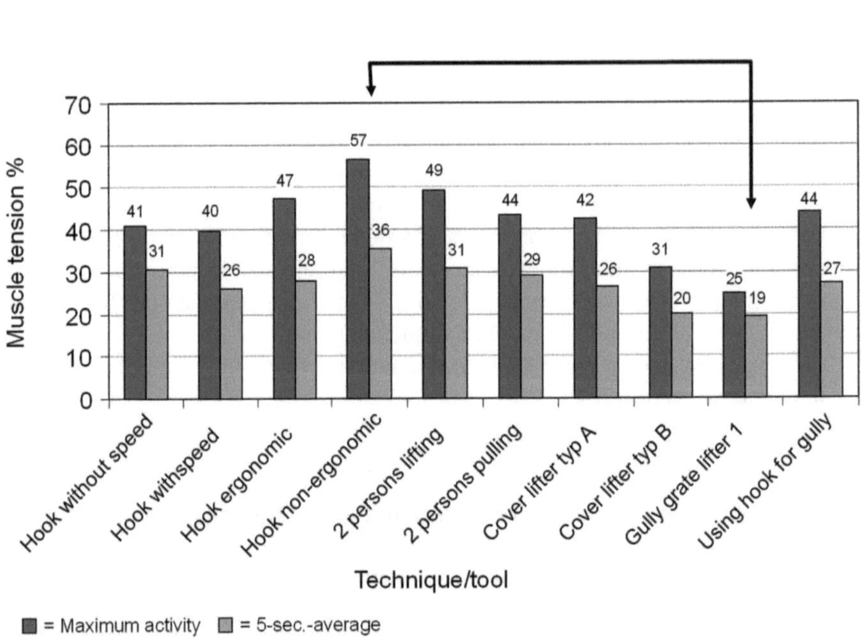

Fig. 6 Muscle activity while applying different techniques for open manhole covers

ascertained that they bypassed them because they preferred to retain a familiar working partner, care should be taken to shift teams rather than individuals.

One finding points to the usefulness of information exchange between individuals and teams. For some of the operations related to manhole reconstruction, we found large interindividual differences in scores that could not be attributed to the ages of the participants. A detailed analysis of the work steps used by each worker showed that one of the teams had developed and used a much gentler technique for accomplishing their tasks. To better communicate such innovations, we proposed regular visits and work shadowing between teams.

Yet another finding points to the necessity for communication; regular feedback in particular. The burdens classified as harmful in manhole inspection (Fig. 5) were found to be caused by a lack of feedback. The decisions and recommendations resulting from inspection by experienced workers are naturally accompanied by uncertainty: whether the right decision has been made will only become evident during the repair. A feedback loop between manhole inspectors and repair teams would significantly reduce uncertainty-based mental strain.

The heavy special vehicles used for gully clearing and re-opening congested sewers can block narrow streets, which repeatedly leads to severe social strain caused by impatient drivers complaining angrily and at length, and in rare circumstances even physically attacking the sewer workers. Although the employees already undergo conflict resolution training, we further proposed placing a placard at the rear of the vehicles to inform waiting drivers what is occurring, but not visible to them, at the front of the blocking vehicle.

In addition to the above-mentioned and a number of other specific measures, the health reports and data obtained gave rise to some general findings that were not related to specific tasks. The data strongly suggested that overweight employees are less satisfied and employees who actively participate in sports are more satisfied with their state of health. Our recommendations therefore included nutritional advice, optimization of canteen food, and maintaining and further developing measures to promote sports and specific muscle training.

The data raised the possibility of whether older workers require different breaks than their younger colleagues, though this remained unclear. Because this and related questions cannot be answered by questionnaires or in field experiments, the HSE has funded elaborate laboratory experiments that are described elsewhere (Kakarot et al. 2012; Kakarot and Müller 2014).

3.2 An Unexpected Outcome

Because older workers are more experienced in identifying damage to manholes when they inspect them, and because this task was considered the least strenuous of the physical tasks required by the sewage company, it was previously considered a suitable task for older workers. At opening the manhole covers are rather slipped out than lifted. Therefore we expected the force applied being <35 kg. However, our measures showed that manhole inspection is the most demanding of all evaluated tasks. This was confirmed by Scientists of the Institut für Bewegungswissenschaft at Hamburg University thereupon. Sensors applied to the hooks used occasional registered forces near 100 kg. Fortunately, a detailed analysis of the collected data gave rise to various possible means of considerably reducing work-related strain. This demonstrates that a valid data set provides the perfect guide for operational decisions.

3.3 Supplement

Since the study described above was started, some major changes have taken place in the Hamburg Sewage Company. A significant number of employees reached retirement and some new workers had been taken in, thus reducing the average age of the workforce to 48.9 years. At the same time manhole inspection was completely re-organized thereby reducing the number of yearly inspections by about 75 %. The total number of inspections will not increase 70 per team and day. A follow up study on the mid- and long-term effects of the measures proposed is planned and will take place in the very near future.

Acknowledgement We like to thank management, works council and employees of the HSE for perfect cooperation, support, openness and inspiring insights into an fascinating organization.

References

Bartenwerfer H (1969) Einige praktische Konsequenzen aus der Aktivierungstheorie. Z Exp Angew Psychol 16(2):195–221

European Committee for Standardization (2000) Ergonomic principles related to mental workload. Part 1: general terms and definitions EN ISO 10075-1

Heller O (1981) Zur Quantifizierung psychischer Anspannung. In: Tent L (ed) Erkennen, Wollen, Handeln. Hogrefe, Göttingen, pp 213–223

Heller O (1982) Theorie und Praxis des Verfahrens der Kategorienunterteilung (KU). In: Heller O (ed) Forschungsbericht, vol 198. Psychologisches Institut Lehrstuhl für Allgemeine Psychologie, Würzburg, pp 1–15

Kakarot N, Müller F (2014) Assessment of physical strain in younger and older subjects using heart rate and scalings of perceived exertion. Ergonomics 57(7):1052–1067

Kakarot N, Mueller F et al (2012) Activity–rest schedules in physically demanding work and the variation of responses with age. Ergonomics 55(3):282–294

Louhevaara V, Suurnäkki T (1992) OWAS: a method for the evaluation of postural load during work. Institute of Occupational Health and Centre for Occupational Safety, Helsinki

Müller F (1996) Veridical scaling of the intensity of perceptions, feelings and emotions. In: Mital A et al (eds) Advances in occupational ergonomics and safety, vol 2. ISOES, Cincinnati, pp 793–804

Müller F, Neely G, Fichtl E (1995) Scaling of loudness, perceived exertion, and pain intensity: a comparison between the category partitioning (CP) and category ratio (CR-20) scaling procedures. In: Neely G (ed) Category-ratio scaling of sensory magnitude in comparison with other methods. Dissertation, Universität Stockholm

Rohmert W (1984) Das Belastungs-Beanspruchungs-Konzept. Zeitschrift für Arbeitswissenschaft 38(4):193–200

Salewski-Renner M (1998) Belastungsanalysen mit SIGMA: Zusammenfassende Ergebnisse und Konsequenzen für die Verwendbarkeit. In: von Benda H, Bratge D (eds) Psychologie der Arbeitssicherheit, vol 9. Workshop 1997. Asanger, Heidelberg, pp 301–305

Windel A (1998) Entwicklung und Aufbau des Screening-Instruments zur Bewertung und Gestaltung von menschengerechten Arbeitstätigkeiten (SIGMA). In: von Benda H, Bratge D (eds) Psychologie der Arbeitssicherheit, vol 9. Workshop 1997. Asanger, Heidelberg, pp 285–289

Work–Life Balance. A Challenge for Working Moms: Differences and Similarities Between Germany and the USA

Petia Genkova and Michaela Wieser

Abstract

Work–life balance is one of the challenges of a globalized world. The study described in this chapter aimed to identify the factors influencing the work–life balance of working mothers. Additionally, we will describe differences and similarities between the well-being and working styles of German and US-American working mothers. It is particularly difficult for mothers to be successful, because to do so, they must master the double burden of work and family. In this study, 320 working mothers were surveyed, 142 in Germany and 178 in the USA. It was found that the cultural concepts of work–life balance of these working mothers are comparable. Furthermore, differences in working styles were identified: US-Americans valued both well-being and work–life balance more strongly than Germans. In both cultures, younger mothers felt more burdened by work than older mothers. This may indicate that competences are learned with an increasing age, enabling a more successful mastery of challenges, even though an increasing age usually has a negative effect on health. The number of children, family status and education did not influence work–life balance. These findings are further discussed in this chapter.

P. Genkova (✉)
Hochschule Osnabrück, Fakultät für Wirtschafts- und Sozialwissenschaften, Caprivistr.30A, 49076 Osnabrück, Germany
e-mail: P.Genkova@hs-osnabrueck.de

M. Wieser
Universität Passau, BA Medien und Kommunikation, Sarasatestrasse 24, 81247 München, Germany
e-mail: michaela.wieser@gmx.de

M. Wiencke et al. (eds.), *Healthy at Work*, DOI 10.1007/978-3-319-32331-2_26

1 Introduction

Nowadays, to achieve a successful work–life balance is both extremely important and a challenge for our society and for individuals. It is becoming increasingly difficult to accomplish a balance between career and family, especially for working mothers. Because work–life balance is influenced by the social and cultural background, the question arises, whether and how work–life balance differs between countries. Unlike in the USA, the increasing employment of women in Germany has resulted in a very little redistribution of work and stress (Metz-Göckel 2004). According to Metz-Göckel (2004, p. 124), almost all empirical studies have confirmed only minimal changes in intra-family distribution of tasks despite the mothers' participation in the labor force. Thus, the behavior of fathers only changes slightly when their wives work full-time.

In both the USA and Germany, tension between work and family and the problems this causes for couples have led to unstable relationships. As a possible consequence, there are high divorce rates (in 2007, the divorce rate was 10.24 % in Germany and 11.5 % in the USA) and a growing proportion of single-parent and so-called patchwork families (European Commission 2007). Action is required to equilibrate the conflictual tension between work and family in such families.

In Germany, a typical solution for this problem has been for increasing numbers of women to work part-time. Between 2001 and 2006, the part-time rate among women in Germany increased from 39.9 to 45.8 % (European Commission 2007). In 2006, 87 % of all part-time workers were women (IAQ [Institut Arbeit und Qualifikation] 2008). Part-time work creates space for both, paid work and family and thus minimizes or avoids an imbalance in the main living areas. According to Metz-Göckel (2004), this solution to mothers' conflicts is not implemented in the same way in other European countries or in the USA.

According to Krumpholz (1998), years ago career opportunities and financial incentives were of paramount importance. However, nowadays, fun at work, deployment and decision possibilities, responsibility and contact with other people, for example through teamwork, are more important. Goals, needs, and interests of individual employees have diversified. There is a tendency to individualization that can result in an amorphous motivational structure. In addition, there is a tendency for women to prefer to achieve recognition and satisfaction at the work-place rather than in the family. Children and parenting are increasingly considered as work, rather than a rewarding and enriching aspect of life and thus no longer provide a sufficient compensation for a lack of confirmation and satisfaction. A reversal characterized by obtaining gratification not in the family, but at the work-place, has begun to occur. In this context, it is particularly striking that a strong recognition at the work-place can be achieved by applying for more overtime, resulting in even greater gratification, but exacerbates tension at home, because one consequence is that less time is spent with the family. Thus, employment retains positive connotations while the family is negatively affected; leading to a downward spiral in which work–life balance steadily deteriorates.

Thus, work is acquiring more positive connotations whereas family is being negatively affected, resulting in a downward spiral characterized by an intensification of work–life imbalance. Engagement in work or family is more and more analyzed in terms of the cost-benefit ratio: "Investments of time, energy, and money in work and family compete against each other. Thus, the following questions become relevant: to what extent do such investments reap benefits and for whom, who "does more", and who receives more acknowledgement and security" (Kastner 2004, p. 71).

A consequence of the increase in numbers of working women are so-called double earner couples, in which both partners "strive for a career in the narrower sense" (Wiese 2007, p. 247). These couples tend to leave their children alone more often or to have them cared by paid service providers. Thus, there is less shared time with the family and time regimes become increasingly strict. As a compensation, attempts are made to introduce so-called "quality time"; namely, shared time that is intense and of high quality. These attempts are often unsuccessful, resulting, for example, in children becoming alienated from their parents, avoiding them, or showing reluctance to conform with their inflexible time planning. The reduced time spent with the family results in an intensification of conflicts; and home as a comfortable, relaxed place that is preferable to the work-place is becoming a myth. In particular, women who work high in a company's hierarchy find spending time at the work-place more attractive than dealing with the difficult and exhausting problems of their family at home.

According to Metz-Göckel (2004, p. 108), there is "currently a creeping extension of working hours in the USA, Europe and Germany simultaneously accompanied by more or less pronounced unemployment". This extension of working hours affects both sexes. In particular, highly qualified employees are working more and more unpaid hours and the difference between the agreed and actual working time is increasing. In many contracts, the working time is now no longer fixed; rather, employees must complete their workloads and are not paid according to the hours worked. As a result, many employees work more hours per week than they probably would do with contractually agreed arrangements out of the fear to fulfill the requirements.

Especially in the USA, companies see themselves as "family" (Hochschild 2002). For this purpose, they provide a familial atmosphere. Employees are bound to the company by corporate outings, trainings, praise, and recognition. Organizations expect their employees to spend as much time at work as possible; consequently, the time for the family dwindles. The ability to make decisions and take personal responsibility serves as an incentive for co-determination, which enhances self-worth. Hoff et al. (2005) have noted that a unique and clear separation between the work and private roles is becoming more difficult to achieve. To create a balance, organizations implement work–life balance measures, which in turn reinforce the identification with the company. Thus, the time spent in the company increases, which paradoxically leads to an exacerbation of the work–life imbalance (Hochschild 2002).

The company Google is a good example for a company at which employees positively perceive the offering of such work–life balance measures (LFPress 2014). Other companies such as Facebook and Apple offer to assist their female employees to have their ova frozen, thus enabling them to delay to have children and consequently focus on their careers while they are young (TechCrunch 2014).

These findings of Hochschild (2002) about company culture in the USA cannot be generally transferred to Germany; however, it is clear that the culture of family-friendly companies by no means guarantees the achieving of a good work–life balance or the improvement of the employees' family situations.

The policy framework for work–life balance in the USA is dissimilar to the one in Germany. In the USA, parents do not receive child benefits as German parents do; they can only apply for tax refunds, which amount to approximately $83 per month. Furthermore, only a 12 weeks' parental leave is available in the USA. The USA government offers some financial and statutory support. The so-called "work/life programs and policies" developed in the US Office of Personnel Management aimed to create more flexible and responsive work environments that support commitments to community, home, and family. The USA government promotes a public, family-friendly and health-promoting corporate policy that reflects the cultural prestige of work–life balance in the USA.

As to the gender distribution of gainful employment and family-related work, it is striking that in both countries, Germany and the USA, men spend more time at work, whereas women spend more time with the family or household. In contrast, in Sweden, this distribution is more balanced (European Union 2015). One approach to assess work–life balance is to calculate the percentage of time spent on each type of activity. However, this does not address the question of the subjective satisfaction with the time distribution.

In the subsequent empirical part of this chapter, the extent to which well-being and work style differ between German and American working mothers will be further investigated, as well as whether age, marital status and the number of children affect work–life balance. This is presented under the headings of the research questions and hypotheses posed.

2 Questions of Measuring Instruments

According to published studies, sociopolitical conditions and cultural ideas in relation to the understanding of roles are the main factors that determine work–life balance. As already discussed in Sect. 1, there are different approaches of work–life balance in Germany and the USA. In the USA, the mean number of weekly working hours is 39.1, whereas in Germany it is 34; in addition, fewer employed mothers work part-time in the USA. Thus, US-American working mothers face higher loads than their German counterparts.

Thus, the answer to the following question: "Are there any differences in the well-being and working styles of German and US-American working mothers?" is crucial.

Hypothesis 1: US-American and German employed mothers have different styles of work.

Hypothesis 2: US-American employed mothers have a poorer work–life balance than their German counterparts.

That diverse cultural concepts in different countries influence not only the style of work, but also the understanding of well-being, supports these hypotheses. Thus, it is theoretically assumed that Germany and the USA differ in their work styles and well-being as well as in their work–life balance. This assumption will be investigated to ascertain whether there are indeed significant differences between German and US-American working mothers in relation to well-being, work style, and work–life balance.

Theoretically, well-being and work style should differ according to age. Not only do the specific situations and stress factors differ between working mothers, according to Abele (2005) they can only achieve a good work–life balance when their children are no longer infants. For biological reasons, older mothers are likely to have older children than younger mothers. From this, the second question is derived: Does the work–life balance of working mothers differ according to age?

Hypothesis 3: The well-being of working mothers depends on their age.

Hypothesis 4: Younger employed mothers have a poorer work–life balance than older ones.

Older employed mothers likely achieve a better work–life balance because this is a more realistic goal than it is for younger mothers, who have to pay more attention to their children and thus have less time for their paid work. In addition, older employed mothers often have a more established life situation and thus fewer worries about everyday things or their financial situations. Furthermore, they no longer have to prove themselves at work, unlike many younger and less experienced mothers. These hypotheses regarding a possible relationship between work–life balance and the age of working mothers will be investigated.

Working professionals with one or more children have, as already demonstrated, a particular challenge. Reconciling the demands of work, family, and their own leisure activities requires special skills and a great talent for organization. This raises the question about the impact of the number of children on the work–life balance of their mothers. Thus, the question: "Are there differences in work–life balance between working mothers with only one child and those with more than one?" is a further vital question of this survey.

Hypothesis 5: Women with only one child have a better work–life balance than mothers with two or more children.

That women with more than one child must better organize and coordinate their lives than those with only one child supports this hypothesis. It is easier for the latter to create some personal freedom, for example through arranging for their child a visit to a friend. The more children a woman has, the more difficult it will be for her to find time for herself. Our data are therefore examined to determine whether this hypothesis is true.

3 Questionnaire and Sample Description

To answer these questions, the following two measuring instruments were chosen after an extensive search. These measuring instruments measure the construct work–life balance from the perspective of health behavior, life satisfaction (Bongartz 2000) and work satisfaction and organizational behavior (Braun 2000).

The first part of the questionnaire concerns the assessment of well-being, (Bongartz 2000). Here, questions about psychological and physical well-being, satisfaction with life, and health behavior are asked. The second part is the Landauer Questionnaire for work style, which has six sections, each with eight items (48 items in all) (Braun 2000). There are several subscales: clarity of aims, organizational orientation, satisfaction, willingness to change (the job and to leave the organization), generation of intention and planning, and clarity of means for achieving aims.

In this study 320 women, including 142 German and 178 US-American employed mothers, were interviewed.

The sample was acquired through several organizational networks. This was considered vital, because no comparisons can be made when employees of a single organization are interviewed. Working mothers, some of whom were well qualified with tertiary educations, comprised the sample surveyed.

The subjects' age ranged from 27 to 58 years with a mean of 43.87 years. The mean age of the German respondents was 45.57 years and the one of the US-American women was slightly lower at 42.14 years. Overall, the proportions were well balanced, with no particular age group predominating. However, the ratios were slightly different for the two nationalities studied: whereas respondents aged less than 44 years made up 49.5 % of the whole sample, only 39.2 % of the German respondents were in this age group, compared with 60 % of US-American respondents.

As to marital status, 84.2 % of the tests persons were married, 1 % were in committed relationships, 4 % were single, 3 % were widowed, and 7.9 % were divorced or separated. Surprisingly, 15.7 % of German respondents were divorced/separated, whereas none of the US-American subjects were.

Because only mothers were interviewed, all study subjects had children, the mean number was 1.91. Regarding both nationalities, most mothers had two children (German, 43.1 %; US-American, 68.0 %). There was a higher percentage of US-American than German mothers with two children, which is consistent with the different birth rates in these two countries.

Most of the interviewed German and US-American women had completed secondary-level education. Overall, 70.3 % had completed high school and some had college degrees. Of note, 6 % of the US-American and 39.2 % of the German respondents reported that their highest level of education was a lower secondary school or a high-school diploma. This difference may be attributable to the differences between Germany and the USA in definitions of the level of education and structures of educational bodies.

A difference in terms of occupation was also identified. Employees or office workers comprised the biggest proportion of respondents (48.5 %). Most German respondents (80.4 %) were employees, whereas only 16 % of the US-American women held such positions. Most of the surveyed US-American women (72 %) reported that they were self-employed artisans.

4 Results and Discussion

A factorial analysis of variance was used to ascertain the truth of the hypotheses. Mean differences in various factors such as nationality, age, marital status, and number of children were assessed.

To test the first hypothesis, an ANOVA was performed ($F=11.12$, degrees of freedom [df] = 1.318, $p=0.001$). The hypothesis "US-American and German employed mothers have different styles of work" was confirmed by the analysis, the two target groups differ significantly in working styles. The mean value was 2.50 (SD = 0.38) for German respondents and 2.70 (SD = 0.20) for US-American interviewees. The higher the value in this scale, the better is the style of work from a subjective perspective (see Fig. 1).

To ascertain the reasons for this difference, the following individual parts of the work style questionnaire were assessed separately: clarity of objectives, orientation of organization, availability of resources, planning, satisfaction, and trends to change the occupation (see Fig. 2 and Table 1). The values for some of these questionnaire components did not differ significantly; however, there were

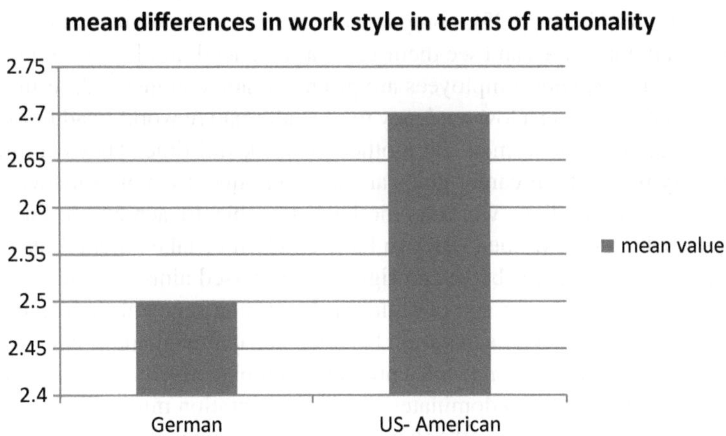

Fig. 1 Mean differences in work style according to nationality

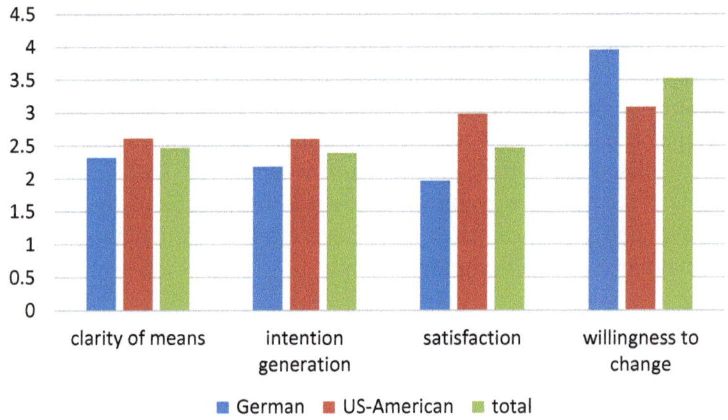

Fig. 2 Average values for work style subscales

significant differences in resource clarity (F = 8.60; df = 1.318; p = 0.004), planning (F = 7.33; df = 1.318; p = 0.008), satisfaction (F = 55.80; df = 1.318; p = 0.000), and trends in change (F = 115.92; df = 3.99; p=0.000).

Thus German (2.32 SD = 0.65) differ from US-American mothers (2.61, SD = 0.25) in resource clarity, satisfaction (German mothers 1.97, SD = 0.80; US-American mothers (2.98 SD = 0.54). German mothers (2.18, SD = 0.83) had lower scores than US-Americans (2.60, SD = 0.72) for planning. The same was true for the change tendency (German mothers 3.96, SD = 0.53; US-American mothers 3.08, SD = 0.23).

On average, the interviewed US-American working mothers scored higher on resources to achieve their work-place goals and on the planning concerning how to fulfill their job requirements. Moreover, they were more satisfied and had a lower tendency to seek alternative work than the German participants.

According to Abele (2005), companies in the USA place great importance on retaining their employees and see themselves as extended families, which results in a good work atmosphere. Employees are promoted and commended, resulting in a high job satisfaction and a low tendency to seek alternative work. In addition, in the USA, it is common for women and mothers to work full-time. They can therefore more readily pursue their career goals and are consequently more satisfied.

Because only childless working mothers are able to achieve a satisfactory balance (Abele 2005), women defer to have children or take parental leave when they do. The discrepancies between originally expressed aims and the actual ones decrease with an increasing age of children. Most men segment their lives in this way, that is to create a clear separation between occupational and private aims and actions, with the aim of permanently protecting an objective balance in which the occupational area of life predominates. Such segmentation minimizes conflicts on the one hand; however, on the other hand, it creates an imbalance between work and family life.

Table 1 Results of ANOVA on selected variables

		Mean value	Standard deviation	Df	F	Significance
Clarity of means	German	2.32	.65	1.318	8.60	0.004
	US-American	2.61	.25			
	Total	2.47	.51			
Intention generation	German	2.18	.83	1.318	7.33	0.008
	US-American	2.60	.72			
	Total	2.39	.80			
Satisfaction	German	1.97	.80	1.318	55.80	0.000
	US-American	2.98	.54			
	Total	2.47	.85			
Willingness to change	German	3.96	.53	1.318	115.92	0.000
	US-American	3.08	.23			
	Total	3.52	.60			

"US-American employed mothers have a poorer work–life balance than their German counterparts". The interviewed employed German and US-American mothers had significantly different working styles, which prompts the question of whether this large difference has an impact on work–life balance in general. The second major subscale of work–life balance is well-being. In this regard, the two groups of interviewees did not differ significantly ($p = 0.856$) from each other.

Next, hypothesis 3: "The well-being of working mothers depends on their age" and hypothesis 4: "Younger employed mothers have a poorer work–life balance than older ones" was examined.

The sample was divided into two age groups, one younger and the other older than the median age of the whole sample. To test the hypotheses, a two-way ANOVA was performed and it was found that only the domains of well-being and health had a relationship with age. Time with family ($r = 0.340^{**}$), good partnership ($r = 0.243^{*}$), sufficient sleep ($r = 0.438^{***}$) and healthy nutrition ($r = 0.421^{***}$) were positively correlated with age, whereas sufficient sleep and healthy nutrition were more positively correlated with well-being. Professional work was negatively related to well-being with a high correlation coefficient ($r = -0.523^{**}$).

Overall, there were no significant differences between younger and older working mothers in well-being ($p = 0.314$) or style of work ($p = 0.778$). Differences between younger and older employed mothers were found only in specific areas, for example vocation-oriented, cultural, and educational work. Looking at health behavior and the well-being domains more differentiated, one can assert that there are no significant differences between both age groups. For both age groups, family and relationships, sleep, and healthy diet often had positive effects on well-being. Professional work had the strongest negative effect on work–life balance.

The analysis showed no significant differences between age groups in terms of work–life balance ($F = 1.024$; $df = 2.318$; $p = 0.314$ for well-being group; $F = 0.080$, $df = 2.318$; $p = 0.778$ for working style group).

To test the hypothesis that women with only one child have a better work–life balance than mothers with two or more children, the sample was divided into two groups: working mothers with only one child (29.3 %) and those with two or more children (72.7 %).

This hypothesis was refuted: there were no significant differences between these two groups in well-being ($F = 0.491$; $df = 3.317$; $p = 0.485$) or work styles ($F = 0.152$; $df = 3.317$; $p = 0.698$). Additionally, when nationality was considered, no significant differences were detected between the groups in terms of well-being or work.

One possible explanation is that the birth of a first child changes a woman's life considerably, affecting her well-being and work style, whereas additional major changes do not occur with the birth of further children because their mother is already familiar with the double burden of work and family life and has already implemented techniques for organizing and coordinating her life. An additional possibility is that the increased responsibilities with additional children are outweighed by the joy that those children give their mothers (see Table 2 and Fig. 3).

This study aimed to highlight that work–life balance is a challenge for working mothers and to identify differences and similarities between well-being, work style, and work–life balance of German and US-American working mothers. The concept of work–life balance is complex and has many different aspects. Our study made clear that a successful work–life balance is especially difficult for mothers to achieve because of the need to master a balancing act between work and family in the face of the reality that children explicitly require care.

Furthermore, we showed that different cultural and sociopolitical conditions may have an impact on achieving a good work–life balance. The quantitative analysis of our data showed differences in work style between the interviewed German and US-American working mothers, which are likely attributable to different organizational and work structures in the two cultures. However, German and

Table 2 Results of ANOVA of work–life balance according to number of children (one versus two or more)

		Df	F	Significance
Well-being scale	Between groups	3	0.491	0.485
	Within groups	317		
	Total	320		
Work style scale	Between groups	3	0.152	0.698
	Within groups	378		
	Total	320		

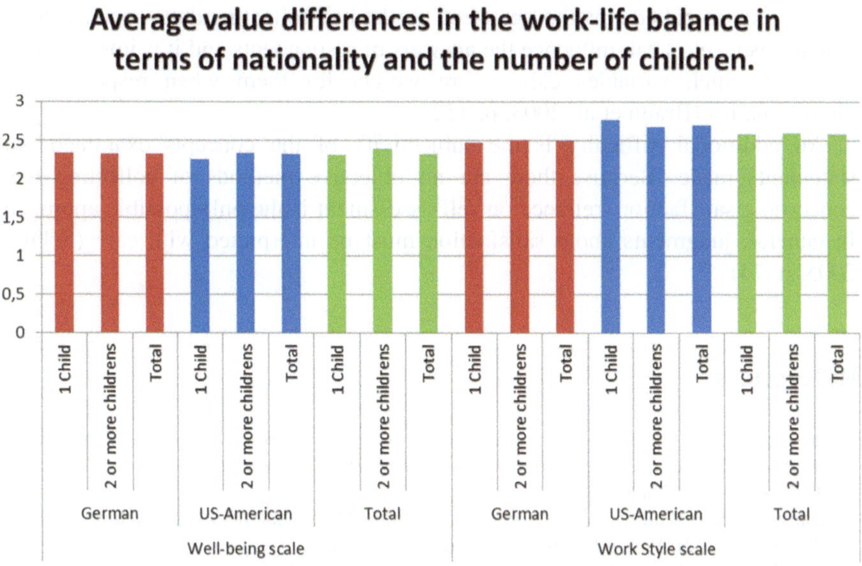

Fig. 3 Average differences in work–life balance according to nationality and number of children

US-American working mothers differed only slightly in well-being. There were small differences in favor of US-American women in work–life balance in general.

Our data showed that there are no significant differences between younger and older working mothers. The well-being of both groups was positively affected by having children/family and a live-in partner, as well as adequate sleep and a healthy diet. "Business-oriented work" often influenced the well-being of younger mothers negatively, which supports the contention that work–life balance is more a challenge for younger working mothers, who tend to be more closely involved with their families than older working mothers. According to the analysis of our data, the number of children does not affect the work–life balance of working mothers.

However, the limitations of our study must also be mentioned. Because many of the variables were not normally distributed, a comparability cannot be guaranteed. Additionally, satisfaction with life is affected by impairment in various domains of satisfaction from the pools "work" (satisfaction with work, profession and success) and "life" (satisfaction with partner, free time, family etc.) (OECD 2007, p. 120). The consideration of the work–life balance concept leads to the notion that a subjective successful work–life balance is distinguished by high satisfaction with various aspects of both work and family life (Guest 2001, p. 256).

The assessment of work–life balance requires satisfaction with certain life domains and often includes investigating a third aspect; namely health or well-being (Stock-Homburg and Bauer 2007). Another limitation of this study is that judgments about satisfaction are easier for individuals to make if they are related to certain moments rather than longer periods of time (Fischer 2006, p. 5). The setting of the survey also has to be scrutinized. Numerous factors (mood, time pressure, locality, the so-called Hawthorne effect, method of collecting data, socially desirable answers, etc.) can influence the answers of respondents and it is impossible to ascertain which variables carry more weight for them when responding to questionnaires (Braun et al. 2003, p. 152).

An additional difficulty is the subjectivity of the concepts examined and statements made. Because there are no objective methods of collecting data concerning satisfaction, reliance on self-assessment is the only possible approach. In general, judgments about satisfaction must be interpreted with care (Schulte 2005, p. 74).

References

Abele AE (2005) Objectives, self-concept and work–life balance in long-term life-shaping—results of the Erlangen longitudinal section study BELA-E with academics. Z Arb Organ 49:176–186 (in German)

Bongartz N (2000) Well-being as a parameter of health. Theory and treatment-oriented diagnosis (Psychologie 32). Verlag Empirische Pädagogik, Landau (in German)

Braun OL (2000) A model of active adaptation: occupational clarity of goals, organizational orientation, clarity of means and generation of intention/planning as precursory conditions of work-satisfaction, tendency to change and performance (Psychologie 36). Verlag Empirische Pädagogik, Landau (in German)

Braun OL, Adjei M, Münch M (2003) Self-management and life-satisfaction. In: Müller GF (ed) Self-actualization in working life. Pabst, Lengerich, pp 151–170 (in German)

Institut Arbeit und Qualifikation (IAQ) an der Universität Duisburg–Essen (2008) IAQ-Report 2008-04: more and more women are employed—but with fewer weekly hours of work. Available from http://www.iaq.uni-due.de/iaq-report/2008/report2008-04.php. Accessed 17 Nov 2014 (in German)

European Commission (2007) Employment in Europe 2007. Office for Official Publications of the European Communities, Luxembourg

European Union (2015) Sweden: successful reconciliation of work and family life. Available from http://europa.eu/epic/countries/sweden/index_en.htm. Accessed 17 Feb 2015

Fischer L (ed) (2006) Work satisfaction. Concepts and empirical results. Hogrefe, Göttingen (in German)

Guest DE (2001) Perspectives on the study of work–life balance. Social Science Information 2. Maison des Sciences de l'Homme, Paris, pp 255–279

Hochschild AR (2002) No time. When the company is becoming home and at home there is only work waiting. Leske, Budrich (in German)

Hoff E-H, Grote S, Dettmar S et al (2005) Work–life-balance: shaping of occupational and private life of highly qualified professional women and men. Z Arb Organ 49:196–207 (in German)

Kastner M (2004) Different approaches to work life balance. In: Kastner M (ed) The future of work life balance: how can work and family, work and free time be arranged? Asanger, Kröning (in German)

Krumpholz D (1998) Gruppendynamik und Organisationsberatung [The consequences of the shift in values for organizations and managers]. Z Angew Sozialpsychologie 29(4):349–358 (in German)

LFPress (2014) Google tops list of best places to work in 2015. Available from http://www.lfpress.com/2014/12/10/glassdoor-names-top-50-best-places-to-work-in-2015. Accessed 17 Feb 2015

Metz-Göckel S (2004) When work eats family: work life balance—a gender problem? In: Kastner M (ed) The future of work life balance: how can work and family, work and free time be arranged? Asanger, Kröning (in German)

OECD (2007) Society at a glance. OECD social indicators. OECD, Paris (in German)
Schulte K (2005) Work-satisfaction over the span of life. Papst, Lengerich (in German)
Stock-Homburg R, Bauer E-M (2007) Work–life-balance in top management. Aussen Politik Zeitgeschichte 34:25–32 (in German)
TechCrunch (2014) Facebook and Apple offer to pay for female employees to freeze their eggs. Available from http://techcrunch.com/2014/10/14/facebook-and-apple-offer-to-pay-for-female-employees-to-freeze-their-eggs/. Accessed 17 Feb 2015
Wiese BS (2007) Work–life balance. In: Moser K (ed) Wirtschaftspsychologie [Business psychology]. Springer, Berlin, pp 245–263 (in German)